Matrix-Based Introduction to Multivariate Data Analysis

Kohei Adachi

Matrix-Based Introduction
to Multivariate Data Analysis

 Springer

Kohei Adachi
Graduate School of Human Sciences
Osaka University
Osaka
Japan

ISBN 978-981-10-9595-5   ISBN 978-981-10-2341-5   (eBook)
DOI 10.1007/978-981-10-2341-5

Printed on acid-free paper

This Springer imprint is published by Springer Nature
The registered company is Springer Nature Singapore Pte Ltd.
The registered company address is: 152 Beach Road, #22-06/08 Gateway East, Singapore 189721, Singapore

Preface

A set of multivariate data can be expressed as a table, i.e., a matrix, of individuals (rows) by variables (columns), with the variables interrelated. Statistical procedures for analyzing such data sets are generally referred to as multivariate data analysis. The demand for this kind of analysis is increasing in a variety of fields. Each procedure in multivariate data analysis features a special purpose. For example, predicting future performance, classifying individuals, visualizing inter-individual relationships, finding a few factors underlying a number of variables, and examining causal relationships among variables are included in the purposes for the procedures.

The aim of this book is to enable readers who may not be familiar with matrix operations to understand major multivariate data analysis procedures in matrix forms. For that aim, this book begins with explaining fundamental matrix calculations and the matrix expressions of elementary statistics, followed by an introduction to popular multivariate procedures, with chapter-by-chapter advances in the levels of matrix algebra. The organization of this book allows readers without knowledge of matrices to deepen their understanding of multivariate data analysis.

Another feature of this book is its emphasis on the model that underlies each procedure and the objective function that is optimized for fitting the model to data. The author believes that the matrix-based learning of such models and objective functions is the shortest way to comprehend multivariate data analysis. This book is also arranged so that readers can intuitively capture for what purposes multivariate analysis procedures are utilized; plain explanations of the purposes with numerical examples precede mathematical descriptions in almost all chapters.

The preceding paragraph featured three key words: purpose, model, and objective function. The author considers that capturing those three points for each procedure suffices to understand it. This consideration implies that the mechanisms behind how objective functions are optimized must not necessarily be understood. Thus, the mechanisms are only described in appendices and some exercises.

This book is written with the following guidelines in mind:

(1) Not using mathematics except matrix algebra
(2) Emphasizing singular value decomposition (SVD)
(3) Preferring a simultaneous solution to a successive one

Although the exceptions to (1) are found in Appendix A6, where differential calculus is used, and in some sections of Part III and Chap. 15, where probabilities are used, those exceptional parts only occupy a limited number of pages; the majority of the book is matrix-intensive. Matrix algebra is also exclusively used for formulating the optimization of objective functions in Appendix A4. For matrix-intensive formulations, ten Berge's (1983, 1993) theorem is considered to be the best starting fact, as found in Appendix A4.1.

Guideline (2) is due to the fact that SVD can be defined for any matrix, and a number of important properties of matrices are easily derived from SVD. In the former point, SVD is more general than eigenvalue decomposition (EVD), which is only defined for symmetric matrices. Thus, EVD is only mentioned in Sect. 6.2. Further, SVD takes on an important role in optimizing trace and least squares functions of matrices: The optimization problems are formulated with the combination of SVD and ten Berge's (1983, 1993) theorem, as found in Appendix A4.2 and Appendix A4.3.

Guideline (3) is particularly concerned with principal component analysis (PCA), which can be formulated as minimizing $\|\mathbf{X} - \mathbf{FA'}\|^2$ over PC score matrix $\mathbf{F}$ and loading matrix $\mathbf{A}$ for a data matrix $\mathbf{X}$. In some of the literature, PCA is described as obtaining the first component, the second, and the remaining components in turn (i.e., per column of $\mathbf{F}$ and $\mathbf{A}$). This can be called a successive solution. On the other hand, PCA can be described as obtaining $\mathbf{F}$ and $\mathbf{A}$ matrix-wise, which can be called a simultaneous solution. This is preferred in this book, as the above formulation is actually made matrix-wise and the simultaneous solution facilitates understanding PCA as a reduced rank approximation of $\mathbf{X}$.

This book is appropriate for undergraduate students who have already learned introductory statistics, as the author has used preliminary versions of the book in a course for such students. It is also useful for graduate students and researchers who are not familiar with the matrix-intensive formulations of multivariate data analysis.

I owe this book to the people who can be called the "matricians" in statistics, more exactly, the ones taking matrix-intensive approaches for formulating and developing data analysis procedures. Particularly, I have been influenced by the Dutch psychometricians, as found above, in that I emphasize the theorem by Jos M.F. ten Berge (Professor Emeritus, University of Groningen). Yutaka Hirachi of Springer has been encouraging me since I first considered writing this book. I am most grateful to him. I am also thankful to the reviewers who read through drafts of this book. Finally, I must show my gratitude to Yoshitaka Shishikura of the publisher Nakanishiya Shuppan, as he readily agreed to the use of the numerical examples in this book, which I had originally used in that publisher's book.

Kyoto, Japan   Kohei Adachi
May 2016

Contents

Part I
Elementary Statistics with Matrices

This part begins with introducing elementary matrix operations, followed by explanations of fundamental statistics with their matrix expressions. These initial chapters serve as preparation for learning the multivariate data analysis procedures that are described in Part II and thereafter.

Chapter 1
Elementary Matrix Operations

The mathematics for studying the properties of matrices is called *matrix algebra* or *linear algebra*. This chapter treats the introductory part of matrix algebra required for learning multivariate data analysis. I begin by explaining what a matrix is, in order to describe elementary matrix operations.

In later chapters, more advanced properties of matrices are described, where necessary, with reference to Appendices for more detailed explanations.

1.1 Matrices

Let us note that Table 1.1 is a 6 teams × 4 items table. When such a table (i.e., a two-way array) is treated as a unit entity and expressed as

$$\mathbf{X} = \begin{bmatrix} 0.617 & 731 & 140 & 3.24 \\ 0.545 & 680 & 139 & 4.13 \\ 0.496 & 621 & 143 & 3.68 \\ 0.493 & 591 & 128 & 4.00 \\ 0.437 & 617 & 186 & 4.80 \\ 0.408 & 615 & 184 & 4.80 \end{bmatrix},$$

it is called a 6 (rows) × 4 (columns) *matrix*, or a matrix of 6 rows by 4 columns. "*Matrices*" is the plural of "matrix." Here, a horizontal array and a vertical one are called a *row* and a *column*, respectively. For example, the fifth row of $\mathbf{X}$ is "0.437, 617, 0.260, 4.80," while the third column is "140, 139, 143, 128, 186, 184." Further, the cell at which the 5th row and 3rd column intersect is occupied by 186, which is called "the (5,3)" *element*. Rewriting the rows of a matrix as columns (or its columns as rows) is referred to as a *transpose*. The transpose of $\mathbf{X}$ is denoted as $\mathbf{X}'$:

© Springer Nature Singapore Pte Ltd. 2016
K. Adachi, *Matrix-Based Introduction to Multivariate Data Analysis*,
DOI 10.1007/978-981-10-2341-5_1

Table 1.1 Six teams × four items matrix for the averages

Team	Item			
	Win (%)	Runs	HR	ERA
Tigers	0.617	731	140	3.24
Dragons	0.545	680	139	4.13
BayStars	0.496	621	143	3.68
Swallows	0.493	591	128	4.00
Giants	0.437	617	186	4.80
Carp	0.408	615	184	4.80

$$\mathbf{X}' = \begin{bmatrix} 0.617 & 0.545 & 0.496 & 0.493 & 0.437 & 0.408 \\ 731 & 680 & 621 & 591 & 617 & 615 \\ 140 & 139 & 143 & 128 & 186 & 184 \\ 3.24 & 4.13 & 3.68 & 4.00 & 4.80 & 4.80 \end{bmatrix}$$

Let us describe a matrix in a generalized setting. The array of a_{ij} ($i = 1, \ldots, n$; $j = 1, \ldots, m$) arranged in n rows and m columns, i.e.,

$$\mathbf{A} = \begin{bmatrix} a_{11} & a_{12} & \cdots & a_{1m} \\ a_{21} & a_{22} & \cdots & a_{2m} \\ \vdots & \vdots & \vdots & \vdots \\ a_{n1} & a_{n2} & \cdots & a_{nm} \end{bmatrix}, \tag{1.1}$$

is called an $n \times m$ *matrix* with a_{ij} its (i, j) *element*. The transpose of $\mathbf{A}$ is an $m \times n$ matrix

$$\mathbf{A}' = \begin{bmatrix} a_{11} & a_{21} & \cdots & a_{n1} \\ a_{12} & a_{22} & \cdots & a_{n2} \\ \vdots & \vdots & \vdots & \vdots \\ a_{1m} & a_{2m} & \cdots & a_{nm} \end{bmatrix}. \tag{1.2}$$

The transpose of a transposed matrix is obviously the original matrix, with $(\mathbf{A}')' = \mathbf{A}$.

The expression of matrix $\mathbf{A}$ as the right-hand side in (1.1) takes a large amount of space. For economy of space, matrix $\mathbf{A}$ in (1.1) is also expressed as:

$$\mathbf{A} = (a_{ij}), \tag{1.3}$$

using the general expression a_{ij} for the elements of $\mathbf{A}$. The statement "We define an $n \times m$ matrix $\mathbf{A} = (a_{ij})$" stands for the matrix $\mathbf{A}$ being expressed as (1.1).

1.2 Vectors

A vertical array,

$$\mathbf{a} = \begin{bmatrix} a_1 \\ a_2 \\ \vdots \\ a_n \end{bmatrix}, \tag{1.4}$$

is called a *column vector* or simply a *vector*. In exactness, (1.4) is said to be an $n \times 1$ *vector*, since it contains n elements. Vectors can be viewed as a special case of matrices; (1.4) can also be called an $n \times 1$ matrix. Further, a *scalar* is a 1×1 matrix. The right side of (1.4) is vertically long, and for the sake of the economy of space, (1.4) is often expressed as:

$$\mathbf{a} = [a_1, a_2, \ldots, a_n]' \text{ or } \mathbf{a}' = [a_1, a_2, \ldots, a_n], \tag{1.5}$$

using a transpose. A horizontal array as $\mathbf{a}'$ is called a *row vector*.

We can use vectors to express a matrix: By using $n \times 1$ vectors $\mathbf{a}_j = [a_{1j}, a_{2j}, \ldots, a_{nj}]'$, with $j = 1, 2, \ldots, m$, or by using $m \times 1$ vectors $\tilde{\mathbf{a}}_i = [a_{i1}, a_{i2}, \ldots, a_{im}]'$, with $i = 1, 2, \ldots, n$, matrix (1.1) or (1.3) is expressed as:

$$\mathbf{A} = [\mathbf{a}_1, \mathbf{a}_2, \ldots, \mathbf{a}_m] = \begin{bmatrix} \tilde{\mathbf{a}}'_1 \\ \tilde{\mathbf{a}}'_2 \\ \vdots \\ \tilde{\mathbf{a}}'_n \end{bmatrix} = [\tilde{\mathbf{a}}_1, \tilde{\mathbf{a}}_2, \ldots, \tilde{\mathbf{a}}_n]'. \tag{1.6}$$

In this book, a bold uppercase letter such as $\mathbf{X}$ is used for denoting a *matrix*, a bold lowercase letter such as $\mathbf{x}$ is used for a *vector*, and an *italic* letter (not bold) such as x is used for a *scalar*. Though a *series of integers* has so far been expressed as $i = 1, 2, \ldots, n$, this may be rewritten as $i = 1, \ldots, n$, omitting 2 when it obviously follows 1. With this notation, (1.1) or (1.6) is rewritten as

$$\mathbf{A} = \begin{bmatrix} a_{11} & \cdots & a_{1m} \\ \vdots & \vdots & \vdots \\ a_{n1} & \cdots & a_{nm} \end{bmatrix} = [\mathbf{a}_1, \ldots, \mathbf{a}_m] = [\tilde{\mathbf{a}}_1, \cdots \tilde{\mathbf{a}}_n]'.$$

1.3   Sum of Matrices and Their Multiplication by Scalars

The sum of matrices can be defined when they are of the *same size*. Let matrices
A and **B** be equivalently $n \times m$. Their *sum* **A** + **B** yields the $n \times m$ matrix, each of
whose *elements is the sum of the corresponding ones* of $\mathbf{A} = (a_{ij})$ and $\mathbf{B} = (b_{ij})$: The
sum is defined as:

$$\mathbf{A} + \mathbf{B} = (a_{ij} + b_{ij}), \tag{1.7}$$

using the notation in (1.3). For example, when $\mathbf{X} = \begin{bmatrix} 3 & -2 & 6 \\ 8 & 0 & -2 \end{bmatrix}$ and
$\mathbf{Y} = \begin{bmatrix} 2 & 1 & -9 \\ -7 & 2 & -3 \end{bmatrix}$,

$$\mathbf{X} + \mathbf{Y} = \begin{bmatrix} 3+2 & -2+1 & 6-9 \\ 8-7 & 0+2 & -2-3 \end{bmatrix} = \begin{bmatrix} 5 & -1 & -3 \\ 1 & 2 & -5 \end{bmatrix}.$$

The multiplication of matrix $\mathbf{A} = (a_{ij})$ by scalar s is defined as *all elements* of
A being *multiplied* by s:

$$s\mathbf{A} = (s \times a_{ij}), \tag{1.8}$$

using the notation in (1.3). For example, when $\mathbf{Z} = \begin{bmatrix} 8 & -2 & 6 \\ -5 & 0 & -3 \end{bmatrix}$

$$-0.1\mathbf{Z} = \begin{bmatrix} -0.1 \times 8 & -0.1 \times (-2) & -0.1 \times 6 \\ -0.1 \times (-5) & -0.1 \times 0 & -0.1 \times (-3) \end{bmatrix}$$
$$= \begin{bmatrix} -0.8 & 0.2 & -0.6 \\ 0.5 & 0 & 0.3 \end{bmatrix}.$$

The *sum of the matrices multiplied by scalars* is defined simply as the combination of (1.7) and (1.8):

$$v\mathbf{A} + w\mathbf{B} = (va_{ij} + wb_{ij}). \tag{1.9}$$

For example, when $\mathbf{X} = \begin{bmatrix} 4 & -2 & 6 \\ 8 & 0 & -2 \end{bmatrix}$ and $\mathbf{Y} = \begin{bmatrix} 2 & 1 & -9 \\ -7 & 2 & -3 \end{bmatrix}$,

$$0.5\mathbf{X} + (-2)\mathbf{Y} = \begin{bmatrix} 2-4 & -1-2 & 3+18 \\ 4+14 & 0-4 & -1+6 \end{bmatrix} = \begin{bmatrix} -2 & -3 & 21 \\ 18 & -4 & 5 \end{bmatrix}.$$

Obviously, setting $v = 1$ and $w = -1$ in (1.9) leads to the definition of the matrix
difference **A** − **B**.

The above definition is generalized as

$$\sum_{k=1}^{K} v_k \mathbf{A}_k = v_1 \mathbf{A}_1 + \cdots + v_K \mathbf{A}_K = \left(\sum_{k=1}^{K} v_k a_{ijk} \right), \tag{1.10}$$

where $\mathbf{A}_1, \ldots, \mathbf{A}_K$ are of the same size and a_{ijk} is the (i, j) element of $\mathbf{A}_k$ $(k = 1, \ldots, K)$.

1.4   Inner Product and Norms of Vectors

The *inner product* of the vectors $\mathbf{a} = [a_1, \ldots, a_m]'$ and $\mathbf{b} = [b_1, \ldots, b_m]'$ is defined as:

$$\mathbf{a}'\mathbf{b} = \mathbf{b}'\mathbf{a} = [a_1, \ldots, a_m] \begin{bmatrix} b_1 \\ \vdots \\ b_m \end{bmatrix} = a_1 b_1 + \cdots + a_m b_m = \sum_{k=1}^{m} a_k b_k. \tag{1.11}$$

Obviously, this can be defined only for the vectors of the same size. The inner product is expressed as $\mathbf{a}'\mathbf{b}$ or $\mathbf{b}'\mathbf{a}$, i.e., the form of a *transposed column vector* (i.e., *row vector*) followed by a *column vector*, so as to be congruous to the matrix product introduced in the next section.

The inner product of the identical vectors $\mathbf{a}$ and $\mathbf{a}$ is in particular called the *squared norm* of $\mathbf{a}$ and denoted as $\|\mathbf{a}\|^2$:

$$\|\mathbf{a}\|^2 = \mathbf{a}'\mathbf{a} = [a_1, \ldots, a_m]' \begin{bmatrix} a_1 \\ \vdots \\ a_m \end{bmatrix} = a_1^2 + \cdots + a_m^2 = \sum_{k=1}^{m} a_k^2. \tag{1.12}$$

The square root of $\|\mathbf{a}\|^2$, that is, $\|\mathbf{a}\|$, is simply called the *norm* of the vector $\mathbf{a} = [a_1, \ldots, a_m]'$ with

$$\|\mathbf{a}\| = \sqrt{a_1^2 + \cdots + a_m^2}. \tag{1.13}$$

It is also called the *length* of $\mathbf{a}$, for the following reason: If $m = 3$ with $\mathbf{a} = [a_1, a_2, a_3]'$ and $\mathbf{a}$ is viewed as the line extending from the origin to the point whose coordinate is $[a_1, a_2, a_3]'$, as illustrated in Fig. 1.1, (1.13) expresses the length of the line. It also holds for $m = 1, 2$. If $m > 3$, the line cannot be depicted or seen by those of us who live in three-dimensional world, but the length of $\mathbf{a}$ is also defined as (1.13) for $m > 3$ in mathematics. (In it, the entities that do not exist in the real world are also considered, if they are treated logically.)

Fig. 1.1 Graphical
representation of a vector

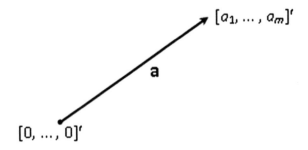

1.5   Product of Matrices

Let $n \times m$ and $m \times p$ matrices be defined as:

$$\mathbf{A} = \begin{bmatrix} \mathbf{a}_1' \\ \vdots \\ \mathbf{a}_n' \end{bmatrix} = \begin{bmatrix} a_{11} & \cdots & a_{1m} \\ \vdots & \vdots & \vdots \\ a_{n1} & \cdots & a_{nm} \end{bmatrix} \quad \text{and} \quad \mathbf{B} = [\mathbf{b}_1 \cdots \mathbf{b}_p] = \begin{bmatrix} b_{11} & \cdots & b_{1p} \\ \vdots & \vdots & \vdots \\ b_{m1} & \cdots & b_{mp} \end{bmatrix},$$

respectively. Then, the *post-multiplication* of $\mathbf{A}$ by $\mathbf{B}$ is defined as:

$$\mathbf{AB} = \begin{bmatrix} \mathbf{a}_1'\mathbf{b}_1 & \cdots & \mathbf{a}_1'\mathbf{b}_p \\ \vdots & \cdots & \vdots \\ \mathbf{a}_n'\mathbf{b}_1 & \cdots & \mathbf{a}_n'\mathbf{b}_p \end{bmatrix} = (\mathbf{a}_i'\mathbf{b}_j), \tag{1.14}$$

using the *inner products* of the *row vectors of the preceding matrix* $\mathbf{A}$ and *the column vectors of the following matrix* $\mathbf{B}$. The resulting matrix $\mathbf{AB}$ is the $n \times p$ matrix whose (i, j) element is the inner product of the ith row of $\mathbf{A}$ and the jth column of $\mathbf{B}$:

$$\mathbf{a}_i'\mathbf{b}_j = a_{i1}b_{1j} + \cdots + a_{im}b_{mj} = \sum_{k=1}^{m} a_{ik}b_{kj}. \tag{1.15}$$

For example, if $\mathbf{A} = \begin{bmatrix} \mathbf{a}_1' \\ \mathbf{a}_2' \end{bmatrix} = \begin{bmatrix} 2 & -4 \\ 1 & 7 \end{bmatrix}$, $\mathbf{B} = [\mathbf{b}_1 \quad \mathbf{b}_2] = \begin{bmatrix} -3 & 1 \\ 2 & -5 \end{bmatrix}$, then

$$\begin{aligned} \mathbf{AB} &= \begin{bmatrix} \mathbf{a}_1'\mathbf{b}_1 & \mathbf{a}_2'\mathbf{b}_2 \\ \mathbf{a}_2'\mathbf{b}_1 & \mathbf{a}_2'\mathbf{b}_2 \end{bmatrix} = \begin{bmatrix} 2 \times (-3) + (-4) \times 2 & 2 \times 1 + (-4) \times (-5) \\ 1 \times (-3) + 7 \times 2 & 1 \times 1 + 7 \times (-5) \end{bmatrix} \\ &= \begin{bmatrix} -14 & 22 \\ 11 & -34 \end{bmatrix}. \end{aligned}$$

As found above, the matrix product **AB** is defined only when the following holds:

$$\text{The number of columns in } \mathbf{A} = \text{the number of rows in } \mathbf{B}. \tag{1.16}$$

The resulting matrix **AB** is:

$$(\text{the number of rows in } \mathbf{A}) \times (\text{the number of columns in } \mathbf{B}). \tag{1.17}$$

Thus, the product is sometimes expressed as:

$$\underset{n \times m}{\mathbf{A}} \underset{m \times p}{\mathbf{B}} = \underset{n \times p}{\mathbf{C}}, \text{ or, more simply, } {}_n\mathbf{A}_m\mathbf{B}_p = {}_n\mathbf{C}_p, \tag{1.18}$$

with which we can easily verify (1.16) and (1.17). If $n = p$, we can define products **AB** and **BA**. Here, we should note

$$\mathbf{AB} \neq \mathbf{BA}, \tag{1.19}$$

except for special **A** and **B**, which is different from the product of scalars with $st = ts$, the inner product (1.11), and that of scalar s and matrix **A** with

$$s\mathbf{A} = \mathbf{A} \times s. \tag{1.20}$$

For this reason, we call **AB** "the *post-multiplication* of **A** by **B**" or "the premultiplication of **B** by **A**," so as to clarify the order of the matrices.

Here, four examples of matrix products are presented:

Ex. 1 For $\mathbf{X} = \begin{bmatrix} 2 & 3 & -1 \\ -2 & 0 & 4 \end{bmatrix}$ and $\mathbf{Y} = \begin{bmatrix} 3 & 5 & 4 \\ -1 & 0 & -2 \\ 0 & 6 & 0 \end{bmatrix}$,

$$\begin{aligned}
\mathbf{XY} &= \begin{bmatrix} 2 \times 3 + 3 \times (-1) + (-1) \times 0 & 2 \times 5 + 3 \times 0 + (-1) \times 6 & 2 \times 4 + 3 \times (-2) + (-1) \times 0 \\ -2 \times 3 + 0 \times (-1) + 4 \times 0 & -2 \times 5 + 0 \times 0 + 4 \times 6 & -2 \times 4 + 0 \times (-2) + 4 \times 0 \end{bmatrix} \\
&= \begin{bmatrix} 3 & 4 & 2 \\ -6 & 14 & -8 \end{bmatrix}.
\end{aligned}$$

Ex. 2 For $\mathbf{F} = \begin{bmatrix} 2 & -1 \\ -3 & 0 \\ 1 & 3 \\ -2 & -3 \end{bmatrix}$ and $\mathbf{A} = \begin{bmatrix} -4 & 1 \\ 6 & -3 \\ 2 & 5 \end{bmatrix}$,

$$\mathbf{FA'} = \begin{bmatrix} 2 & -1 \\ -3 & 0 \\ 1 & 3 \\ -2 & -3 \end{bmatrix} \begin{bmatrix} -4 & 6 & 2 \\ 1 & -3 & 5 \end{bmatrix}$$

$$= \begin{bmatrix} 2 \times (-4) + (-1) \times 1 & 2 \times 6 + (-1) \times (-3) & 2 \times 2 + (-1) \times 5 \\ -3 \times (-4) + 0 \times 1 & -3 \times 6 + 0 \times (-3) & -3 \times 2 + 0 \times 5 \\ 1 \times (-4) + 3 \times 1 & 1 \times 6 + 3 \times (-3) & 1 \times 2 + 3 \times 5 \\ -2 \times (-4) + (-3) \times 1 & -2 \times 6 + (-3) \times (-3) & -2 \times 2 + (-3) \times 5 \end{bmatrix}$$

$$= \begin{bmatrix} -9 & 11 & -1 \\ 12 & -18 & -6 \\ -1 & -3 & 17 \\ 5 & -3 & -19 \end{bmatrix},$$

where it should be noted that $\mathbf{A}$ has been transposed in the product.

Ex. 3 In statistics, the product of a matrix and its transpose is often used. For $\mathbf{A} = \begin{bmatrix} -4 & 1 \\ 6 & -3 \\ 2 & 5 \end{bmatrix}$, the post-multiplication of $\mathbf{A}$ by $\mathbf{A'}$, which we denote by $\mathbf{S}$, is

$$\mathbf{S} = \mathbf{AA'} = \begin{bmatrix} (-4)^2 + 1^2 & -4 \times 6 + 1 \times (-3) & -4 \times 2 + 1 \times 5 \\ 6 \times (-4) + (-3) \times 1 & 6^2 + (-3)^2 & 6 \times 2 + (-3) \times 5 \\ 2 \times (-4) + 5 \times 1 & 2 \times 6 + 5 \times (-3) & 2^2 + 5^2 \end{bmatrix}$$

$$= \begin{bmatrix} 17 & -27 & -3 \\ -27 & 45 & -3 \\ -3 & -3 & 29 \end{bmatrix}.$$

The premultiplication of $\mathbf{A}$ by $\mathbf{A'}$, which we denote by $\mathbf{T}$, is

$$\mathbf{T} = \mathbf{A'A} = \begin{bmatrix} (-4)^2 + 6^2 + 2^2 & (-4) \times 1 + 6 \times (-3) + 2 \times 5 \\ 1 \times (-4) + (-3) \times 6 + 5 \times 2 & 1^2 + (-3)^2 + 5^2 \end{bmatrix}$$

$$= \begin{bmatrix} 56 & -12 \\ -12 & 35 \end{bmatrix}.$$

Ex. 4 The product of vectors is a special case of that of matrices:

$$\text{For } \mathbf{u} = \begin{bmatrix} 2 \\ -1 \\ 3 \end{bmatrix} \text{ and } \mathbf{v} = \begin{bmatrix} -2 \\ 3 \\ -4 \end{bmatrix},$$

and the inner product yields a scalar as:

$$\mathbf{u}'\mathbf{v} = 2 \times (-2) + (-1) \times 3 + 3 \times (-4) = -19,$$

but the post-multiplication of 3×1 vector $\mathbf{u}$ by 1×3 $\mathbf{v}'$ gives a 3×3 matrix with

$$\mathbf{u}\mathbf{v}' = \begin{bmatrix} 2 \\ -1 \\ 3 \end{bmatrix} \begin{bmatrix} -2 & 3 & -4 \end{bmatrix} = \begin{bmatrix} 2 \times (-2) & 2 \times 3 & 2 \times (-4) \\ (-1) \times (-2) & (-1) \times 3 & (-1) \times (-4) \\ 3 \times (-2) & 3 \times 3 & 3 \times (-4) \end{bmatrix}.$$

1.6 Two Properties of Matrix Products

The *transposed product* of matrices satisfies

$$(\mathbf{AB})' = \mathbf{B}'\mathbf{A}', \quad (\mathbf{ABC})' = \mathbf{C}'\mathbf{B}'\mathbf{A}' \tag{1.21}$$

Let $\mathbf{A}$ and $\mathbf{B}$ be matrices of size $n \times m$; let $\mathbf{C}$ and $\mathbf{D}$ be those of $m \times l$. Then, the *product of their sums multiplied by scalars* s, t, u, and v satisfies

$$(s\mathbf{A} + t\mathbf{B})(u\mathbf{C} + v\mathbf{D}) = su\mathbf{AC} + sv\mathbf{AD} + tu\mathbf{BC} + tv\mathbf{BD}. \tag{1.22}$$

1.7 Trace Operator and Matrix Norm

A matrix with the number of rows equivalent to that of columns is said to be *square*. For a square matrix $\mathbf{S} = \begin{bmatrix} s_{11} & s_{12} & \cdots & s_{1n} \\ s_{21} & s_{22} & \cdots & s_{2n} \\ \vdots & \vdots & \ddots & \vdots \\ s_{n1} & s_{n2} & \cdots & s_{nn} \end{bmatrix}$, the elements on the diagonal, i.e., $s_{11}, \ldots, s_{nn}$, are called the *diagonal elements* of $\mathbf{S}$. Their sum is called a *trace* and is denoted as:

$$\mathrm{tr}\mathbf{S} = s_{11} + s_{22} + \cdots + s_{nn}. \tag{1.23}$$

Obviously,

$$\mathrm{tr}\mathbf{S}' = \mathrm{tr}\mathbf{S}. \tag{1.24}$$

The trace fulfills important roles when it is defined for the matrix product. Let us

consider matrix $\mathbf{A} = [\mathbf{a}_1 \cdots \mathbf{a}_m] = \begin{bmatrix} \tilde{\mathbf{a}}_1' \\ \vdots \\ \tilde{\mathbf{a}}_n' \end{bmatrix} = \begin{bmatrix} a_{11} & \cdots & a_{1m} \\ \vdots & \vdots & \vdots \\ a_{n1} & \cdots & a_{nm} \end{bmatrix}$ and an

$m \times n$ matrix $\mathbf{B} = [\mathbf{b}_1 \cdots \mathbf{b}_n] = \begin{bmatrix} \tilde{\mathbf{b}}_1' \\ \vdots \\ \tilde{\mathbf{b}}_m \end{bmatrix} = \begin{bmatrix} b_{11} & \cdots & b_{1n} \\ \vdots & \vdots & \vdots \\ b_{m1} & \cdots & b_{mn} \end{bmatrix}$. Then, $\mathbf{AB}$ and $\mathbf{BA}$

are $n \times n$ and $m \times m$ square matrices, respectively, for which traces can be defined, with

$$\mathbf{AB} = \begin{bmatrix} \tilde{\mathbf{a}}_1' \mathbf{b}_1 & & \# \\ & \ddots & \\ \# & & \tilde{\mathbf{a}}_n' \mathbf{b}_n \end{bmatrix} \quad \text{and} \quad \mathbf{BA} = \begin{bmatrix} \tilde{\mathbf{b}}_1' \mathbf{a}_1 & & \# \\ & \ddots & \\ \# & & \tilde{\mathbf{b}}_m' \mathbf{a}_m \end{bmatrix}.$$

Here, # is used for all elements other than the diagonal ones. In this book, the matrix product precedes the trace operation:

$$\text{tr}\mathbf{AB} = \text{tr}(\mathbf{AB}). \tag{1.25}$$

Thus,

$$\text{tr}\mathbf{AB} = \sum_{i=1}^{n} \tilde{\mathbf{a}}_i' \mathbf{b}_i = \sum_{i=1}^{n} (a_{i1}b_{1i} + \cdots + a_{im}b_{mi}) = \sum_{i=1}^{n} \sum_{j=1}^{m} a_{ij}b_{ji}, \tag{1.26}$$

$$\text{tr}\mathbf{BA} = \sum_{j=1}^{m} \tilde{\mathbf{b}}_j' \mathbf{a}_j = \sum_{j=1}^{m} (b_{j1}a_{1j} + \cdots + b_{jn}a_{nj}) = \sum_{j=1}^{m} \sum_{i=1}^{n} b_{ji}a_{ij} = \sum_{i=1}^{n} \sum_{j=1}^{m} a_{ij}b_{ji}. \tag{1.27}$$

Both are found to be equivalent, i.e.,

$$\text{tr}\mathbf{AB} = \text{tr}\mathbf{BA}, \tag{1.28}$$

and express the *sum* of $a_{ij}b_{ji}$ over all pairs *of* i and j.

It is an important property of the trace that (1.28) implies

$$\text{tr}\mathbf{ABC} = \text{tr}\mathbf{CAB} = \text{tr}\mathbf{BCA}; \quad \text{tr}\mathbf{ABCD} = \text{tr}\mathbf{BCDA} = \text{tr}\mathbf{CDAB} = \text{tr}\mathbf{DABC}. \tag{1.29}$$

Using (1.21), (1.28), and (1.29), we also have

$$\text{tr}(\mathbf{AB})' = \text{tr}\mathbf{B}'\mathbf{A}' = \text{tr}\mathbf{A}'\mathbf{B}'; \quad \text{tr}(\mathbf{ABC})' = \text{tr}\mathbf{C}'\mathbf{B}'\mathbf{A}' = \text{tr}\mathbf{A}'\mathbf{C}'\mathbf{B}' = \text{tr}\mathbf{B}'\mathbf{A}'\mathbf{C}'. \tag{1.30}$$

Substituting $\mathbf{A}'$ into $\mathbf{B}$ in (1.25), we have $\text{tr}\mathbf{A}\mathbf{A}' = \text{tr}\mathbf{A}'\mathbf{A} = \sum_{i=1}^{n} \sum_{j=1}^{m} a_{ij}^2$ which is the sum of the squared elements of $\mathbf{A}$. This is called the *squared norm* of $\mathbf{A}$, i.e., the matrix version of (1.12), and is denoted as $\|\mathbf{A}\|^2$:

$$\|\mathbf{A}\|^2 = \mathrm{tr}\mathbf{A}\mathbf{A}' = \mathrm{tr}\mathbf{A}'\mathbf{A} = \sum_{i=1}^{n}\sum_{j=1}^{m} a_{ij}^2. \tag{1.31}$$

The squared norm of the sum of matrices weighted by scalars is expanded as:

$$
\begin{aligned}
\|s\mathbf{X}+t\mathbf{Y}\|^2 &= \mathrm{tr}(s\mathbf{X}+t\mathbf{Y})'(s\mathbf{X}+t\mathbf{Y}) \\
&= \mathrm{tr}(s^2\mathbf{X}'\mathbf{X}+st\mathbf{X}'\mathbf{Y}+st\mathbf{X}'\mathbf{Y}+t^2\mathbf{Y}'\mathbf{Y}) \\
&= s^2\mathrm{tr}\mathbf{X}'\mathbf{X}+st\mathrm{tr}\mathbf{X}'\mathbf{Y}+st\mathrm{tr}\mathbf{X}'\mathbf{Y}+t^2\mathrm{tr}\mathbf{Y}'\mathbf{Y} \\
&= s^2\mathrm{tr}\mathbf{X}'\mathbf{X}+2st\mathrm{tr}\mathbf{X}'\mathbf{Y}+t^2\mathrm{tr}\mathbf{Y}'\mathbf{Y} \\
&= s^2\|\mathbf{X}\|^2+2st\mathrm{tr}\mathbf{X}'\mathbf{Y}+t^2\|\mathbf{Y}\|^2.
\end{aligned}
\tag{1.32}
$$

1.8   Vectors and Matrices Filled with Ones or Zeros

A *zero vector* refers to a vector filled with zeros. The $p \times 1$ zero vector is denoted as $\mathbf{0}_p$, using the boldfaced zero:

$$\mathbf{0}_p = \begin{bmatrix} 0 \\ 0 \\ \vdots \\ 0 \end{bmatrix}. \tag{1.33}$$

A zero matrix refers to a matrix whose elements are all zeros. In this book, the $n \times p$ zero matrix is denoted as ${}_n\mathbf{O}_p$, using the boldfaced "O":

$$_n\mathbf{O}_p = \begin{bmatrix} 0 & \cdots & 0 \\ \vdots & \cdots & \vdots \\ 0 & \cdots & 0 \end{bmatrix}. \tag{1.34}$$

A *vector of ones* refers to a vector filled with ones. The $n \times 1$ vector of ones is denoted as $\mathbf{1}_n$, with the boldfaced one:

$$\mathbf{1}_n = \begin{bmatrix} 1 \\ 1 \\ \vdots \\ 1 \end{bmatrix}. \tag{1.35}$$

The $n \times p$ matrix filled with ones is given by:

$$1_n 1_p' = \begin{bmatrix} 1 & \cdots & 1 \\ \vdots & \cdots & \vdots \\ 1 & \cdots & 1 \end{bmatrix}.$$
(1.36)

1.9 Special Square Matrices

A square matrix $\mathbf{S} = (s_{ij})$ satisfying

$$\mathbf{S} = \mathbf{S}', \quad \text{i.e.,} \quad s_{ij} = s_{ji}$$
(1.37)

is said to be *symmetric*. An example of a 3×3 symmetric matrix is
$\mathbf{S} = \begin{bmatrix} 2 & -4 & 9 \\ -4 & 6 & -7 \\ 9 & -7 & 3 \end{bmatrix}$. The products of a matrix $\mathbf{A}$ and its transpose, $\mathbf{AA}'$ and $\mathbf{A}'\mathbf{A}$,
are symmetric; using (1.21), we have

$$(\mathbf{AA}')' = (\mathbf{A}')'\mathbf{A}' = \mathbf{AA}' \quad \text{and} \quad (\mathbf{A}'\mathbf{A})' = \mathbf{A}'(\mathbf{A}')' = \mathbf{A}'\mathbf{A}.$$
(1.38)

This has already been exemplified in Ex. 3 (Sect. 1.5).

The elements of $\mathbf{A} = (a_{ij})$ with $i \neq j$ are called the off-diagonal elements of $\mathbf{A}$. A square matrix $\mathbf{D}$ whose off-diagonal elements are all zeros,

$$\mathbf{D} = \begin{bmatrix} d_1 & 0 & \cdots & 0 \\ 0 & d_2 & 0 & \vdots \\ \vdots & 0 & \ddots & 0 \\ 0 & \cdots & 0 & d_p \end{bmatrix},$$
(1.39)

is called a *diagonal matrix*. The products of two diagonal matrices are easily obtained as:

$$\begin{bmatrix} c_1 & 0 & \cdots & 0 \\ 0 & c_2 & 0 & \vdots \\ \vdots & 0 & \ddots & 0 \\ 0 & \cdots & 0 & c_p \end{bmatrix} \begin{bmatrix} d_1 & 0 & \cdots & 0 \\ 0 & d_2 & 0 & \vdots \\ \vdots & 0 & \ddots & 0 \\ 0 & \cdots & 0 & d_p \end{bmatrix} = \begin{bmatrix} c_1 d_1 & 0 & \cdots & 0 \\ 0 & c_2 d_2 & 0 & \vdots \\ \vdots & 0 & \ddots & 0 \\ 0 & \cdots & 0 & c_p d_p \end{bmatrix},$$
(1.40)

$$\mathbf{D}^t = \mathbf{DD} \ldots \mathbf{D} = \begin{bmatrix} d_1^t & 0 & \cdots & 0 \\ 0 & d_2^t & 0 & \vdots \\ \vdots & 0 & \ddots & 0 \\ 0 & \cdots & 0 & d_p^t \end{bmatrix}, \tag{1.41}$$

where $\mathbf{D}^t$ denotes the matrix obtained by multiplying $\mathbf{D}$ t times. Thus, we use the following expression:

$$\mathbf{D}^{-t} = \begin{bmatrix} d_1^{-t} & 0 & \cdots & 0 \\ 0 & d_2^{-t} & 0 & \vdots \\ \vdots & 0 & \ddots & 0 \\ 0 & \cdots & 0 & d_p^{-t} \end{bmatrix}. \tag{1.42}$$

when $-t = -1/2$, the diagonal elements of $\mathbf{D}^{-1/2}$ are $1/\sqrt{d_1}, \ldots, 1/\sqrt{d_p}$.

The *identity matrix* refers to the diagonal matrix whose diagonal elements are all ones. The $p \times p$ identity matrix is denoted as $\mathbf{I}_p$, using the boldfaced "I":

$$\mathbf{I}_p = \begin{bmatrix} 1 & 0 & \cdots & 0 \\ 0 & 1 & 0 & \vdots \\ \vdots & 0 & \ddots & 0 \\ 0 & \cdots & 0 & 1 \end{bmatrix}. \tag{1.43}$$

For example, $\mathbf{I}_3 = \begin{bmatrix} 1 & 0 & 0 \\ 0 & 1 & 0 \\ 0 & 0 & 1 \end{bmatrix}$. An important property of the identity matrix is:

$$\mathbf{AI}_p = \mathbf{A} \quad \text{and} \quad \mathbf{I}_p\mathbf{B} = \mathbf{B}. \tag{1.44}$$

The identity matrix of 1×1 is $\mathbf{I}_1 = 1$, with $s \times 1 = 1 \times s = s$. That is, $\mathbf{I}_p$ is a generalization of one (or unit).

1.10 Bibliographical Notes

Matrix operations, which are necessary for describing multivariate data analysis, but have not been treated in this chapter, are introduced in the following chapters.

In this section, books dealing with matrix algebra are introduced that are useful for statistics. As in the present book, introductory matrix operations are treated intelligibly in Carroll et al. (1997), where geometric illustrations are emphasized.

Banerjee and Roy (2014) and Schott (2005) are among the textbooks recommended for those who finished reading the present book.

Advanced and exhaustive descriptions are found in Harville (1997) and Seber (2008). Though differential calculus for matrices is mainly treated in Magnus and Neudecker (1991), its chapters one to three are useful for capturing matrix algebra. Computational aspects are specially emphasized in Eldén (2007) and Gentle (2007). Formulas for matrix operations are exhaustively listed in Lütkepohl (1996).

Horn and Johnson (2013) is among the most advanced books of matrix algebra, and Golub and van Loan (1996) is among those for matrix computations.

Exercises

1.1   Let $\mathbf{X} = (x_{ij})$ be an $n \times p$ matrix. Express $\mathbf{X}$ using $n \times 1$ vectors $\mathbf{x}_j = [x_{1j}, \ldots, x_{nj}]'$, $j = 1, \ldots, p$, and express $\mathbf{X}$ using $p \times 1$ vectors $\tilde{\mathbf{x}}_i = [x_{i1}, \ldots, x_{ip}]'$, $i = 1, \ldots, n$.

1.2   Let $\mathbf{A} = \begin{bmatrix} -2 & 3 & 9 \\ 1 & -6 & -5 \\ 8 & 2 & 0 \\ -4 & 6 & -3 \end{bmatrix}$ and $\mathbf{B} = \begin{bmatrix} 5 & 7 \\ 6 & -8 \\ -2 & 1 \end{bmatrix}$. Compute $\mathbf{AB}$, $\mathbf{B'B}$, $\mathbf{BB'}$, $\mathbf{A'A}$, and $\mathbf{AA'}$.

1.3   Let $\mathbf{A}_1 = \begin{bmatrix} -2 & 3 & 9 \\ 1 & -6 & -5 \\ 8 & 2 & 0 \\ -4 & 6 & -3 \end{bmatrix}$, $\mathbf{A}_2 = \begin{bmatrix} 7 & -1 & -5 \\ -2 & -2 & 3 \\ 0 & 3 & 9 \\ 6 & -4 & 0 \end{bmatrix}$, $\mathbf{B}_1 = \begin{bmatrix} 2 & -3 \\ -9 & 6 \\ 1 & -7 \end{bmatrix}$,

$\mathbf{B}_2 = \begin{bmatrix} 5 & 7 \\ 6 & -8 \\ -2 & 1 \end{bmatrix}$, $s_1 = -5$, $s_2 = 7$, $t_1 = 3$, and $t_2 = -2$. Compute $(s_1\mathbf{A}_1 + s_2\mathbf{A}_2)(t_1\mathbf{B}_1 + t_2\mathbf{B}_2)$.

1.4   Let $\mathbf{B} = [\mathbf{b}_1, \ldots, \mathbf{b}_m]$. Show $\mathbf{AB} = [\mathbf{Ab}_1, \ldots, \mathbf{Ab}_m]$.

1.5   Prove $\mathrm{tr}\mathbf{ABCDE} = \mathrm{tr}\mathbf{C'B'A'E'D'}$.

1.6. Let $\mathbf{W} = [\mathbf{w}_1, \ldots, \mathbf{w}_p]$. Show that the (j, k) element of $\mathbf{W'X'XW}$ is $\mathbf{w}_j'\mathbf{X'Xw}_k$ and $\mathrm{tr}\mathbf{W'X'XW} = \sum_{j=1}^{p} \mathbf{w}_j'\mathbf{X'Xw}_j$.

1.7   Let a matrix $\mathbf{F}$ satisfy $\frac{1}{n}\mathbf{F'F} = \mathbf{I}_m$. Show $\|\mathbf{X} - \mathbf{FA'}\|^2 = \|\mathbf{X}\|^2 - 2\mathrm{tr}\mathbf{F'XA} + n\|\mathbf{A}\|^2$.

1.8   Compute $\mathbf{1}_4'[4, 2, 6, 1]'$ and $\mathbf{1}_4[4, 2, 6, 1]$.

1.9   Prove $\left(\mathbf{I}_n - \frac{1}{n}\mathbf{1}_n\mathbf{1}_n'\right)'\left(\mathbf{I}_n - \frac{1}{n}\mathbf{1}_n\mathbf{1}_n'\right) = \mathbf{I}_n - \frac{1}{n}\mathbf{1}_n\mathbf{1}_n'$.

1.10 If $\mathbf{F} = [\mathbf{f}_1, \mathbf{f}_2, \mathbf{f}_3]$ with $\|\mathbf{f}_j\|^2 = n$ ($j = 1, 2, 3$) and $\mathbf{f}_j'\mathbf{f}_k = 0$ for $j \neq k$, show $\frac{1}{n}\mathbf{F'F} = \mathbf{I}_3$.

Chapter 2
Intravariable Statistics

This chapter begins with expressing data sets by matrices. Then, we introduce two statistics (statistical indices): average and variance, where the *average* is an index value that represents scores and the *variance* stands for how widely scores disperse. Further, how the original scores are transformed into *centered* and *standard scores* using the average and variance is described.

As the statistics in this chapter summarize the scores *within* a variable, the chapter is named *intra*variable statistics, in contrast to the immediately following chapter entitled *inter*-variable statistics, where the statistics *between* variables would be treated.

2.1 Data Matrices

A *multivariate data* set refers to a set of values arranged in a table whose rows and columns are individuals and variables, respectively. This is illustrated in each panel of Table 2.1. Here, the term "*individuals*" implies the sources from which data are obtained; for example, individuals are participants, cities, and baseball teams, respectively, in panels (A), (B), and (C) of Table 2.1. On the other hand, the term "*variables*" refers to the indices or items for which individuals are measured; for example, variables are Japanese, mathematics, English, and sciences in Table 2.1 (A). By attaching "multi" to "variate," which is a synonym of "variable," we use the adjective "*multivariate*" for the data sets with multiple variables, as shown in Table 2.1. On the other hand, data with a single variable are called "*univariate data*".

© Springer Nature Singapore Pte Ltd. 2016
K. Adachi, *Matrix-Based Introduction to Multivariate Data Analysis*,
DOI 10.1007/978-981-10-2341-5_2

Table 2.1 Three examples of multivariate data

(A) Test scores (Artificial example)

Participant	Item			
	Japan	Mathematics	English	Science
1	82	70	70	76
2	96	65	67	71
3	84	41	54	65
4	90	54	66	80
5	93	76	74	77
6	82	85	60	89

(B) Weather in cities in January (http://www2m.biglobe.ne.jp/ZenTech/world/kion/Japan/Japan.htm)

City	Weather		
	Min °C	Max °C	Precipitation
Sapporo	−7.7	−0.9	110.7
Tokyo	2.1	9.8	48.6
.	.	.	.
.	.	.	.
.	.	.	.
Naha	14.3	19.1	114.5

(C) Team scores (2005 in Japan) (http://npb.jp/bis/2005/stats/)

Team	Averages				
	Win %	Runs	HR	Avg.	ERA
Tigers	0.617	731	140	0.274	3.24
Dragons	0.545	680	139	0.269	4.13
BayStars	0.496	621	143	0.265	3.68
Swallows	0.493	591	128	0.276	4.00
Giants	0.437	617	186	0.260	4.80
Carp	0.408	615	184	0.275	4.80

Let us express a data set as an n-individuals $\times$ p-variables matrix

$$
\mathbf{X} = \begin{bmatrix} x_{11} & \cdots & x_{1j} & \cdots & x_{ip} \\ \vdots & & \vdots & & \vdots \\ x_{i1} & \cdots & x_{ij} & \cdots & x_{ip} \\ \vdots & & \vdots & & \vdots \\ x_{n1} & \cdots & x_{nj} & \cdots & x_{np} \end{bmatrix} = \begin{bmatrix} \mathbf{x}_1, \ldots, \mathbf{x}_j, \ldots, \mathbf{x}_p \end{bmatrix}, \tag{2.1}
$$

whose jth column

Table 2.2   Raw, centered, and standard scores of tests with their averages, variances, and standard deviations (SD) (artificial example)

Student	(A) Raw		(B) Centered		(C) Standard	
	History	Mathematics	History	Mathematics	History	Mathematics
1	66	74	5	−3	0.52	−0.20
2	72	98	11	21	1.15	1.43
3	44	62	−17	−15	−1.78	−1.02
4	58	88	−3	11	−0.31	0.75
5	70	56	9	−21	0.94	−1.43
6	56	84	−5	7	−0.52	0.48
Average	61.0	77.0	0	0	0	0
Variance	91.67	214.33	91.67	214.33	1.00	1.00
SD	9.57	14.64	9.57	14.64	1.00	1.00

$$\mathbf{x}_j = \begin{bmatrix} x_{1j} \\ \vdots \\ x_{nj} \end{bmatrix} = \begin{bmatrix} x_{1j}, \ldots, x_{nj} \end{bmatrix}' \tag{2.2}$$

stands for the *j*th *variable*. Examples of (2.1) are given in Table 2.1(A), (B), (C). A different example is presented in Table 2.2(A), where *n*-individuals and *p*-variables are 6 students and 2 items, respectively, with x_{ij} the score of individual *i* for item *j* and $\mathbf{x}_j$ the 6 × 1 vector containing the scores on the *j*th variable:

$$\underset{6 \times 2}{\mathbf{X}} = \begin{bmatrix} 66 & 74 \\ 72 & 98 \\ \vdots & \vdots \\ 56 & 84 \end{bmatrix} \quad \text{with} \quad \mathbf{x}_1 = \begin{bmatrix} 66 \\ 72 \\ \vdots \\ 56 \end{bmatrix} \quad \text{and} \quad \mathbf{x}_2 = \begin{bmatrix} 74 \\ 98 \\ \vdots \\ 84 \end{bmatrix}.$$

The scores in Table 2.2(B) and (C) will be explained later, in Sects. 2.4 and 2.6.

2.2   Distributions

The distribution of the six students' scores for each variable in Table 2.2(A) is graphically depicted in Fig. 2.1, where those scores are plotted on lines extending from 0 to 100. The distributions allow us to intuitively recognize that [1] their scores in history are lower on *average* than those in mathematics, and [2] the scores *disperse* more widely in mathematics than in history. The statistics related to [1] and [2] are introduced in Sects. 2.3 and 2.5, respectively.

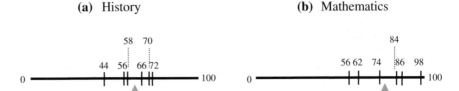

(a) History   **(b)** Mathematics

Fig. 2.1 Distributions of the test scores in Table 2.2(A)

2.3   Averages

Let us consider summarizing n scores into a single statistic. The most popular statistic for the summary is the *average*, which is defined as:

$$\bar{x}_j = \frac{1}{n}(x_{1j} + \cdots + x_{nj}) = \frac{1}{n}\sum_{i=1}^{n} x_{ij} \tag{2.3}$$

for variable j, i.e., the jth column of **X**. For example, the average score in mathematics ($j = 2$) in Table 2.2(A) is $\bar{x}_2 = (74 + 98 + 62 + 88 + 56 + 84)/6 = 77.0$. The average can be rewritten, using the $n \times 1$ *vector of ones* $\mathbf{1}_n = [1,1,\ldots, 1]'$ defined in (1.35): The *sum* $x_{1j} + \cdots + x_{nj}$ is expressed as:

$$\mathbf{1}'_n\mathbf{x}_j = [1, \ldots, 1]\begin{bmatrix} x_{1j} \\ \vdots \\ x_{nj} \end{bmatrix}, \tag{2.4}$$

thus, the *average* (2.3) is also simply expressed as:

$$\bar{x}_j = \frac{1}{n}\mathbf{1}'_n\mathbf{x}_j, \tag{2.5}$$

without using the complicated "Sigma" symbol. For example, the average score in history ($j = 1$) in Table 2.2(A) is expressed as $6^{-1}\mathbf{1}_6'\mathbf{x}_1$ with $\mathbf{x}_1 = [66, 72, 44, 58, 70, 56]'$. The resulting average is $6^{-1}\mathbf{1}_6'\mathbf{x}_1 = 61.0$.

2.4   Centered Scores

The raw scores minus their average are called *centered scores* or *deviations from average*. Let the centered score vector for variable j be denoted as $\mathbf{y}_j = [y_{1j}, \ldots, y_{nj}]'$ ($n \times 1$), which is expressed as:

$$\mathbf{y}_j = \begin{bmatrix} y_{1j} \\ \vdots \\ y_{nj} \end{bmatrix} = \begin{bmatrix} x_{1j} - \bar{x}_j \\ \vdots \\ x_{nj} - \bar{x}_j \end{bmatrix} = \begin{bmatrix} x_{1j} \\ \vdots \\ x_{nj} \end{bmatrix} - \begin{bmatrix} \bar{x}_j \\ \vdots \\ \bar{x}_j \end{bmatrix} = \mathbf{x}_j - \begin{bmatrix} \bar{x}_j \\ \vdots \\ \bar{x}_j \end{bmatrix}. \qquad (2.6)$$

In Table 2.2(B), the centered data for (A) are shown: The centered scores [5, 11, …, −5]′ for history are given by subtracting 61 from all elements of [66, 72, …, 56]′ and the centered scores for mathematics are given by subtracting 77 in a parallel manner.

Here, we rewrite (2.6) in a simpler form. First, let us note that all elements of the subtracted vector $[\bar{x}_j, \dots, \bar{x}_j]'$ in (2.6) are equal to an average $\bar{x}_j$, and thus, that vector can be written as:

$$\begin{bmatrix} \bar{x}_j \\ \vdots \\ \bar{x}_j \end{bmatrix} = \bar{x}_j \mathbf{1}_n = \mathbf{1}_n \times \bar{x}_j, \qquad (2.7)$$

where we have used (1.20). Substituting (2.5) into $\bar{x}_j$ in (2.7), it is rewritten as:

$$\begin{bmatrix} \bar{x}_j \\ \vdots \\ \bar{x}_j \end{bmatrix} = \mathbf{1}_n \times \left(\frac{1}{n} \mathbf{1}'_n \mathbf{x}_j \right) = \frac{1}{n} \mathbf{1}_n \left(\mathbf{1}'_n \mathbf{x}_j \right) = \frac{1}{n} \mathbf{1}_n \mathbf{1}'_n \mathbf{x}_j \qquad (2.8)$$

Here, we have made use of the fact that "× scalar (n^{-1})" can be moved and $\mathbf{A}(\mathbf{BC}) = \mathbf{ABC}$ generally holds for matrices $\mathbf{A}$, $\mathbf{B}$, and $\mathbf{C}$, which implies $\mathbf{1}_n(\mathbf{1}'_n \mathbf{x}_j) = \mathbf{1}_n \mathbf{1}'_n \mathbf{x}_j$. Using (2.8) in (2.6) and noting property (1.44) for an identity matrix, the *centered score vector* (2.6) can be rewritten as:

$$\mathbf{y}_j = \begin{bmatrix} x_{1j} - \bar{x}_j \\ \vdots \\ x_{nj} - \bar{x}_j \end{bmatrix} = \mathbf{x}_j - \begin{bmatrix} \bar{x}_j \\ \vdots \\ \bar{x}_j \end{bmatrix} = \mathbf{I}_n \mathbf{x}_j - \frac{1}{n} \mathbf{1}_n \mathbf{1}'_n \mathbf{x}_j = \left(\mathbf{I}_n - \frac{1}{n} \mathbf{1}_n \mathbf{1}'_n \right) \mathbf{x}_j = \mathbf{J} \mathbf{x}_j,$$

$$(2.9)$$

where $\mathbf{J} = \mathbf{I}_n - n^{-1} \mathbf{1}_n \mathbf{1}'_n$ and we have made use of the fact that $\mathbf{BC} + \mathbf{EC} = (\mathbf{B} + \mathbf{E})\mathbf{C}$ holds for matrices $\mathbf{B}$, $\mathbf{C}$, and $\mathbf{E}$. The matrix $\mathbf{J}$ has a special name and important properties:

Note 2.1. Centering Matrix
It is defined as

$$\mathbf{J} = \mathbf{I}_n - \frac{1}{n}\mathbf{1}_n\mathbf{1}'_n. \tag{2.10}$$

The centering matrix has the following properties:

$$\mathbf{J} = \mathbf{J}' \text{ (symmetric)} \tag{2.11}$$

$$\mathbf{J}^2 = \mathbf{JJ} = \mathbf{J} \tag{2.12}$$

$$\mathbf{1}'_n\mathbf{J} = \mathbf{0}'_n \tag{2.13}$$

Equation (2.11) can easily be found. Equations (2.12) and (2.13) can be proved as:

$$\mathbf{JJ} = (\mathbf{I}_n - n^{-1}\mathbf{1}_n\mathbf{1}'_n)(\mathbf{I}_n - n^{-1}\mathbf{1}_n\mathbf{1}'_n) = \mathbf{I}_n - n^{-1}\mathbf{1}_n\mathbf{1}'_n - n^{-1}\mathbf{1}_n\mathbf{1}'_n + n^{-2}\mathbf{1}_n\mathbf{1}'_n\mathbf{1}_n\mathbf{1}'_n$$
$$= \mathbf{I}_n - n^{-1}\mathbf{1}_n\mathbf{1}'_n - n^{-1}\mathbf{1}_n\mathbf{1}'_n + n^{-2}\mathbf{1}_n(n)\mathbf{1}'_n = \mathbf{I}_n - n^{-1}\mathbf{1}_n\mathbf{1}'_n$$

and

$$\mathbf{1}'_n(\mathbf{I}_n - n^{-1}\mathbf{1}_n\mathbf{1}'_n) = \mathbf{1}'_n - n^{-1}\mathbf{1}'_n\mathbf{1}_n\mathbf{1}'_n = \mathbf{1}'_n - n^{-1}(n)\mathbf{1}'_n = \mathbf{0}'_n,$$

respectively, where $\mathbf{1}'_n\mathbf{1}_n = n$ has been used.

Equations (2.12) and (2.13) further lead to the following important facts:

Note 2.2. Matrices Premultiplied by the Centering Matrix
A matrix $s\mathbf{JA}$ with $\mathbf{A}$ an $n \times p$ matrix and s a scalar satisfies:

$$\mathbf{1}'_n(s\mathbf{JA}) = s\mathbf{1}'_n\mathbf{JA} = \mathbf{0}'_p, \tag{2.14}$$

$$\mathbf{J}(s\mathbf{JA}) = s\mathbf{JJA} = s\mathbf{JA}. \tag{2.15}$$

When $\mathbf{A}$ is an $n \times 1$ vector $\mathbf{a}$, those equations are rewritten as $\mathbf{1}'_n(s\mathbf{Ja}) = 0$ and $\mathbf{J}(s\mathbf{Ja}) = s\mathbf{Ja}$, respectively.

Comparing (2.9) with (2.14), we can find that the sum and average of centered scores are always *zero*:

$$\mathbf{1}_n'\mathbf{y}_j = \frac{1}{n}\mathbf{1}_n'\mathbf{y}_j = 0. \tag{2.16}$$

This is shown in the row named "Average" in Table 2.1(B). Figure 2.3(B) (*on a later page*) illustrates (2.16); the centered scores are distributed with their average being the zero which is a *center between negative and positive values*. This property provides the name "centered scores," and the transformation of raw scores into centered ones is called *centering*. Comparing (2.9) with (2.15), we also find:

$$\mathbf{J}\mathbf{y}_j = \mathbf{y}_j. \tag{2.17}$$

The centered score vector, premultiplied by the centering matrix, remains unchanged.

2.5   Variance and Standard Deviation

The locations of averages in the distributions of scores are indicated by triangles in Fig. 2.1, which do not stand for how widely scores disperse. The most popular statistic for indicating dispersion is *variance*. It is defined using the *sum of squared distances* between *scores* and *their average*, which is illustrated in Fig. 2.2. Denoting the variance for variable j as v_{jj}, it is formally expressed as:

$$\begin{aligned} v_{jj} &= \frac{1}{n}\left\{|x_{1j} - \bar{x}_j|^2 + \cdots + |x_{nj} - \bar{x}_j|^2\right\} \\ &= \frac{1}{n}\left\{(x_{1j} - \bar{x}_j)^2 + \cdots + (x_{nj} - \bar{x}_j)^2\right\} = \frac{1}{n}\sum_{i=1}^{n}(x_{ij} - \bar{x}_j)^2, \end{aligned} \tag{2.18}$$

where the same subscript j is used twice as v_{jj}, for the sake of accordance with the related statistic introduced in the next chapter. The variance of the scores for mathematics in Table 2.2(A) is obtained as $6^{-1}\{(74 - 77)^2 + (98 - 77)^2 + \cdots + (84 - 77)^2\} = 214.33$, for example.

Fig. 2.2 Distances of scores to their average, which are squared, summed, and divided by n, to give the variance of the mathematics scores in Table 2.2(A)

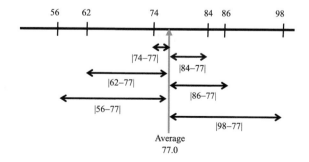

To express (2.18) in *vector* form, we should note that it can be rewritten as:

$$v_{jj} = \frac{1}{n} [x_{1j} - \bar{x}_j, \ldots, x_{nj} - \bar{x}_j] \begin{bmatrix} x_{1j} - \bar{x}_j \\ \vdots \\ x_{nj} - \bar{x}_j \end{bmatrix}. \tag{2.19}$$

Comparing (2.19) with $\begin{bmatrix} x_{1j} - \bar{x}_j \\ \vdots \\ x_{nj} - \bar{x}_j \end{bmatrix} = \mathbf{Jx}_j$ in (2.9), the *variance* (2.18) or (2.19) is

expressed as:

$$v_{jj} = \frac{1}{n} (\mathbf{Jx}_j)' \mathbf{Jx}_j = \frac{1}{n} ||\mathbf{Jx}_j||^2 = \frac{1}{n} \mathbf{x}_j' \mathbf{J}' \mathbf{Jx}_j = \frac{1}{n} \mathbf{x}_j' \mathbf{JJx}_j = \frac{1}{n} \mathbf{x}_j' \mathbf{Jx}_j, \tag{2.20}$$

where (1.12), (2.11), and (2.12) have been used. Further, we can use (2.9) in (2.20) to rewrite it as:

$$v_{jj} = \frac{1}{n} \mathbf{x}_j' \mathbf{Jx}_j = \frac{1}{n} \mathbf{x}_j' \mathbf{J}' \mathbf{Jx}_j = \frac{1}{n} \mathbf{y}_j' \mathbf{y}_j = \frac{1}{n} ||\mathbf{y}_j||^2. \tag{2.21}$$

The variance of raw scores is expressed using their *centered score vector* simply as $n^{-1}||\mathbf{y}_j||^2$. We can also find in (2.20) and (2.21) that the variance is the squared length of vector $\mathbf{y}_j = \mathbf{Jx}_j$ divided by n.

How is the variance of the centered scores (rather than raw scores) expressed? To find this, we substitute the centered score vector $\mathbf{y}_j$ for $\mathbf{x}_j$ in the variance (2.20). Then, we use (2.17) and (2.9) to get:

$$\frac{1}{n} \mathbf{y}_j' \mathbf{J}' \mathbf{Jy}_j = \frac{1}{n} \mathbf{y}_j' \mathbf{y}_j = \frac{1}{n} \mathbf{x}_j' \mathbf{J}' \mathbf{Jx}_j, \tag{2.22}$$

which is equal to (2.20); the variance of the centered scores equals that for their raw scores.

The square root of variance

$$\sqrt{v_{jj}} = \sqrt{\frac{1}{n} \mathbf{x}'_j \mathbf{Jx}_j} = \frac{1}{\sqrt{n}} ||\mathbf{Jx}_j|| = \frac{1}{\sqrt{n}} ||\mathbf{y}_j|| \tag{2.23}$$

is called the *standard deviation*, which is also used for reporting the dispersion of data. We can find in (2.23) that the standard deviation is the length of vector $\mathbf{y}_j = \mathbf{Jx}_j$ divided by $n^{1/2}$.

2.6 Standard Scores

The centered scores (i.e., the raw scores minus their average) divided by their standard deviation are called *standard scores* or *z-scores*. Let the standard score vector for variable j be denoted by $\mathbf{z}_j = [z_{1j}, \ldots, z_{nj}]'$, which is expressed as:

$$\mathbf{z}_j = \begin{bmatrix} (x_{1j} - \bar{x}_j)/\sqrt{v_{jj}} \\ \vdots \\ (x_{nj} - \bar{x}_j)/\sqrt{v_{jj}} \end{bmatrix} = \frac{1}{\sqrt{v_{jj}}} \begin{bmatrix} x_{1j} - \bar{x}_j \\ \vdots \\ x_{nj} - \bar{x}_j \end{bmatrix} = \frac{1}{\sqrt{v_{jj}}} \mathbf{J}\mathbf{x}_j = \frac{1}{\sqrt{v_{jj}}} \mathbf{y}_j, \qquad (2.24)$$

where we have used (2.9). In Table 2.2(C), the standard scores for (A) are shown; the standard scores $[-0.2, \ldots, 0.48]'$ for mathematics are given by dividing its centered scores (B) by 14.64. Transforming raw scores into standard ones is called *standardization*.

Standard scores have two important properties. One is that the sum and average of standard scores are always *zero*, as are those of centered scores:

$$\mathbf{1}'_n \mathbf{z}_j = \frac{1}{n} \mathbf{1}'_n \mathbf{z}_j = 0, \qquad (2.25)$$

which follows from (2.16) and (2.24). The other property is that the variance of standard scores is always *one*, which is shown as follows: The substitution of $\mathbf{z}_j$ into $\mathbf{x}_j$ in (2.20) leads to the variance of standard scores being expressed as $n^{-1}\mathbf{z}_j'\mathbf{J}'\mathbf{J}\mathbf{z}_j = n^{-1}\mathbf{z}_j'\mathbf{z}_j$, where we have used $\mathbf{z}_j = \mathbf{J}\mathbf{z}_j$, following from the use of (2.17) in (2.24). Further, the variance can be rewritten, using (1.12), (2.21), and (2.24), as:

$$\frac{1}{n}\mathbf{z}_j'\mathbf{J}'\mathbf{J}\mathbf{z}_j = \frac{1}{n}\mathbf{z}_j'\mathbf{z}_j = \frac{1}{n}\|\mathbf{z}_j\|^2 = \frac{1}{nv_{jj}}\mathbf{y}_j'\mathbf{y}_j = \frac{n}{n\|\mathbf{y}_j\|^2}\mathbf{y}_j'\mathbf{y}_j = 1. \qquad (2.26)$$

This also implies that the length of every standard score vector is always $\|\mathbf{z}_j\| = n^{1/2}$.

2.7 What Centering and Standardization Do for Distributions

The properties of centered and standard scores shown with (2.16), (2.22), (2.25), and (2.26) are summarized in Table 2.3.

Table 2.3 Averages and variances of centered and standard scores

	Average	Variance
Centered scores	0	Variance of raw scores
Standard scores	0	1

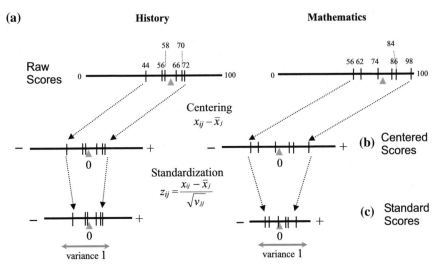

Fig. 2.3 Distributions of raw, centered, and standard scores in Table 2.2

Figure 2.3 illustrates the roles that centering and standardization (i.e., transforming raw scores into centered and standard ones) perform for the distributions of data: *Centering* simply *moves the distributions* of raw scores so that the average of the moved distributions is zero, and *standardization* further *accommodates the scale* of the moved distributions so that their variances are equal to one. The standard scores are unified among different variables so that the averages and variances are zero and one, respectively; thus, the greatness/smallness of the standard scores can be compared reasonably between variables.

2.8   Matrix Representation

We will now introduce a basic formula in matrix algebra:

Note 2.3. A Property of Matrix Product
If $\mathbf{A}$ is a matrix of $n \times m$ and $\mathbf{b}_1, \ldots, \mathbf{b}_K$ are $m \times 1$ vectors, then

$$[\mathbf{Ab}_1, \ldots, \mathbf{Ab}_K] = \mathbf{A}[\mathbf{b}_1, \ldots, \mathbf{b}_K]. \tag{2.27}$$

Using this and (2.5), the $1 \times p$ row vector containing the *averages* of p-variables is expressed as:

$$\left[\bar{x}_1, \ldots, \bar{x}_p\right] = \left[\frac{1}{n}\mathbf{1}'_n\mathbf{x}_1, \ldots, \frac{1}{n}\mathbf{1}'_n\mathbf{x}_p\right] = \frac{1}{n}\mathbf{1}'_n\left[\mathbf{x}_1, \ldots, \mathbf{x}_p\right] = \frac{1}{n}\mathbf{1}'_n\mathbf{X}. \qquad (2.28)$$

For example, when $\mathbf{X}$ consists of the six students' scores in Table 2.2(A), $6^{-1}\mathbf{1}'_6\mathbf{X} = [61.0, 71.0]$.

Let $\mathbf{Y} = [\mathbf{y}_1, \ldots, \mathbf{y}_p]$ denote the $n \times p$ matrix of *centered scores* whose jth column is defined as (2.9) for the corresponding column of $\mathbf{X}$. Then, we can use (2.9) and (2.27) to express $\mathbf{Y}$ as:

$$\mathbf{Y} = \left[\mathbf{J}\mathbf{x}_1, \ldots, \mathbf{J}\mathbf{x}_p\right] = \mathbf{J}\left[\mathbf{x}_1, \ldots, \mathbf{x}_p\right] = \mathbf{J}\mathbf{X}, \qquad (2.29)$$

an example of which is presented in Table 2.2(B).

Let $\mathbf{Z} = [\mathbf{z}_1, \ldots, \mathbf{z}_p]$ denote the $n \times p$ matrix of *standard scores* whose jth column is defined as (2.24) for the corresponding columns of $\mathbf{X}$ and $\mathbf{Y}$. Then, $\mathbf{Z}$ is expressed as:

$$\mathbf{Z} = \left[\frac{1}{\sqrt{v_{11}}}\mathbf{y}_1, \ldots, \frac{1}{\sqrt{v_{pp}}}\mathbf{y}_p\right] = \left[\mathbf{y}_1, \ldots, \mathbf{y}_p\right]\begin{bmatrix} \frac{1}{\sqrt{v_{11}}} & & \\ & \ddots & \\ & & \frac{1}{\sqrt{v_{pp}}} \end{bmatrix} = \mathbf{Y}\mathbf{D}^{-1}. \quad (2.30)$$

Here, the blanks in $\begin{bmatrix} \frac{1}{\sqrt{v_{11}}} & & \\ & \ddots & \\ & & \frac{1}{\sqrt{v_{pp}}} \end{bmatrix}$ stand for the corresponding elements being

zeros and $\mathbf{D} = \begin{bmatrix} \sqrt{v_{11}} & & \\ & \ddots & \\ & & \sqrt{v_{pp}} \end{bmatrix}$ is the $p \times p$ diagonal matrix whose diagonal

elements are the standard deviations for p variables: We should recall (1.42) to notice that $\mathbf{D}^{-1}$ is the diagonal matrix whose diagonal elements are the reciprocals of the standard deviations. Those readers who have difficulties in understanding (2.30) should note the following simple example with $\mathbf{Y}$ being 3×2:

$$\mathbf{Y}\mathbf{D}^{-1} = \begin{bmatrix} y_{11} & y_{12} \\ y_{21} & y_{22} \\ y_{31} & y_{32} \end{bmatrix}\begin{bmatrix} \frac{1}{\sqrt{v_{11}}} & \\ & \frac{1}{\sqrt{v_{22}}} \end{bmatrix} = \begin{bmatrix} y_{11}/\sqrt{v_{11}} & y_{12}/\sqrt{v_{22}} \\ y_{21}/\sqrt{v_{11}} & y_{22}/\sqrt{v_{22}} \\ y_{31}/\sqrt{v_{11}} & y_{23}/\sqrt{v_{22}} \end{bmatrix}, \qquad (2.31)$$

which illustrates the equalities in (2.30) in the reverse order. The standard score matrix $\mathbf{Z}$ can also be expressed as:

$$\mathbf{Z} = \mathbf{J}\mathbf{X}\mathbf{D}^{-1}, \qquad (2.32)$$

using (2.29) in (2.30).

Table 2.4 Data matrix $\mathbf{X}$ of 5 persons × 3 variables

Person	Height	Weight	Sight
Bill	172	64	0.8
Brian	168	70	1.4
Charles	184	80	1.2
Keith	176	64	0.2
Michael	160	62	1.0

2.9   Bibliographical Notes

Carroll et al. (1997, Chap. 3) and Reyment and Jöreskog (1996, Chap. 2) are among the literature in which the matrix expressions of intravariable statistics are intelligibly treated.

Exercises

2.1. Compute $\mathbf{J} = \mathbf{I}_5 - 5^{-1}\mathbf{1}_5\mathbf{1}_5'$ and obtain the centered score matrix $\mathbf{Y} = \mathbf{J}\mathbf{X}$ for the 5 × 3 matrix $\mathbf{X}$ in Table 2.4.

2.2. Compute the variance $v_{jj} = 5^{-1}\mathbf{x}_j'\mathbf{J}\mathbf{x}_j$ $(j = 1, 2, 3)$, the diagonal matrix

$$\mathbf{D}^{-1} = \begin{bmatrix} \frac{1}{\sqrt{v_{11}}} & & \\ & \frac{1}{\sqrt{v_{22}}} & \\ & & \frac{1}{\sqrt{v_{33}}} \end{bmatrix}, \text{ and the standard score matrix } \mathbf{Z} = \mathbf{J}\mathbf{X}\mathbf{D}^{-1} \text{ for}$$

$\mathbf{X} = [\mathbf{x}_1, \mathbf{x}_2, \mathbf{x}_3]$ (5 × 3) in Table 2.4.

2.3. Discuss the benefits of standardizing the data in Table 2.4.

2.4. If the average for each column of $\mathbf{Y}$ $(n \times p)$ is zero, show that the average for each column of $\mathbf{YA}$ is also zero.

2.5. Let $\mathbf{Z}$ be an n-individuals × p-variables matrix containing standard scores. Show that $\|\mathbf{Z}\|^2 = \mathrm{tr}\mathbf{Z}'\mathbf{Z} = \mathrm{tr}\mathbf{Z}\mathbf{Z}' = np$.

2.6. Let $\mathbf{x}$ be an $n \times 1$ vector with $v = n^{-1}\mathbf{x}'\mathbf{J}\mathbf{x}$ the variance of the elements in $\mathbf{x}$. Show that the variance of the elements in $b\mathbf{x} + c\mathbf{1}_n$ is b^2v.

2.7. Let $\mathbf{y} = [y_1, ..., y_n]'$ contain centered scores. Show that the average of the elements in $-\mathbf{y} + c\mathbf{1}_n = [-y_1 + c, ..., -y_n + c]'$ is c, and their variance is equivalent to that for $\mathbf{y}$.

2.8. Let $\mathbf{z} = [z_1, ..., z_n]'$ contain standard scores. Show that the average of the elements in $b\mathbf{z} + c\mathbf{1}_n = [bz_1 + c, ..., bz_n + c]'$ is c, and their standard deviation is b.

Chapter 3
Inter-variable Statistics

In the previous chapter, we described the two statistics, average and variance, which summarize the distribution of scores *within* a variable. In this chapter, we introduce *covariance* and the *correlation coefficient*, which are the *inter*-variable statistics indicating the relationships *between* two variables. Finally, the rank of a matrix, an important notion in linear algebra, is introduced.

3.1 Scatter Plots and Correlations

As in the previous chapter, we consider an n-individuals $\times$ p-variables data matrix (2.1), i.e., $\mathbf{X} = [\mathbf{x}_1, \ldots, \mathbf{x}_j, \ldots, \mathbf{x}_p]$. An example of $\mathbf{X}$ is presented in Table 3.1(A). There, n-individuals are 10 kinds of *foods* ($n = 10$) and p-variables are their *sweetness*, degree of *spice*, and *sales* ($p = 3$). The relationship between two variables, j and k ($j, k = 1, \ldots, p$), which is called a *correlation*, can be captured by the *scatter plot* in which n-individuals are plotted as points with their coordinates [x_{ij}, x_{ik}], $i = 1, \ldots, n$, where x_{ij} and x_{ik} are the scores of individual i for variables j and k, respectively. The plots for Table 3.1(A) are shown in Fig. 3.1a–c. For example, Fig. 3.1c is the scatter plot for *sweetness* and *sales*, where 10 foods are plotted as points with their coordinates [x_{i1}, x_{i3}], $i = 1, \ldots, 10$, i.e., [32, 62] [28, 83], ..., [22, 63].

In Fig. 3.1, distributions are easily captured by the ellipses roughly surrounding the points. The slope of the ellipse in Fig. 3.1c shows that *sales* are positively proportional to *sweetness*. Two variables with such a proportional relation are said to have a *positive correlation*. The inverse relationship is found between *spice* and *sweetness* in Fig. 3.1a; the former tends to decrease with an increase in the latter, which is expressed as the variables having a *negative correlation*. On the other hand, the ellipse in Fig. 3.1b is not sloped; *no correlation* is found between *spice* and *sales*.

© Springer Nature Singapore Pte Ltd. 2016
K. Adachi, *Matrix-Based Introduction to Multivariate Data Analysis*,
DOI 10.1007/978-981-10-2341-5_3

Table 3.1 Data matrices of 10 individuals × 3 variables (artificial example)

Food	(A) Raw score: **X**			(B) Centered scores: **Y**			(C) Standard scores: **Z**		
	$\mathbf{x}_1$	$\mathbf{x}_2$	$\mathbf{x}_3$	$\mathbf{y}_1$	$\mathbf{y}_2$	$\mathbf{y}_3$	$\mathbf{z}_1$	$\mathbf{z}_2$	$\mathbf{z}_3$
	Sweet	*Spice*	*Sales*	*Sweet*	*Spice*	*Sales*	*Sweet*	*Spice*	*Sales*
1	32	10	62	3.5	−7.7	−5.5	0.69	−1.77	−0.35
2	28	20	83	−0.5	2.3	15.5	−0.10	0.53	0.98
3	20	19	34	−8.5	1.3	−33.5	−1.68	0.30	−2.11
4	34	21	91	5.5	3.3	23.5	1.09	0.76	1.48
5	25	16	53	−3.5	−1.7	−14.5	−0.69	−0.39	−0.91
6	35	14	70	6.5	−3.7	2.5	1.28	−0.85	0.16
7	25	20	62	−3.5	2.3	−5.5	−0.69	0.53	−0.35
8	30	18	73	1.5	0.3	5.5	0.30	0.07	0.35
9	34	13	84	5.5	−4.7	16.5	1.09	−1.08	1.04
10	22	26	63	−6.5	8.3	−4.5	−1.28	1.90	−0.28
Average	28.5	17.7	67.5	0.00	0.00	0.00	0.00	0.00	0.00
Variance	25.65	19.01	251.45	25.65	19.01	251.45	1.00	1.00	1.00
SD	5.06	4.36	15.86	5.06	4.36	15.86	1.00	1.00	1.00

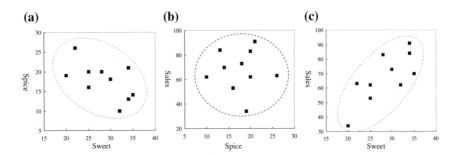

Fig. 3.1 Scatter plots for the pairs of the variables in Table 3.1(A)

3.2   Covariance

The correlation between two variables j and k can be indicated by a *covariance*, which is defined as:

$$v_{jk} = \frac{1}{n}\sum_{i=1}^{n}(x_{ij} - \bar{x}_j)(x_{ik} - \bar{x}_k), \tag{3.1}$$

the average of the product $(x_{ij} - \bar{x}_j) \times (x_{ik} - \bar{x}_k)$ over $i = 1, \ldots, n$, with $x_{ij} - \bar{x}_j$ and $x_{ik} - \bar{x}_k$ being the centered scores for variables j and k, respectively. It takes a *positive* value when variables j and k have a positive correlation, while v_{jk} shows a *negative* value when the variables have a negative correlation, and v_{jk} is close to

Table 3.2 The covariance matrix for Table 3.1(A)

Variable	Sweet	Spice	Sales
Sweet	25.65	−12.65	60.15
Spice	−12.65	19.01	0.15
Sales	60.15	0.15	251.45

zero when j and k have no correlation. This property can be verified as follows: The covariance between the variables *sweet* and *spice* in Table 3.1(A) is computed as:

$$v_{12} = \frac{1}{10} \{3.5 \times (-7.7) + (-0.5) \times 2.3 + \cdots + (-6.5) \times 8.3\} = -12.65, \quad (3.2)$$

using the centered scores in (B). Those variables are found negatively correlated in Fig. 3.1a, and their covariance (3.2) is also negative. In a parallel manner, we can have the positive $v_{13} = 60.15$, which is the covariance between *sweet* and *sales* correlated positively, as in Fig. 3.1b, while we can find $v_{23} = 0.15$ closing to zero, which is the covariance between *spices* and *sales* uncorrelated, as in Fig. 3.1c. Those covariances are summarized in Table 3.2.

To express (3.1) in a *vector* form, (3.1) can be rewritten as:

$$v_{jk} = \frac{1}{n} \begin{bmatrix} x_{1j} - \bar{x}_j, \ldots, x_{nj} - \bar{x}_j \end{bmatrix} \begin{bmatrix} x_{1k} - \bar{x}_k \\ \vdots \\ x_{nk} - \bar{x}_k \end{bmatrix}. \quad (3.3)$$

Here, (2.9) should be recalled, noticing that the right matrix in (3.3) can be

expressed as $\begin{bmatrix} x_{1k} - \bar{x}_k \\ \vdots \\ x_{nk} - \bar{x}_k \end{bmatrix} = \mathbf{y}_k = \mathbf{Jx}_k$ by replacing the subscript j in (2.9) by k, with

$\mathbf{y}_k$ the $n \times 1$ vector of centered scores corresponding to the raw scores $\mathbf{x}_k$ and $\mathbf{J}$ the $n \times n$ centering matrix defined in (2.10). Thus, (3.3) is rewritten as:

$$v_{jk} = \frac{1}{n} (\mathbf{Jx}_j)' \mathbf{Jx}_k = \frac{1}{n} \mathbf{x}_j' \mathbf{J}' \mathbf{Jx}_k = \frac{1}{n} \mathbf{x}_j' \mathbf{JJx}_k = \frac{1}{n} \mathbf{x}_j' \mathbf{Jx}_k = \frac{1}{n} \mathbf{y}_j' \mathbf{y}_k, \quad (3.4)$$

in which (2.9), (2.11) and (2.12) have been used. That is, the covariance between variables j and k is the *inner product* of the *centered score vectors* $\mathbf{y}_j = \mathbf{Jx}_j$ and $\mathbf{y}_k = \mathbf{Jx}_k$ divided by n.

A p-variables × p-variables matrix containing covariances, as in Table 3.2, is called a *covariance matrix*. Each of its diagonal elements expresses the covariance for the same variable.

$$v_{jj} = \frac{1}{n} \mathbf{x}_j' \mathbf{Jx}_j = \frac{1}{n} \mathbf{y}_j' \mathbf{y}_j, \quad (3.5)$$

which equals (2.21), i.e., the *variance* for that variable. This implies that covariance is an extension of the concept of variance for two variables. We should thus consider *covariance* as including *variance* as a *special case*.

Let us substitute the centered score vector $\mathbf{y}_j$ for $\mathbf{x}_j$ in covariance (3.4). Then, we have

$$v_{jk} = \frac{1}{n}\left(\mathbf{Jy}_j\right)'\mathbf{Jy}_k = \frac{1}{n}\mathbf{y}_j'\mathbf{Jy}_k = \frac{1}{n}\mathbf{y}_j'\mathbf{y}_k = \frac{1}{n}\mathbf{x}_j'\mathbf{J}'\mathbf{Jx}_k, \tag{3.6}$$

which equals (3.4); the covariance of centered scores equals that of their raw scores.

Though the covariance is a theoretically important statistic, an inconvenient property of the covariance is that its value does not allow us to easily capture *how strong* the positive/negative *correlations* between variables are. For example, (3.2) shows that *sweet* and *spice* are negatively correlated with its sign (negative), but its absolute value of 12.65 does not show to what degree those variables are negatively correlated. This problem can easily be dealt with by modifying the covariance into a correlation coefficient, as described next.

3.3 Correlation Coefficient

A *correlation coefficient* between variables j and k is given by dividing the covariance (3.1) or (3.4) by the square roots of the variances of variables j and k (i.e., by the standard deviations of j and k). That is, the coefficient is defined using (2.23) as:

$$r_{jk} = \frac{v_{jk}}{\sqrt{v_{jj}}\sqrt{v_{kk}}} = \frac{\frac{1}{n}\mathbf{x}_j'\mathbf{Jx}_k}{\sqrt{\frac{1}{n}\mathbf{x}_j'\mathbf{Jx}_j}\sqrt{\frac{1}{n}\mathbf{x}_k'\mathbf{Jx}_k}}. \tag{3.7}$$

Here, it should be noted that n^{-1} and the two square roots of n^{-1} in the right-hand side can be canceled out; (3.7) is rewritten as:

$$r_{jk} = \frac{\mathbf{x}_j'\mathbf{Jx}_k}{\sqrt{\mathbf{x}_j'\mathbf{Jx}_j}\sqrt{\mathbf{x}_k'\mathbf{Jx}_k}} = \frac{\mathbf{x}_j'\mathbf{Jx}_k}{\sqrt{(\mathbf{Jx}_j)'\mathbf{Jx}_j}\sqrt{(\mathbf{Jx}_k)'\mathbf{Jx}_k}} = \frac{(\mathbf{Jx}_j)'\mathbf{Jx}_k}{\|\mathbf{Jx}_j\|\|\mathbf{Jx}_k\|} = \frac{\mathbf{y}_j'\mathbf{y}_k}{\|\mathbf{y}_j\|\|\mathbf{y}_k\|} \tag{3.8}$$

which shows that the correlation coefficient is defined as the *inner product* of the *centered score* vectors $\mathbf{y}_j = \mathbf{Jx}_j$ and $\mathbf{y}_k = \mathbf{Jx}_k$ divided by their *lengths*. The coefficient (3.7) or (3.8) is also called Pearson's product–moment correlation coefficient, named after Karl Pearson (1857–1936, British statistician), who derived the coefficient.

The correlation coefficient r_{jk} between variables j and k has the following properties:

[1] Its absolute value cannot exceed one with $-1 \leq r_{jk} \leq 1$.
[2] It takes a *positive value* when j and k have a *positive correlation*.
[3] It takes a *negative value* when j and k have a *negative correlation*.
[4] It approximates *zero* when j and k have *no correlation*.

Property [1], which is not possessed by covariances, allows us to easily capture the strength of the correlation, as illustrated in the following paragraph. Before that, we will show some numerical examples. The coefficient between *sweet* and *spice* can be obtained as:

$$r_{12} = \frac{v_{12}}{\sqrt{v_{11}}\sqrt{v_{22}}} = \frac{-12.65}{\sqrt{25.65}\sqrt{19.01}} = -0.57, \tag{3.9}$$

using (3.2) and v_{jj} (Table 3.2) in (3.7). The value from (3.9) shows that *sweetness* is negatively correlated with *spice*. In a parallel manner, the coefficient between *spice* and *sales* is computed as:

$$r_{23} = \frac{v_{23}}{\sqrt{v_{22}}\sqrt{v_{33}}} = \frac{0.15}{\sqrt{19.01}\sqrt{251.45}} = 0.00, \tag{3.10}$$

indicating that *spice* and *sales* have no correlation, while r_{13} is found to be 0.75, which shows that *sweetness* is positively correlated with *sales*.

The *upper limit* $r_{jk} = 1$, shown in Property [1], is attained for $\mathbf{y}_j = a\mathbf{y}_k$ with $a > 0$; its substitution in (3.8) leads to $r_{jk} = a\mathbf{y}_j'\mathbf{y}_j/(\|\mathbf{y}_j\| \times a\|\mathbf{y}_j\|) = 1$. On the other hand, the *lower limit* $r_{jk} = -1$ is attained when $\mathbf{y}_j = b\mathbf{y}_k$ with $b < 0$. The scatter plots of the variables with $r_{jk} = 1$ and $r_{jk} = -1$ are presented at the far left and right in Fig. 3.2, respectively. In each plot, all points are located on a straight line. Any r_{jk} takes a value between the two extremes ± 1, as shown in Fig. 3.2. There, we can find that the *strength of a positive or negative correlation* is captured by noting to what degree r_{jk} is *far from the 0 point* corresponding to no correlation and *close*

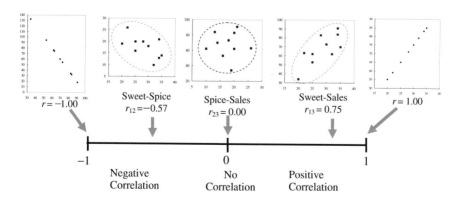

Fig. 3.2 Scatter plots and the corresponding correlation coefficients with their locations on the scale ranging from -1 to 1

Table 3.3 The correlation matrix for Table 3.1

Variable	Sweet	Spice	Sales
Sweet	1.00	−0.57	0.75
Spice	−0.57	1.00	0.00
Sales	0.75	0.00	1.00

to ± 1. For example, $r_{13} = 0.75$ is close to 1, which indicates *sweetness* and *sales* are highly positively correlated. On the other hand, $r_{12} = -0.57$ is located a little to the left of the midpoint between 0 and -1, which indicates that *sweetness* and *spice* have a fairly negative correlation, though the correlation might not be said to be high.

The correlation coefficients among the variables in Table 3.1(A) are shown in Table 3.3. Such a *p*-variables × *p*-variables matrix is called a *correlation matrix*. As found in its diagonal elements, the correlation for the same variable is always one: $r_{jj} = \mathbf{y}_j'\mathbf{y}_j/(\|\mathbf{y}_j\|\|\mathbf{y}_j\|) = 1$.

3.4   Variable Vectors and Correlations

In this section, vector $\mathbf{y}_j = [y_{1j}, \ldots, y_{nj}]'$ is regarded as the line extending from $[0, \ldots, 0]'$ to $[y_{1j}, \ldots, y_{nj}]'$. As explained in "Appendix A.1.1" with ("A.1.3"), the cosine of the angle between two vectors is equal to the division of their inner product by the product of their lengths. Thus, (3.8) shows that the correlation coefficient equals the *cosine of the angle* θ_{jk} between vectors $\mathbf{y}_j = \mathbf{J}\mathbf{x}_j$ and $\mathbf{y}_k = \mathbf{J}\mathbf{x}_k$ with

$$r_{jk} = \frac{\mathbf{y}_j'\mathbf{y}_k}{\|\mathbf{y}_j\|\|\mathbf{y}_k\|} = \cos\theta_{jk} \begin{cases} = 1 & \text{if} & \theta_{jk} = 0° \\ > 0 & \text{if} & \theta_{jk} < 90° \\ = 0 & \text{if} & \theta_{jk} = 90° \\ < 0 & \text{if} & \theta_{jk} > 90° \\ = -1 & \text{if} & \theta_{jk} = 180° \end{cases} \tag{3.11}$$

Here, the right-hand side shows the relationships of θ_{jk} to $\cos\theta_{jk}$. In (3.11), we can find that the angles between the vectors of *positively correlated* variables are *less than 90°*, while the angles between *negatively correlated* ones are *more than 90°*, and the angle between *uncorrelated* variable vectors is *90°*. When the angles between two vectors are 90°, they are said to be *orthogonal*. Using (3.11), we can find that $r_{12} = -0.57$, $r_{13} = 0.75$, and $r_{23} = 0.00$ lead to $\theta_{12} = 125°$, $\theta_{13} = 41°$, and $\theta_{23} = 90°$, respectively, which allows us to visually illustrate the variable vectors as in Fig. 3.3.

Fig. 3.3 Illustration of correlations with variable vectors in a three-dimensional space

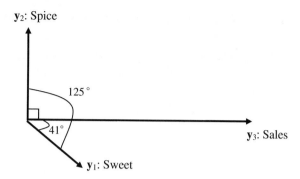

3.5 Covariances and Correlations for Standard Scores

Let us recall (2.24), i.e., the definition of standard scores. By substituting (2.24) into x_j in (3.5), the covariance between standard score vectors z_j and z_k is expressed as:

$$v_{jk}^{[z]} = \frac{1}{n}\left(\mathbf{J}\mathbf{z}_j\right)'\mathbf{J}\mathbf{z}_k = \frac{1}{n}\mathbf{z}_j'\mathbf{z}_k = \frac{1}{n}\left(\frac{1}{\sqrt{v_{jj}}}\mathbf{y}_j\right)'\frac{1}{\sqrt{v_{kk}}}\mathbf{y}_k = \frac{\frac{1}{n}\mathbf{y}_j'\mathbf{y}_k}{\sqrt{v_{jj}}\sqrt{v_{kk}}} = \frac{v_{jk}}{\sqrt{v_{jj}}\sqrt{v_{kk}}}.$$

$$(3.12)$$

Here, $z_j = \mathbf{J}z_k$ has been used. We can find (3.12) be equal to (3.7); the covariance of standard scores is equivalent to the correlation coefficient of raw scores.

The correlation coefficient between standard score vectors z_j and z_k is expressed as $r_{jk}^{[z]} = v_{jk}^{[z]}/(\sqrt{v_{jj}^{[z]}}\sqrt{v_{kk}^{[z]}})$ by replacing v_{jk} in (3.7) by the covariance (3.12). Here, the variances $v_{jj}^{[z]}$ and $v_{kk}^{[z]}$ for standard scores are equal to one, as found in Table 2.3 ; thus, $r_{jk}^{[z]} = v_{jk}^{[z]}$ or (3.12); the correlation coefficient of standard scores equals the correlation coefficients of the raw scores.

Table 3.4 summarizes the properties of the covariances and correlation coefficients for standard and centered scores. The correlation coefficients for centered/standard scores and the covariances for standard scores equal the correlation coefficients of their original raw scores, and the covariances for centered scores equal those of the raw scores.

Table 3.4 Covariances and correlations of centered and standard scores

	Covariance	Correlation coefficient
Centered scores	Covariance for raw scores	Correlation coefficient for raw scores
Standard scores	Correlation coefficient for raw scores	

3.6   Matrix Expressions of Covariances and Correlations

Using (3.6) with (2.27), the p-variables $\times$ p-variables *covariance matrix* $\mathbf{V} = (v_{jk})$ for the data matrix $\mathbf{X}$ (2.1) can be expressed as:

$$
\begin{aligned}
\mathbf{V} &= \frac{1}{n}
\begin{bmatrix}
\mathbf{x}_1' \mathbf{J}' \mathbf{J} \mathbf{x}_1 & \cdots & \mathbf{x}_1' \mathbf{J}' \mathbf{J} \mathbf{x}_k & \cdots & \mathbf{x}_1' \mathbf{J}' \mathbf{J} \mathbf{x}_p \\
\vdots & \cdots & \cdots & \cdots & \vdots \\
\mathbf{x}_j' \mathbf{J}' \mathbf{J} \mathbf{x}_1 & \cdots & \mathbf{x}_j' \mathbf{J}' \mathbf{J} \mathbf{x}_k & \cdots & \mathbf{x}_j' \mathbf{J}' \mathbf{J} \mathbf{x}_p \\
\vdots & \vdots & \vdots & \cdots & \vdots \\
\mathbf{x}_p' \mathbf{J}' \mathbf{J} \mathbf{x}_1 & \cdots & \mathbf{x}_p' \mathbf{J}' \mathbf{J} \mathbf{x}_k & \cdots & \mathbf{x}_p' \mathbf{J}' \mathbf{J} \mathbf{x}_p
\end{bmatrix} \\[2mm]
&= \frac{1}{n}
\begin{bmatrix}
\mathbf{x}_1' \mathbf{J}' \\
\vdots \\
\mathbf{x}_j' \mathbf{J}' \\
\vdots \\
\mathbf{x}_p' \mathbf{J}'
\end{bmatrix}
\begin{bmatrix} \mathbf{J} \mathbf{x}_1 \cdots \mathbf{J} \mathbf{x}_k \cdots \mathbf{J} \mathbf{x}_p \end{bmatrix} \\[2mm]
&= \frac{1}{n}
\begin{bmatrix}
\mathbf{x}_1' \\
\vdots \\
\mathbf{x}_j' \\
\vdots \\
\mathbf{x}_p'
\end{bmatrix}
\mathbf{J}' \mathbf{J} \begin{bmatrix} \mathbf{x}_1 \cdots \mathbf{x}_k \cdots \mathbf{x}_p \end{bmatrix} = \frac{1}{n} \mathbf{X}' \mathbf{J}' \mathbf{J} \mathbf{X} = \frac{1}{n} \mathbf{X}' \mathbf{J} \mathbf{X},
\end{aligned}
\tag{3.13}
$$

in which (2.11) and (2.12) have been used. We can use (2.29) to rewrite the covariance matrix (3.13) simply as:

$$
\mathbf{V} = \frac{1}{n} \mathbf{Y}' \mathbf{Y}.
\tag{3.14}
$$

Let $\mathbf{R} = (r_{jk})$ denote the p-variables $\times$ p-variables *correlation matrix* $\mathbf{R} = (r_{jk})$ for $\mathbf{X}$. Since the covariance for the standard scores is equal to the correlation coefficient for the raw scores, as shown in Table 3.4, we can use an n-individuals $\times$ p-variables *standard score* matrix $\mathbf{Z} = [\mathbf{z}_1, \ldots, \mathbf{z}_p]$ to express $\mathbf{R}$ as:

$$
\mathbf{R} = \frac{1}{n}
\begin{bmatrix}
\mathbf{z}_1'\mathbf{J}\mathbf{z}_1 & \cdots & \mathbf{z}_1'\mathbf{J}\mathbf{z}_k & \cdots & \mathbf{z}_1'\mathbf{J}\mathbf{z}_p \\
\vdots & \cdots & \cdots & \cdots & \vdots \\
\mathbf{z}_j'\mathbf{J}\mathbf{z}_1 & \cdots & \mathbf{z}_j'\mathbf{J}\mathbf{z}_k & \cdots & \mathbf{z}_j'\mathbf{J}\mathbf{z}_p \\
\vdots & \vdots & \vdots & \cdots & \vdots \\
\mathbf{z}_p'\mathbf{J}\mathbf{z}_1 & \cdots & \mathbf{z}_p'\mathbf{J}\mathbf{z}_k & \cdots & \mathbf{z}_p'\mathbf{J}\mathbf{z}_p
\end{bmatrix}
= \frac{1}{n}
\begin{bmatrix}
\mathbf{z}_1'\mathbf{z}_1 & \cdots & \mathbf{z}_1'\mathbf{z}_k & \cdots & \mathbf{z}_1'\mathbf{z}_p \\
\vdots & \cdots & \cdots & \cdots & \vdots \\
\mathbf{z}_j'\mathbf{z}_1 & \cdots & \mathbf{z}_j'\mathbf{z}_k & \cdots & \mathbf{z}_j'\mathbf{z}_p \\
\vdots & \vdots & \vdots & \cdots & \vdots \\
\mathbf{z}_p'\mathbf{z}_1 & \cdots & \mathbf{z}_p'\mathbf{z}_k & \cdots & \mathbf{z}_p'\mathbf{z}_p
\end{bmatrix}
$$

$$
= \frac{1}{n}\mathbf{Z}'\mathbf{Z},
$$

$$(3.15)$$

in which we have used (3.12). Using (2.32) in (3.15), $\mathbf{R}$ is also expressed as:

$$
\mathbf{R} = \frac{1}{n}\mathbf{D}^{-1}\mathbf{X}'\mathbf{J}'\mathbf{J}\mathbf{X}\mathbf{D}^{-1} = \frac{1}{n}\mathbf{D}^{-1}\mathbf{X}'\mathbf{J}\mathbf{X}\mathbf{D}^{-1},
\tag{3.16}
$$

with $\mathbf{D} = \begin{bmatrix} \sqrt{v_{11}} & & \\ & \ddots & \\ & & \sqrt{v_{pp}} \end{bmatrix}$, as defined in Sect. 2.8. Further, if we compare
(3.16) with (3.13), we have

$$
\mathbf{R} = \mathbf{D}^{-1}\mathbf{V}\mathbf{D}^{-1}.
\tag{3.17}
$$

3.7 Unbiased Covariances

For covariances (and variances), a definition exits that is different from (3.4). In this
definition, $\mathbf{x}_j'\mathbf{J}\mathbf{x}_k$ is divided by $n-1$ in place of n; the covariance matrix for $\mathbf{X}$ may
be defined as:

$$
\mathbf{V}^{\#} = \frac{1}{n-1}\mathbf{X}'\mathbf{J}\mathbf{X} = \frac{n}{n-1}\mathbf{V}.
\tag{3.18}
$$

It is called an *unbiased covariance* matrix. Its off-diagonal and diagonal elements
are called unbiased covariances and unbiased variances, respectively, for distinguishing (3.18) from (3.13); one may use either equation. In this book, we refrain
from explaining why two types of definition exist, and (3.13) is used throughout.

Though the covariance is defined in the two above manners, the correlation
coefficient is defined uniquely, i.e., in a single way, as follows: If we use covariance
(3.18), the correlation in (3.7) is rewritten as:

$$r_{jk} = \frac{v_{jk}}{\sqrt{v_{jj}}\sqrt{v_{kk}}} = \frac{\frac{1}{n-1}\mathbf{x}_j'\mathbf{J}\mathbf{x}_k}{\sqrt{\frac{1}{n-1}\mathbf{x}_j'\mathbf{J}\mathbf{x}_k}\sqrt{\frac{1}{n-1}\mathbf{x}_j'\mathbf{J}\mathbf{x}_k}} = \frac{\mathbf{x}_j'\mathbf{J}\mathbf{x}_k}{\sqrt{\mathbf{x}_j'\mathbf{J}\mathbf{x}_k}\sqrt{\mathbf{x}_j'\mathbf{J}\mathbf{x}_k}}. \tag{3.19}$$

Here, $n - 1$ in the numerator and denominator are canceled out so that (3.19) becomes equivalent to (3.8), i.e., (3.7).

3.8   Centered Matrix

When a data matrix $\mathbf{X}$ contains centered scores, i.e., is already centered, with

$$\mathbf{1}_n'\mathbf{X} = \mathbf{0}_p', \text{ or, equivalently, } \mathbf{X} = \mathbf{J}\mathbf{X}, \tag{3.20}$$

$\mathbf{X}$ is said to be *centered*. The equivalence of the two equations in (3.20) will now be proved.

> **Note 3.1. Two Expressions of Zero Average**
> The sum and average of the elements in each column of an $n \times m$ vector $\mathbf{F}$ being zero are equivalent to the premultiplication of $\mathbf{F}$ by the centering matrix being $\mathbf{F}$:
>
> $$\mathbf{1}_n'\mathbf{F} = \frac{1}{n}\mathbf{1}_n'\mathbf{F} = \mathbf{0}_m' \text{ is equivalent to } \mathbf{J}\mathbf{F} = \mathbf{F}. \tag{3.21}$$
>
> This is proved by showing both [1] $\mathbf{J}\mathbf{F} = \mathbf{F} \Rightarrow \mathbf{1}_n'\mathbf{F} = \mathbf{0}_m$ and [2] $\mathbf{1}_n'\mathbf{F} = \mathbf{0}_m' \Rightarrow \mathbf{J}\mathbf{F} = \mathbf{F}$. [1] is easily found by using (2.13) in $\mathbf{1}_n'\mathbf{F} = \mathbf{1}_n'\mathbf{J}\mathbf{F}$, while [2] follows from premultiplying of both sides of $\mathbf{1}_n'\mathbf{F} = \mathbf{0}_m'$ by $-n^{-1}\mathbf{1}_n$ yields $-n^{-1}\mathbf{1}_n\mathbf{1}_n'\mathbf{F} = {}_n\mathbf{O}_m$, to whose both sides we can add $\mathbf{F}$ so as to have $\mathbf{F} - n^{-1}\mathbf{1}_n\mathbf{1}_n'\mathbf{F} = \mathbf{F}$, i.e., $(\mathbf{I}_n - n^{-1}\mathbf{1}_n\mathbf{1}_n')\mathbf{F} = \mathbf{F}$.

When $\mathbf{X}$ is centered, (3.13) and (3.16) are expressed as:

$$\mathbf{V} = \frac{1}{n}\mathbf{X}'\mathbf{X}, \tag{3.22}$$

$$\mathbf{R} = \frac{1}{n}\mathbf{D}^{-1}\mathbf{X}'\mathbf{X}\mathbf{D}^{-1}, \tag{3.23}$$

respectively, simply without using $\mathbf{J}$.

So far, the covariance matrix that has been treated contains covariances among the variables in a single data matrix $\mathbf{X}$. Now, let us consider the $p \times m$ matrix containing covariances between the variables in $\mathbf{X}$ $(n \times p)$ and those corresponding to the columns $\mathbf{F}$ $(n \times m)$. The covariance matrix is expressed as:

$$\mathbf{V}_{XF} = \frac{1}{n}\mathbf{X}'\mathbf{JF}. \tag{3.24}$$

If both $\mathbf{X}$ and $\mathbf{F}$ are centered with $\mathbf{X} = \mathbf{JX}$ and $\mathbf{F} = \mathbf{JF}$, (3.24) is simplified as:

$$\mathbf{V}_{XF} = \frac{1}{n}\mathbf{X}'\mathbf{F}. \tag{3.25}$$

Further, when both $\mathbf{X}$ and $\mathbf{F}$ are centered and contain standard scores, (3.25) also expresses a correlation matrix.

3.9 Ranks of Matrices: Intuitive Introduction

For every matrix, its *rank* is given as an integer. It is an important number that stands for a property of the matrix and is used in the following chapters. In this section, we introduce rank so that it can be *intuitively* captured using the four 5×3 data matrices in Table 3.5.

First, note the data matrix in Table 3.5(A). The values seem to be different among the three variables. Indeed, no relationships exist between $\mathbf{x}_1$, $\mathbf{x}_2$, and $\mathbf{x}_3$. That is, those three variables are regarded as, respectively, conveying three different kinds of information. Such a matrix is said to be one whose rank is three. Next, note (B), whose third column is the same as the first one; though the matrix has three columns, it conveys to us only two kinds of information. The rank of this matrix is two.

The third column in Table 3.5(C) is different from the first one, but multiplication of the latter by -3 gives the third column. Its elements can be considered as expressing the same information as those in the first column, except that the signs of the values are reversed and their scales differ. The rank of this matrix is also two, not three.

Finally, let us note Table 3.5(D). Though the three columns seem to mutually differ, $\mathbf{x}_2 = 1.5\mathbf{x}_1 - 3\mathbf{x}_3$. The rank of this matrix is also two, in that the information conveyed by the second column can be found by knowing that found in the other two.

In the next section, the rank of a matrix is precisely defined.

Table 3.5 Four matrices for illustrating their ranks

	(A)			(B)			(C)			(D)		
	$\mathbf{x}_1$	$\mathbf{x}_2$	$\mathbf{x}_3$	$\mathbf{x}_1$	$\mathbf{x}_2$	$\mathbf{x}_3$	$\mathbf{x}_1$	$\mathbf{x}_2$	$\mathbf{x}_3$	$\mathbf{x}_1$	$\mathbf{x}_2$	$\mathbf{x}_3$
1	2	3	-2	2	3	2	2	3	-6	2	9	-2
2	4	5	9	4	5	4	4	5	-12	4	-21	9
3	-1	7	3	-1	7	-1	-1	7	3	-1	-10.5	3
4	-5	0	3	-5	0	-5	-5	0	15	-5	-16.5	3
5	7	5	2	7	5	7	7	5	-21	7	4.5	2

3.10   Ranks of Matrices: Mathematical Definition

A sum of the vectors $\mathbf{h}_1$, ..., $\mathbf{h}_p$ multiplied by scalars

$$b_1\mathbf{h}_1 + \cdots + b_p\mathbf{h}_p = \mathbf{Hb} \qquad (3.26)$$

is called the *linear combination* of $\mathbf{h}_1$, ..., $\mathbf{h}_p$. Here, $\mathbf{H} = [\mathbf{h}_1, ..., \mathbf{h}_p]$ is an $n \times p$ matrix and $\mathbf{b} = [b_1, ..., b_p]'$ is a $p \times 1$ vector. Before defining the rank of $\mathbf{H}$, we introduce the following two notions:

Note 3.2. Linear Independence
The set of vectors $\mathbf{h}_1$, ..., $\mathbf{h}_p$ is said to be *linearly independent*, if

$$b_1\mathbf{h}_1 + \cdots + b_p\mathbf{h}_p = \mathbf{Hb} = \mathbf{0}_p \text{ implies } \mathbf{b} = \mathbf{0}_p. \qquad (3.27)$$

The inverse of the above is defined as follows:

Note 3.3. Linear Dependence
The set of vectors $\mathbf{h}_1$, ..., $\mathbf{h}_p$ is said to be *linearly dependent*, if $\mathbf{Hb} = \mathbf{0}_p$ does not imply $\mathbf{b} = \mathbf{0}_p$, that is, if

$$b_1\mathbf{h}_1 + \cdots + b_p\mathbf{h}_p = \mathbf{Hb} = \mathbf{0}_p \text{ holds,}$$
$$\text{with at least } b_J (1 \leq J \leq p) \text{ not being zero.} \qquad (3.28)$$

This implies that $b_J\mathbf{h}_J = -\sum_{j \neq J} b_j\mathbf{h}_j$ and we can divide both sides by b_J to have:

$$\mathbf{h}_J = -\sum_{j \neq J} \frac{b_j}{b_J}\mathbf{h}_j. \qquad (3.29)$$

The vector $\mathbf{h}_J$ is a linear combination of the other vectors with coefficient $-b_j/b_J$. Here, $\sum_{j \neq J} a_j$ denotes the sum of a_j over j excluding a_J. When $j = 1$, 2, 3, $\sum_{j \neq 2} a_j = a_1 + a_3$, for example.

The *rank* of **H**, which we denote as rank(**H**), is defined as follows:

Note 3.4. Rank of a Matrix

rank ($\mathbf{H}$) = the maximum number of linearly independent columns in **H**

(3.30)

For illustrating the definition of rank, we present the following three examples:

[1] Let $\mathbf{P} = [\mathbf{p}_1, \mathbf{p}_2, \mathbf{p}_3] = \begin{bmatrix} 1 & 1 & 9 \\ 2 & 2 & 6 \\ 1 & 3 & 4 \\ 2 & 4 & 7 \end{bmatrix}$. Then, rank(**P**) = 3,

since $b_1\mathbf{p}_1 + b_2\mathbf{p}_2 + b_3\mathbf{p}_3 = \mathbf{0}_3$ implies $b_1 = b_2 = b_3 = 0$; we cannot find nonzero b_J with $\mathbf{p}_J = -\Sigma_{j\neq J}b_j/b_J\,\mathbf{p}_j$.

[2] Let $\mathbf{Q} = [\mathbf{q}_1, \mathbf{q}_2, \mathbf{q}_3] = \begin{bmatrix} 1 & 2 & 3 \\ 2 & 4 & 6 \\ 1 & 2 & 3 \\ 2 & 4 & 6 \end{bmatrix}$. Then, rank(**Q**) = 1,

since $\mathbf{q}_2 = 2\mathbf{q}_1$ and $\mathbf{q}_3 = 3\mathbf{q}_1$; the linearly independent vector sets are $\{\mathbf{q}_1\}$, $\{\mathbf{q}_2\}$, and $\{\mathbf{q}_3\}$, each of which consists of a single vector.

[3] Let $\mathbf{R} = [\mathbf{r}_1, \mathbf{r}_2, \mathbf{r}_3] = \begin{bmatrix} 1 & 1 & 3 \\ 2 & 2 & 6 \\ 1 & 3 & 5 \\ 2 & 4 & 8 \end{bmatrix}$. Then, rank(**R**) = 2,

since $\mathbf{r}_3 = 2\mathbf{r}_1 + \mathbf{r}_2$; thus, rank(**R**) < 3, but the set of $\mathbf{r}_1$ and $\mathbf{r}_2$ is linearly independent.

It is difficult to find the rank of a matrix by glancing at it, but we can easily find the rank through the extended version of the *singular value decomposition* introduced in Appendix A.3.1.

The rank of an $n \times p$ matrix **A** satisfies:

$$\text{rank}(\mathbf{A}) = \text{rank}(\mathbf{A}'),$$ (3.31)

which implies that the "columns" in (3.30) may be replaced by "rows." Further, (3.31) implies

$$\text{rank}(\mathbf{A}) \leq \min(n, p).$$ (3.32)

The following properties are also used in the remaining chapters:

$$\text{rank}(\mathbf{BA}) \leq \min(\text{rank}(\mathbf{A}), \text{rank}(\mathbf{B})) \tag{3.33}$$

$$\mathbf{A}'\mathbf{A} = \mathbf{I}_p \text{ implies rank}(\mathbf{A}) = p. \tag{3.34}$$

3.11 Bibliographical Notes

Carroll et al. (1997, Chap. 3), Rencher and Christensen (2012, Chap. 3), and Reyment and Jöreskog (1996, Chap. 2) are among the literatures in which matrix expressions of inter-variable statistics are intelligibly treated. The rank of a matrix is detailed in those books introduced in Sect. 1.9.

Exercises

3.1. Prove the *Cauchy–Schwarz inequality*

$$(\mathbf{a}'\mathbf{b})^2 \leq \|\mathbf{a}\|^2 \|\mathbf{b}\|^2 \tag{3.35}$$

by defining matrix $\mathbf{C} = \mathbf{ab}' - \mathbf{ba}'$ and using $\|\mathbf{C}\|^2 \geq 0$.

3.2. Use (3.35) to show

$$\mathbf{x}_1'\mathbf{Jx}_2 \leq \|\mathbf{Jx}_1\| \|\mathbf{Jx}_2\|, \tag{3.36}$$

with $\mathbf{x}_1$ and $\mathbf{x}_2$ $n \times 1$ vectors and $\mathbf{J} = \mathbf{I}_n - n^{-1}\mathbf{1}_n\mathbf{1}_n'$ the centering matrix.

3.3. Use (3.36) to show that the correlation coefficient takes a value within the range from -1 to 1.

3.4. Let $\mathbf{x} = [x_1, \ldots, x_n]'$ and $\mathbf{y} = [y_1, \ldots, y_n]'$, with $v = n^{-1}\mathbf{x}'\mathbf{Jy}$ the covariance between $\mathbf{x}$ and $\mathbf{y}$. Show that the covariance between $a\mathbf{x} + c\mathbf{1}_n = \begin{bmatrix} ax_1 + c \\ \vdots \\ ax_n + c \end{bmatrix}$

and $b\mathbf{y} + d\mathbf{1}_n = \begin{bmatrix} by_1 + d \\ \vdots \\ by_n + d \end{bmatrix}$ is given by $abv = n^{-1}ab\mathbf{x}'\mathbf{Jy}$.

3.5. Let r denote the correlation coefficient between vectors $\mathbf{x} = [x_1, \ldots, x_n]'$ and $\mathbf{y} = [y_1, \ldots, y_n]'$. Show that the correlation coefficient between $a\mathbf{x} + c\mathbf{1}_n$ and $b\mathbf{y} + d\mathbf{1}_n$ is also r for $ab > 0$, but is $-r$ for $ab < 0$, with the coefficient not defined for $ab = 0$.

3.6. Let $\mathbf{X}$ and $\mathbf{Y}$ be the matrices containing n rows, with $\mathbf{V}_{XY} = n^{-1}\mathbf{X}'\mathbf{JY}$ the covariance matrix between the columns of $\mathbf{X}$ and those of $\mathbf{Y}$. Show that $\mathbf{A}'\mathbf{V}_{XY}\mathbf{B}$ gives the covariance matrix between the columns of $\mathbf{XA}$ and those of $\mathbf{YB}$.

3.7. Let $\mathbf{X}$ and $\mathbf{Y}$ be the matrices containing n rows, $\mathbf{D}_X$ be the diagonal matrix whose jth diagonal element is the standard deviation of the elements in the

jth column of $\mathbf{X}$, and $\mathbf{D_Y}$ be defined for $\mathbf{Y}$ in a parallel manner. Show that $\mathbf{R_{XY}} = n^{-1}\mathbf{D_X^{-1}X'JYD_Y^{-1}}$ gives the correlation matrix between the columns of $\mathbf{X}$ and those of $\mathbf{Y}$.

3.8. Consider the matrices defined in Exercise 3.7. Show that $\mathbf{R_{XY}} = n^{-1}\mathbf{D_X^{-1}X'YD_Y^{-1}}$ gives the correlation matrix between the columns of $\mathbf{X}$ and $\mathbf{Y}$, when they are centered.

3.9. Let $\mathbf{A} = \begin{bmatrix} 1 & 0 & 1 \\ 0 & 2 & 2 \\ 3 & 0 & 3 \\ 0 & 4 & 4 \end{bmatrix}$. Show rank$(\mathbf{A}) = 2$ by noting the columns of $\mathbf{A}$ and rank$(\mathbf{A'}) = 2$ by noting the rows of $\mathbf{A}$.

3.10. Let $\mathbf{G}$ be $p \times q$ and $\mathbf{H}$ be $q \times r$, with $q \geq p \geq r$. Show rank$(\mathbf{GH}) \leq r$.

3.11. Let $\mathbf{F}$ be $n \times m$ and $\mathbf{A}$ be $p \times m$, with $m \leq \min(n, p)$. Show rank$(\mathbf{FA'}) \leq m$.

3.12. Show rank$(\mathbf{I}_n) = n$, with $\mathbf{I}_n$ the $n \times n$ identity matrix.

3.13. Show that rank$(\mathbf{JX}) \leq \min(n - 1, p)$, with $\mathbf{X}$ an $n \times p$ matrix and $\mathbf{J} = \mathbf{I}_n - n^{-1}\mathbf{1}_n\mathbf{1}_n'$ the centering matrix.

Part II
Least Squares Procedures

Regression, principal component, and cluster analyses are introduced as least squares procedures. Here, principal component analysis is treated in two chapters, as it can be described in various ways. The three analysis procedures are formulated as minimizing least squares functions, though other formulations are also possible.

Chapter 4
Regression Analysis

In the previous two chapters, we expressed elementary statistics in matrix form as preparation for introducing multivariate analysis procedures. The introduction to those procedures begins in this chapter. Here, we treat regression analysis, whose purpose is to predict or explain a variable from a set of other variables. The origin of regression analysis is found in the studies of Francis Golton (1822–1911, British scientist) on heredity stature in the mid-1880s. The history of developments in regression analysis is well summarized in Izenman (2008, pp. 107–108).

4.1 Prediction of a Dependent Variable by Explanatory Variables

In Table 4.1, we show a 50 products × 4 variables (*quality*, *price*, *appearance*, and *sales*) data matrix. Let us consider predicting or explaining the *sales* of products by their *quality*, *price*, and *appearance*, with the formula:

$$sales = b_1 \times quality + b_2 \times price + b_3 \times appearance + c + error. \qquad (4.1)$$

Here, the term "*error*" must be attached to the right-hand side, because a perfectly exact prediction of *sales* is impossible.

Let us use x_{i1}, x_{i2}, x_{i3}, and y_i for the *quality*, *price*, *appearance*, and *sales* of the ith product in Table 4.1, respectively. Then, (4.1) is rewritten as:

$$y_i = b_1 x_{i1} + b_2 x_{i2} + b_3 x_{i3} + c + e_i, \qquad (4.2)$$

with e_i the error value for product i. Since (4.2) is supposed for all products, $i = 1, \ldots, 50$ in Table 4.1. Thus, we have

© Springer Nature Singapore Pte Ltd. 2016
K. Adachi, *Matrix-Based Introduction to Multivariate Data Analysis*,
DOI 10.1007/978-981-10-2341-5_4

Table 4.1 Data matrix for the *quality*, *price*, *appearance*, and *sales* of products, which is an artificial example found in Adachi (2006)

Product	Quality	Price	Appearance	Sales
1	10	1800	2.6	48
2	5	1550	4.2	104
3	5	1250	3.0	122
4	5	1150	1.0	104
5	6	1700	7.0	125
6	6	1550	4.0	105
7	5	1200	3.6	135
8	3	1000	1.8	128
9	3	1300	5.8	145
10	5	1300	3.0	124
11	6	1550	5.8	99
12	9	1800	4.2	102
13	8	1400	4.4	146
14	6	1300	3.0	138
15	5	1400	3.8	122
16	10	1950	3.0	13
17	4	1550	5.2	103
18	2	1300	4.0	86
19	7	1800	6.8	109
20	4	1300	3.4	103
21	6	1350	4.0	113
22	9	1450	1.8	100
23	5	1300	4.2	111
24	6	1450	4.0	138
25	8	1750	4.0	101
26	4	1500	4.2	126
27	3	1700	4.6	29
28	6	1500	2.2	73
29	4	1250	3.4	129
30	9	1650	3.2	77
31	5	1500	3.4	84
32	4	1350	3.8	103
33	4	1350	3.8	112
34	3	1550	4.6	77
35	3	1200	3.6	135
36	1	1450	6.0	112
37	4	1600	4.8	106
38	5	1600	3.8	99
39	1	1100	4.2	143
40	6	1600	3.8	54
41	4	1450	6.6	139

(continued)

Table 4.1 (continued)

Product	Quality	Price	Appearance	Sales
42	2	1300	1.6	90
43	4	1200	5.2	203
44	3	1150	2.4	96
45	7	1350	3.2	125
46	7	1200	1.2	107
47	5	1550	5.0	130
48	5	1600	4.2	72
49	7	1400	3.8	137
50	7	1600	5.4	106

$$\begin{bmatrix} 48 \\ 104 \\ \vdots \\ 106 \end{bmatrix} = b_1 \begin{bmatrix} 10 \\ 5 \\ \vdots \\ 7 \end{bmatrix} + b_2 \begin{bmatrix} 1800 \\ 1550 \\ \vdots \\ 1600 \end{bmatrix} + b_3 \begin{bmatrix} 2.6 \\ 4.2 \\ \vdots \\ 5.4 \end{bmatrix} + \begin{bmatrix} c \\ c \\ \vdots \\ c \end{bmatrix} + \begin{bmatrix} e_1 \\ e_2 \\ \vdots \\ e_{50} \end{bmatrix}. \tag{4.3}$$

Further, it is rewritten as:

$$\begin{bmatrix} 48 \\ 104 \\ \vdots \\ 106 \end{bmatrix} = \begin{bmatrix} 10 & 1800 & 2.6 \\ 5 & 1550 & 4.2 \\ \vdots & \vdots & \vdots \\ 7 & 1600 & 5.4 \end{bmatrix} \begin{bmatrix} b_1 \\ b_2 \\ b_3 \end{bmatrix} + c \begin{bmatrix} 1 \\ 1 \\ \vdots \\ 1 \end{bmatrix} + \begin{bmatrix} e_1 \\ e_2 \\ \vdots \\ e_{50} \end{bmatrix} \tag{4.4}$$

by summarizing the vectors for *quality*, *price*, and *appearance* into a matrix. Expressing this matrix as **X** and using **y** for the *sales* vector, (4.4) can expressed as:

$$\mathbf{y} = \mathbf{X}\mathbf{b} + c\mathbf{1}_n + \mathbf{e}, \tag{4.5}$$

with $\mathbf{b} = [b_1, b_2, b_3]'$, $\mathbf{e} = [e_1, \ldots, e_{50}]'$, and $\mathbf{1}_n$ the $n \times 1$ vector of ones ($n = 50$ in this example). *Regression analysis* refers to a procedure for obtaining the *optimal* values of c and the elements of **b** from data **y** and **X**. Though **y** was used for a centered score vector in the last two chapters, it is not so in this chapter.

Hereafter, we generally describe **X** as an *n*-individuals $\times$ *p*-variables matrix, which implies that **y** and **e** are $n \times 1$ vectors and **b** is a $p \times 1$ vector. The model (4.5) for regression analysis is thus expressed as:

$$\underset{\mathbf{y}}{\begin{bmatrix} y_1 \\ \vdots \\ y_i \\ \vdots \\ y_n \end{bmatrix}} = \underset{\mathbf{X}}{\begin{bmatrix} x_{11} & \cdots & x_{1j} & \cdots & x_{1p} \\ \vdots & & \vdots & & \vdots \\ x_{i1} & \cdots & x_{ij} & \cdots & x_{ip} \\ \vdots & & \vdots & & \vdots \\ x_{n1} & \cdots & x_{nj} & \cdots & x_{np} \end{bmatrix}} \underset{\mathbf{b}}{\begin{bmatrix} b_1 \\ \vdots \\ b_j \\ \vdots \\ b_p \end{bmatrix}} + c \underset{\mathbf{1}_n}{\begin{bmatrix} 1 \\ \vdots \\ 1 \\ \vdots \\ 1 \end{bmatrix}} + \underset{\mathbf{e}}{\begin{bmatrix} e_1 \\ \vdots \\ e_i \\ \vdots \\ e_n \end{bmatrix}} \tag{4.6}$$

The term *model* refers to a formula that expresses the idea underlying an analysis procedure.

In this paragraph, we explain the terms used in regression analysis. The predicted or explained vector, i.e., $\mathbf{y}$, is called a *dependent variable*, while the columns of $\mathbf{X}$ are called *explanatory variables*. On the other hand, the elements of $\mathbf{b}$ are referred to as *regression coefficients*, and c is called an *intercept*. In particular, regression analysis with $p = 1$, i.e., a single exploratory variable, is called *simple regression* analysis, while the procedure with $p \geq 2$ is called *multiple regression* analysis; (4.6) is its model.

The terms generally used in analysis procedures are described next:

Note 4.1. Data Versus Parameters
In contrast to $\mathbf{y}$ and $\mathbf{X}$ given as *data* beforehand, the values of $\mathbf{b}$ and c are unknown before regression analysis is performed. Such entities as $\mathbf{b}$ and c, whose values are estimated by an analysis procedure, are generally called *parameters*. When one sees symbols in equations, it is very important to note whether the symbols express *data* or *parameters*.

4.2   Least Squares Method

Parameters $\mathbf{b}$ and c can be estimated using a *least squares method*. It generally refers to a class of the procedures for obtaining parameter values that *minimize* the *sum of squared errors*. This sum for (4.5) is expressed as:

$$\|\mathbf{e}\|^2 = e_1^2 + \cdots + e_n^2 = \|\mathbf{y} - \mathbf{X}\mathbf{b} - c\mathbf{1}_n\|^2, \tag{4.7}$$

since (4.5) is rewritten as $\mathbf{e} = \mathbf{y} - \mathbf{X}\mathbf{b} - c\mathbf{1}_n$. Thus, regression analysis is formulated as:

$$\text{minimizing } f(\mathbf{b}, c) = \|\mathbf{e}\|^2 = \|\mathbf{y} - \mathbf{X}\mathbf{b} - c\mathbf{1}_n\|^2 \text{ over } \mathbf{b} \text{ and } c, \tag{4.8}$$

which can be restated as obtaining the optimal $\mathbf{b}$ and c (i.e., their solutions) that minimize (4.7). Let us express the *solutions* of $\mathbf{b}$ and c for (4.8) as $\hat{\mathbf{b}}$ and $\hat{c}$, respectively, which are given as described in the following paragraphs.

It is known that $\hat{c}$ must satisfy:

$$\hat{c} = \frac{1}{n}\mathbf{1}_n'\mathbf{y} - \frac{1}{n}\mathbf{1}_n'\mathbf{X}\mathbf{b}. \tag{4.9}$$

This result can be derived as follows: We can define $\mathbf{h} = \mathbf{y} - \mathbf{Xb}$ to rewrite (4.7) as $\|\mathbf{h} - c\mathbf{1}_n\|^2$, which is minimized for $c = n^{-1}\mathbf{1}_n'\mathbf{h}$, as shown with (A.2.2) in Appendix A.2.1. The use of $\mathbf{h} = \mathbf{y} - \mathbf{Xb}$ in $\hat{c} = n^{-1}\mathbf{1}_n'\mathbf{h}$ leads to (4.9).

Substituting (4.9) into c in $\mathbf{e} = \mathbf{y} - \mathbf{Xb} - \mathbf{1}_n \times c$, which follows from (4.5), we have:

$$
\begin{aligned}
\mathbf{e} &= \mathbf{y} - \mathbf{Xb} - \left(\frac{1}{n}\mathbf{1}_n\mathbf{1}_n'\mathbf{y} - \frac{1}{n}\mathbf{1}_n\mathbf{1}_n'\mathbf{Xb}\right) \\
&= \left(\mathbf{y} - \frac{1}{n}\mathbf{1}_n\mathbf{1}_n'\mathbf{y}\right) - \left(\mathbf{Xb} - \frac{1}{n}\mathbf{1}_n\mathbf{1}_n'\mathbf{Xb}\right) = \mathbf{Jy} - \mathbf{JXb},
\end{aligned}
\tag{4.10}
$$

with $\mathbf{J} = \mathbf{I}_n - n^{-1}\mathbf{1}_n\mathbf{1}_n'$ the *centering matrix* defined in (2.10). Thus, (4.7) is also rewritten as:

$$\|\mathbf{e}\|^2 = \|\mathbf{Jy} - \mathbf{JXb}\|^2. \tag{4.11}$$

This is minimized when $\mathbf{b}$ is:

$$\hat{\mathbf{b}} = (\mathbf{X}'\mathbf{JX})^{-1}\mathbf{X}'\mathbf{Jy}, \tag{4.12}$$

as shown with (A.2.16) in Appendix A.2.2. Here, $(\mathbf{X}'\mathbf{JX})^{-1}$ is the *inverse matrix* of $\mathbf{X}'\mathbf{JX}$, which is introduced below.

Note 4.2. Inverse Matrix

A $p \times p$ square matrix $\mathbf{H}$ is said to be *nonsingular* if

$$\mathrm{rank}(\mathbf{H}) = p; \tag{4.13}$$

otherwise, $\mathbf{H}$ is said to be singular. If $\mathbf{H}$ is nonsingular, the $p \times p$ matrix $\mathbf{H}^{-1}$ exists that satisfies:

$$\mathbf{H}^{-1}\mathbf{H} = \mathbf{HH}^{-1} = \mathbf{I}_p. \tag{4.14}$$

The matrix $\mathbf{H}^{-1}$ is called the *inverse matrix* of $\mathbf{H}$. This matrix does not exist if $\mathbf{H}$ is singular.

For example,

$$
\mathbf{H} = \begin{bmatrix} 3 & -1 & 2 \\ -4 & 6 & -3 \\ 1 & 0 & 5 \end{bmatrix} \text{ is nonsingular, and}
$$

$$
\mathbf{H}^{-1} = \begin{bmatrix} 0.49 & 0.08 & -0.15 \\ 0.28 & 0.21 & 0.02 \\ -0.10 & -0.02 & 0.23 \end{bmatrix}.
$$

Two basic properties of inverse matrices are

$$(\mathbf{H}')^{-1} = (\mathbf{H}^{-1})', \tag{4.15}$$

$$(\mathbf{GH})^{-1} = \mathbf{H}^{-1}\mathbf{G}^{-1}, \tag{4.16}$$

which includes $(s\mathbf{H})^{-1} = s^{-1}\mathbf{H}^{-1}$ as a special case with $s \neq 0$ a scalar. The inverse matrix of a symmetric matrix $\mathbf{S}$ is also symmetric:

$$\mathbf{S}^{-1} = \mathbf{S}^{-1\prime}. \tag{4.17}$$

As found in the note, we suppose that $\mathbf{X}'\mathbf{JX}$ is nonsingular in (4.12). Actually, the data set in Table 4.1 gives such a $\mathbf{X}'\mathbf{JX}$.

Thus, the solution of regression analysis is given by obtaining (4.12) and substituting $\hat{\mathbf{b}}$ into $\mathbf{b}$ in (4.9). The solution (4.12) for $\mathbf{b}$ is also geometrically derived, as explained later, in Sect. 4.7.

4.3   Predicted and Error Values

The solutions (4.9) and (4.12) for the data set in Table 4.1 are shown in Table 4.2 (A); $\hat{\mathbf{b}} = [\,7.61, \quad -0.18, \quad 18.23\,]'$ and $\hat{c} = 256.4$. Substituting these values into (4.1), we have:

$$sales = 7.61 \times quality - 0.18 \times price + 18.13 \times appearance + 256.4 + error. \tag{4.18}$$

This equation is useful for *predicting* the *future sales* of a product, which is not included in Table 4.1. For example, let us suppose that a product has not yet been sold, but its *quality*, *price*, and *appearance* have been found to be 6, 1500, and 4. We can substitute those values into (4.18) to predict the *sales* as follows:

$$sales = 7.61 \times 6 - 0.18 \times 1500 + 18.13 \times 4 + 256.4 + error = 105 + error. \tag{4.19}$$

That is, *future sales* can be counted as 105, although any *future error* is unknown. However, *errors* that do *not* exist in the *future* can be assessed as described in the following paragraph.

Table 4.2 Results of regression analysis for the data in Table 4.1

Solution	Regression coefficient			Intercept $\hat{c}$	Variance explained	Multiple correlation
	$\hat{b}_1$: quality	$\hat{b}_2$: price	$\hat{b}_3$: appearance			
(A) Unstandardized	7.61	−0.18	18.23	256.4	0.73	0.85
(B) Standardized	0.51	−1.17	0.77	0.0		

Let us consider substituting the solutions $\hat{\mathbf{b}}$ and $\hat{c}$ into (4.5), giving $\mathbf{y} = \mathbf{X}\hat{\mathbf{b}} + \hat{c}\mathbf{1}_n + \hat{\mathbf{e}}$, which is rewritten as:

$$\mathbf{y} = \hat{\mathbf{y}} + \hat{\mathbf{e}}, \quad \text{i.e.,} \quad \hat{\mathbf{e}} = \mathbf{y} - \hat{\mathbf{y}} \tag{4.20}$$

by defining a *predicted value vector* as:

$$\hat{\mathbf{y}} = \mathbf{X}\hat{\mathbf{b}} + \hat{c}\mathbf{1}_n. \tag{4.21}$$

In (4.20), we have attached the "hat" mark to the **e** in (4.5), i.e., replaced it with $\hat{\mathbf{e}}$, in order to emphasize that the error vector **e**, which had been unknown before analysis, becomes known, as shown next: Using $\hat{\mathbf{b}} = [7.61, \quad -0.18, \quad 18.23]'$, $\hat{c} = 256.4$, and **X** in Table 4.1, the values in (4.21) are given by:

$$\hat{\mathbf{y}} = \begin{bmatrix} 10 & 1800 & 2.6 \\ 5 & 1550 & 4.2 \\ \vdots & \vdots & \vdots \\ 7 & 1600 & 5.4 \end{bmatrix} \begin{bmatrix} 7.61 \\ -0.18 \\ 18.23 \end{bmatrix} + 256.4 \begin{bmatrix} 1 \\ 1 \\ \vdots \\ 1 \end{bmatrix} = \begin{bmatrix} 56.0 \\ 92.1 \\ \vdots \\ 120.2 \end{bmatrix}, \tag{4.22}$$

which is used in (4.20) to provide

$$\hat{\mathbf{e}} = \mathbf{y} - \hat{\mathbf{y}} = \begin{bmatrix} 48 \\ 104 \\ \vdots \\ 106 \end{bmatrix} - \begin{bmatrix} 56.0 \\ 92.1 \\ \vdots \\ 120.2 \end{bmatrix} = \begin{bmatrix} -8.0 \\ 11.9 \\ \vdots \\ -14.2 \end{bmatrix}. \tag{4.23}$$

Its squared norm

$$\|\hat{\mathbf{e}}\|^2 = (-8.0)^2 + 11.9^2 + \cdots + (-14.2)^2 \tag{4.24}$$

indicates the largeness of errors.

4.4   Proportion of Explained Variance and Multiple Correlation

The purpose of this section is to introduce a statistic that indicates how *successful* the results of regression analysis are, using (4.24) and the three properties for $\hat{\mathbf{y}}$ and $\hat{\mathbf{e}}$ described in the following paragraph.

The first property is:

$$\mathbf{J}\hat{\mathbf{y}} = \mathbf{J}\mathbf{X}\hat{\mathbf{b}}, \tag{4.25}$$

which is derived as follows: (4.21) implies $\mathbf{J}\hat{\mathbf{y}} = \mathbf{J}\mathbf{X}\hat{\mathbf{b}} + \mathbf{J}(\hat{c}\mathbf{1}_n)$, with $\mathbf{J}(\hat{c}\mathbf{1}_n) = \mathbf{0}_n$ following from (2.13). The second property is:

$$\mathbf{J}\hat{\mathbf{e}} = \hat{\mathbf{e}}, \tag{4.26}$$

which follows from the use of (2.12) in (4.10). Property (4.26) shows that the *average of an error vector* is always *zero*; $n^{-1}\mathbf{1}'_n\hat{\mathbf{e}} = n^{-1}\mathbf{1}'_n\mathbf{J}\hat{\mathbf{e}} = 0$, because of (2.13). The third property is that errors are *uncorrelated* with predicted values with their covariance $n^{-1}\hat{\mathbf{e}}'\mathbf{J}\hat{\mathbf{y}} = 0$, i.e.,

$$\hat{\mathbf{e}}'\mathbf{J}\hat{\mathbf{y}} = 0. \tag{4.27}$$

Readers interested in the proof of (4.27) should see the following note:

Note 4.3. No Correlation Between Errors and Predictive Values
The use of (4.21) and $\mathbf{J}(\hat{c}\mathbf{1}_n) = \mathbf{0}_n$ in (4.20) leads to $\mathbf{J}\hat{\mathbf{e}} = \mathbf{J}\mathbf{y} - \mathbf{J}\mathbf{X}\hat{\mathbf{b}}$. Substituting this and (4.25) in $\hat{\mathbf{e}}'\mathbf{J}\hat{\mathbf{y}} = \hat{\mathbf{e}}'\mathbf{J}'\mathbf{J}\hat{\mathbf{y}}$, it is rewritten as:

$$\hat{\mathbf{e}}'\mathbf{J}\hat{\mathbf{y}} = (\mathbf{J}\mathbf{y} - \mathbf{J}\mathbf{X}\hat{\mathbf{b}})'\mathbf{J}\mathbf{X}\hat{\mathbf{b}} = \mathbf{y}'\mathbf{J}'\mathbf{X}\hat{\mathbf{b}} - \hat{\mathbf{b}}'\mathbf{X}'\mathbf{J}\mathbf{X}\hat{\mathbf{b}}, \tag{4.28}$$

where (2.11) and (2.12) have been used. We can further substitute (4.12) in (4.28) to have:

$$\hat{\mathbf{e}}'\mathbf{J}\hat{\mathbf{y}} = \mathbf{y}'\mathbf{J}'\mathbf{X}(\mathbf{X}'\mathbf{J}\mathbf{X})^{-1}\mathbf{X}'\mathbf{J}\mathbf{y} - \mathbf{y}'\mathbf{J}'\mathbf{X}(\mathbf{X}'\mathbf{J}\mathbf{X})^{-1}\mathbf{X}'\mathbf{J}\mathbf{X}(\mathbf{X}'\mathbf{J}\mathbf{X})^{-1}\mathbf{X}'\mathbf{J}\mathbf{y}$$
$$= \mathbf{y}'\mathbf{J}'\mathbf{X}(\mathbf{X}'\mathbf{J}\mathbf{X})^{-1}\mathbf{X}'\mathbf{J}\mathbf{y} - \mathbf{y}'\mathbf{J}\mathbf{X}(\mathbf{X}'\mathbf{J}\mathbf{X})^{-1}\mathbf{X}'\mathbf{J}\mathbf{y} = 0$$

The premultiplication of the first equality in (4.20) by $\mathbf{J}$ leads to $\mathbf{J}\mathbf{y} = \mathbf{J}\hat{\mathbf{y}} + \mathbf{J}\hat{\mathbf{e}} = \mathbf{J}\hat{\mathbf{y}} + \hat{\mathbf{e}}$ because of (4.26). Further, the angle between $\mathbf{J}\hat{\mathbf{y}}$ and $\hat{\mathbf{e}}$ being $90°$ is found in (4.27). This fact implies that $\mathbf{J}\mathbf{y}, \mathbf{J}\hat{\mathbf{y}}$ and $\hat{\mathbf{e}}$ form the right triangle illustrated in Fig. 4.1. We can thus use the *Pythagorean theorem* to have:

Fig. 4.1 Geometric relationship among **e** (errors), **Jy** (centered dependent variable), and **Jŷ** (centered predicted values)

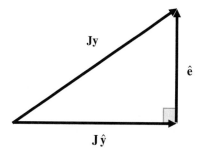

$$\|\mathbf{Jy}\|^2 = \|\mathbf{J\hat{y}}\|^2 + \|\mathbf{\hat{e}}\|^2. \tag{4.29}$$

From (4.29), we can derive a statistic indicating how *successful* the results of regression analysis are, as follows: The division of both sides of (4.29) by $\|\mathbf{Jy}\|^2$ leads to:

$$1 = \frac{\|\mathbf{J\hat{y}}\|^2}{\|\mathbf{Jy}\|^2} + \frac{\|\mathbf{\hat{e}}\|^2}{\|\mathbf{Jy}\|^2}. \tag{4.30}$$

Here, the proportion $\|\mathbf{\hat{e}}\|^2/\|\mathbf{Jy}\|^2$ taking a value within the range [0, 1] stands for the relative largeness of errors; equivalently, one minus that proportion,

$$\frac{\|\mathbf{J\hat{y}}\|^2}{\|\mathbf{Jy}\|^2} = 1 - \frac{\|\mathbf{\hat{e}}\|^2}{\|\mathbf{Jy}\|^2}, \tag{4.31}$$

taking a value within the range [0, 1] indicates the smallness of errors. Statistic (4.31) is called the *proportion of explained variance*, as it can be rewritten as:

$$\frac{\|\mathbf{J\hat{y}}\|^2}{\|\mathbf{Jy}\|^2} = \frac{n^{-1}\|\mathbf{J\hat{y}}\|^2}{n^{-1}\|\mathbf{Jy}\|^2} = \frac{n^{-1}\|\mathbf{JX\hat{b}}\|^2}{n^{-1}\|\mathbf{Jy}\|^2} \tag{4.32}$$

using (4.25). That is, the denominator of (4.32), $n^{-1}\|\mathbf{Jy}\|^2$, is the variance of a dependent variable, while the numerator, $n^{-1}\|\mathbf{JX\hat{b}}\|^2$, is the variance of predicted values based on the explanatory variables in **X**, which implies that (4.32) indicates the *proportion of the variance explained by explanatory variables* in the variance of the dependent variable. The resulting proportion of the explained variance in Table 4.2 is found to be 0.73, which is interpreted to mean that 73 % of the variance of the dependent variable (i.e., how much more/less *sales* are) is explained by *quality*, *price*, and *appearance*. Statistic (4.32) is also called a *coefficient of determination*.

Let us consider the square root of (4.32). It can be rewritten as:

$$\frac{\|\mathbf{J}\hat{\mathbf{y}}\|}{\|\mathbf{J}\mathbf{y}\|} = \frac{\|\mathbf{J}\hat{\mathbf{y}}\|}{\|\mathbf{J}\mathbf{y}\|} \times \frac{\|\mathbf{J}\hat{\mathbf{y}}\|}{\|\mathbf{J}\hat{\mathbf{y}}\|} = \frac{\hat{\mathbf{y}}'\mathbf{J}\hat{\mathbf{y}}}{\|\mathbf{J}\mathbf{y}\| \, \|\mathbf{J}\hat{\mathbf{y}}\|} = \frac{\mathbf{y}'\mathbf{J}\hat{\mathbf{y}}}{\|\mathbf{J}\mathbf{y}\| \, \|\mathbf{J}\hat{\mathbf{y}}\|} = \frac{(\mathbf{J}\mathbf{y})'\mathbf{J}\hat{\mathbf{y}}}{\|\mathbf{J}\mathbf{y}\| \, \|\mathbf{J}\hat{\mathbf{y}}\|} \qquad (4.33)$$

where we have used $\mathbf{y}'\mathbf{J}\hat{\mathbf{y}} = (\hat{\mathbf{y}} + \hat{\mathbf{e}})'\mathbf{J}\hat{\mathbf{y}} = \hat{\mathbf{y}}'\mathbf{J}\hat{\mathbf{y}}$ because of (4.27). Comparing (4.33) with (3.8), we can find (4.33) to be the *correlation coefficient* between $\mathbf{y}$ and $\hat{\mathbf{y}}$. In particular, (4.33) is called the *multiple correlation coefficient* between dependent and explanatory variables, as we can use (4.25) to rewrite (4.33) as:

$$\frac{(\mathbf{J}\mathbf{y})'\mathbf{J}\hat{\mathbf{y}}}{\|\mathbf{J}\mathbf{y}\| \, \|\mathbf{J}\hat{\mathbf{y}}\|} = \frac{(\mathbf{J}\mathbf{y})'\mathbf{J}\mathbf{X}\hat{\mathbf{b}}}{\|\mathbf{J}\mathbf{y}\| \, \|\mathbf{J}\mathbf{X}\hat{\mathbf{b}}\|}, \qquad (4.34)$$

which stands for the relationship of $\mathbf{y}$ to the *multiple* variables in $\mathbf{X}$. Its value, $0.85 = \sqrt{0.73}$ in Table 4.2, is near the upper limit of 1 and indicates a close relationship of *sales* to *quality*, *price*, and *appearance*.

4.5   Interpretation of Regression Coefficients

Simple regression analysis with a single explanatory variable can be formulated in the same manner as the multiple regression analysis described so far, except that p is restricted to one, i.e., $\mathbf{X}$ is set to an $n \times 1$ vector $\mathbf{x}$. *Simple regression* analysis with the model "*sales* = $b \times$ *quality* + c + *error*" for the data in Table 4.1 produces the result:

$$sales = -4.02 \times quality + 128.73 + error. \qquad (4.35)$$

Here, it should be noted that the *regression coefficient* for *quality* is *negative*. The covariance and correlation coefficients between *sales* and *quality* in Table 4.3 are also *negative*. Those negative values show that the products of lower quality tend to sell better. This seems to be *unreasonable*. This is due to the fact that the above coefficients are obtained only from a pair of two variables (*quality* and *sales*) without using the other variables (*prices* and *appearance*), as explained next.

Table 4.3 Covariances and correlations among the four variables in Table 4.1

Variable	V: covariance matrix				R: correlation matrix			
	Quality	Price	Appear[*]	Sales	Quality	Price	Appear[*]	Sales
Quality	4.5				1.00			
Price	245.5	41801.0			0.57	1.00		
Appear[*]	−0.4	104.6	1.8		−0.16	0.39	1.00	
Sales	−18	−3747.7	10.0	985.0	−0.27	−0.58	0.24	1.00

* Appearance

Let us note the positive correlation of *quality* to *price* and the negative one of *price* to *sales* in Table 4.3. We can thus consider the scenario in which the products of *higher quality* tend to have a *higher price*, which tends to *decrease sales*. That is, a *third variable, price, intermediates* between *quality* and *sales*. These may also be intermediated by *appearance*. The effects of *intermediate variables* cannot be considered by the statistics obtained for two variables excluding the intermediate ones.

On the other hand, we can find in Table 4.2 that the coefficient for quality, $\hat{b}_1 = 7.61$, resulting from *multiple regression*, is reasonably *positive*. This is because the other variables are included in the model, as found in (4.1). The coefficient value $\hat{b}_1 = 7.61$ is interpreted as indicating the following relationship: The *sales* increase by 7.61 on average with a unit increase in *quality*, while the values of the other variables are kept fixed. Why this interpretation is derived should be understood in a rather subjective manner: A unit increase in quality with price and appearance fixed in (4.1) can be expressed as:

$$sales^* = b_1 \times (quality + 1) + b_2 \times price + b_3 \times appearance + c + error^*, \quad (4.36)$$

where asterisks have been attached to *sales* and *error*, since an increase in *quality* changes their values from those of the *sales* and *error* in (4.1), i.e.,

$$sales = b_1 \times quality + b_2 \times price + b_3 \times appearance + c + error.$$

The subtraction of both sides of (4.36) from those of (4.1) gives

$$sales^* - sales = b_1 + (error^* - error), \quad (4.37)$$

whose average equals b_1, since the average of errors is zero, i.e., (4.26) leads to $n^{-1}\mathbf{1}_n'\hat{\mathbf{e}} = 0$, implying that the average of $error^* - error$ in (4.37) is zero.

The other coefficients are also interpreted in the same manner. For example, $\hat{b}_2 = -0.18$ in Table 4.2 allows us to consider the following: *sales* tend to *decrease* by 0.18 on average with a unit increase in *price*, while *quality* and *appearance* are fixed.

4.6 Standardization

It is senseless to compare the largeness of the three regression coefficients in Table 4.2(A) ($\hat{b}_1 = 7.61$, $\hat{b}_2 = -0.18$, $\hat{b}_3 = 18.23$), since they are obtained from the raw scores in which the variances (i.e., how widely the values range) differ across variables. For the comparison of coefficients to make sense, regression analysis must be carried out for the standardized data in which the values in all variables have been transformed into standard scores, so that the variances are equivalent among the variables (i.e., all unity). The solutions for standardized data are called *standardized solutions*, while those for raw data, which we have seen so far, are called *unstandardized solutions*. However, the standardized and unstandardized solutions of regression analysis for the same data set can be regarded as the two different expressions of the same solution, as shown next.

The standard score vector for $\mathbf{y}$ is expressed as:

$$\mathbf{y}_S = \frac{1}{\sqrt{v_y}} \mathbf{J} \mathbf{y} \tag{4.38}$$

by substituting $\mathbf{y}$ for $\mathbf{x}_j$ and v_y for v_{jj} in (2.24), where v_y is the variance of the dependent variable; it should be noted that $\mathbf{y}_j$ in (2.24) is different from $\mathbf{y}$ in this chapter. The standard score matrix for $\mathbf{X}$ is expressed as (2.32), i.e.,

$$\mathbf{Z} = \mathbf{J}\mathbf{X}\mathbf{D}^{-1}. \tag{4.39}$$

Here, $\mathbf{D} = \begin{bmatrix} \sqrt{v_{11}} & & \\ & \ddots & \\ & & \sqrt{v_{pp}} \end{bmatrix}$ is the $p \times p$ diagonal matrix, with its jth diagonal element $v_{jj}^{1/2}$ being the standard deviation of the jth explanatory variable, implying that its variance is v_{jj}. Substituting (4.38) and (4.39) into $\mathbf{y}$ and $\mathbf{X}$ in (4.12), respectively, we have the standardized solution of the regression coefficient vector:

$$\begin{aligned}
\hat{\mathbf{b}}_S &= (\mathbf{Z}'\mathbf{J}\mathbf{Z})^{-1}\mathbf{Z}'\mathbf{J}\mathbf{y}_S = \frac{1}{\sqrt{v_y}}\left(\mathbf{D}^{-1}\mathbf{X}'\mathbf{J}\mathbf{X}\mathbf{D}^{-1}\right)^{-1}(\mathbf{D}^{-1}\mathbf{X}'\mathbf{J}')\mathbf{J}\mathbf{y} \\
&= \frac{1}{\sqrt{v_y}}\mathbf{D}(\mathbf{X}'\mathbf{J}\mathbf{X})^{-1}\mathbf{D}\mathbf{D}^{-1}\mathbf{X}'\mathbf{J}\mathbf{y} = \frac{1}{\sqrt{v_y}}\mathbf{D}(\mathbf{X}'\mathbf{J}\mathbf{X})^{-1}\mathbf{X}'\mathbf{J}\mathbf{y} = \frac{1}{\sqrt{v_y}}\mathbf{D}\hat{\mathbf{b}}.
\end{aligned} \tag{4.40}$$

This shows that $\hat{\mathbf{b}}_S$ is easily transformed from the unstandardized solution $\hat{\mathbf{b}}$, i.e., the premultiplication of $\hat{\mathbf{b}}$ by $v_y^{-1/2}\mathbf{D}$. Further, the substitution of (4.38) and (4.39) into $\mathbf{y}$ and $\mathbf{X}$ in (4.9) leads to the standardized solution of the intercept simply being zero:

$$\hat{c}_S = 0. \tag{4.41}$$

Let us substitute (4.38), (4.39), and (4.40) for $\mathbf{y}$, $\mathbf{X}$, and $\hat{\mathbf{b}}$ in (4.32). Then, we have:

$$\frac{n^{-1}\left\|\mathbf{J}\mathbf{Z}\hat{\mathbf{b}}_S\right\|^2}{n^{-1}\left\|\mathbf{J}\mathbf{y}_S\right\|^2} \frac{n^{-1}\left\|\mathbf{J}\mathbf{X}\mathbf{D}^{-1}\left(\frac{1}{\sqrt{v_y}}\mathbf{D}\hat{\mathbf{b}}\right)\right\|^2}{n^{-1}\left\|\frac{1}{\sqrt{v_y}}\mathbf{J}\mathbf{y}\right\|^2} = \frac{n^{-1}\left\|\mathbf{J}\mathbf{X}\hat{\mathbf{b}}\right\|^2}{n^{-1}\left\|\mathbf{J}\mathbf{y}\right\|^2} \tag{4.42}$$

This shows that the *proportion of explained variance* remains equal to (4.32) and its square root (4.33) or (4.34) (multiple correlation coefficient) also remains unchanged, even if the data set is standardized.

Let us see the regression coefficient $\hat{\mathbf{b}}_S$ in the standardized solution, which is called the *standardized regression coefficient*, in Table 4.2(B). A comparison of their values makes sense. We can find that the absolute value of the coefficient for *price* is the largest among the three exploratory variables, showing that the effect of *price* on *sales* is the largest among the three. Further, the coefficient of *price* is negative, implying that *sales* tend to decrease with an increase in *price*. The effect of *quality* is found to be the least among the three variables.

4.7 Geometric Derivation of Regression Coefficients

Those readers uninterested in the subject may skip this section. However, this section can deepen our understanding of regression analysis, although knowledge of the vector space explained in Appendix A.1.3 is necessary here.

The minimization of (4.11) over $\mathbf{b}$ is also restated as minimizing the squared length of the vector (4.10), i.e., $\mathbf{e} = \mathbf{Jy} - \mathbf{JXb}$. To solve this problem, let us consider what $\mathbf{JXb}$ geometrically stands for when the elements of $\mathbf{b}$ take any real values. It can be represented as the grayed plane in Fig. 4.2a. Though it has been depicted as a two-dimensional plane in the figure, the grayed plane indeed represents a *p-dimensional space*, which is formally expressed as:

$$\Xi(\mathbf{JX}) = \{\mathbf{Jx}^* : \mathbf{Jx}^* = \mathbf{JXb} = [\mathbf{Jx}_1, \ldots, \mathbf{Jx}_p]\mathbf{b}$$
$$= b_1\mathbf{Jx}_1 + \cdots + b_p\mathbf{Jx}_p; -\infty < b_j < \infty, \ j = 1, \ldots, p\}, \tag{4.43}$$

with $\mathbf{X} = [\mathbf{x}_1, \ldots, \mathbf{x}_p]$ and $\mathbf{b} = [b_1, \ldots, b_p]$; (4.43) is equivalent to (A.1.12) in Appendix A.1.3, with $\mathbf{H}$ and $\mathbf{h}$ in (A.1.12) replaced by $\mathbf{JX}$ and $\mathbf{Jx}$ in (4.43). We can

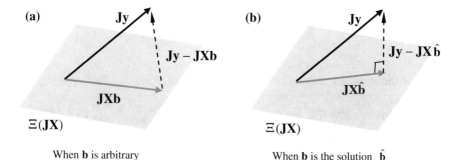

Fig. 4.2 $\mathbf{JXb}$ in space $\Xi(\mathbf{JX})$ with vector $\mathbf{Jy}$

set $b_1, \ldots, b_p$ to any real values so that the terminus of the vector $\mathbf{Jx}^* = \mathbf{JXb}$ moves in the space (4.43), i.e., on the grayed plane in Fig. 4.2a.

The function (4.11) to be minimized is the squared length of the difference vector $\mathbf{e} = \mathbf{Jy} - \mathbf{JXb}$, which is depicted as a dotted line in Fig. 4.2a. It is found to be the shortest, i.e., the minimum, when $\mathbf{e}$ is orthogonal to $\mathbf{JXb}$, as in Fig. 4.2b, that is, when

$$(\mathbf{JXb})'(\mathbf{Jy} - \mathbf{JXb}) = \mathbf{b}'\mathbf{X}'\mathbf{Jy} - \mathbf{b}'\mathbf{X}'\mathbf{JXb} = 0, \tag{4.44}$$

which holds for $\mathbf{b}$ equaling (4.12). It should also be noted that the right triangle found in Fig. 4.2b is the one in Fig. 4.1.

4.8   Bibliographical Notes

There are a number of books in which regression analysis is exhaustively detailed. Among them is Montgomery et al. (2012).

Multivariate data analysis procedures including regression analysis are exhaustively introduced in Lattin et al. (2003) with a number of real data examples. Izenman (2008) and Koch (2014) are examples of advanced books on multivariate data analysis procedures recommended for those who have finished reading the present book.

One topic that has not been mentioned in this chapter is *variable selection*, i.e., the problem of selecting useful exploratory variables and discarding useless ones among the initial set of variables. A modern approach to this problem is simultaneously performing the two tasks: (1) the selection of the regression coefficients to be zero and (2) the estimation of nonzero coefficients' values, which is detailed in Hastie et al. (2015). As described there, this approach has also been used in other procedures besides regression analysis and can generally be called "*sparse analysis*".

Exercises

4.1. Show that

$$\mathbf{A}^{-1} = \frac{1}{a_{11}a_{22} - a_{12}a_{21}} \begin{bmatrix} a_{22} & -a_{12} \\ -a_{21} & a_{11} \end{bmatrix} \tag{4.45}$$

is the inverse matrix of $\mathbf{A} = \begin{bmatrix} a_{11} & a_{12} \\ a_{21} & a_{22} \end{bmatrix}$.

4.2. Let us consider the system of equations $\begin{cases} -6x_1 + 2x_2 = 7 \\ 3x_1 + 9x_2 = -12 \end{cases}$, i.e., $\mathbf{Ax} = \mathbf{c}$,

with $\mathbf{A} = \begin{bmatrix} -6 & 2 \\ 3 & 9 \end{bmatrix}$ and $\mathbf{c} = \begin{bmatrix} 7 \\ -12 \end{bmatrix}$. Compute the solution of $\mathbf{x} = \begin{bmatrix} x_1 \\ x_2 \end{bmatrix}$ for the system using $\mathbf{A}^{-1}$ in (4.45).

4.3. Show $(\mathbf{AB})^{-1'} = \mathbf{A}'^{-1}\mathbf{B}'^{-1}$ with $\mathbf{A}$ and $\mathbf{B}$ being nonsingular.

4.4. Consider the model $y_i = c + e_i (i = 1, \ldots, n)$, i.e., $\mathbf{y} = c\mathbf{1}_n + \mathbf{e}$, for a data vector $\mathbf{y} = [y_1, \ldots, y_n]'$, with $\mathbf{e} = [e_1, \ldots, e_n]'$ containing errors and c the parameter to be obtained. Show that the average $\bar{y} = n^{-1}\sum_{i=1}^{n} y_i$ is the least squares solution of c in the model, i.e., that $f(c) = \|\mathbf{y} - c\mathbf{1}_n\|^2$ is minimized for $c = \bar{y}$, using the facts in Appendix A.2.1.

4.5. Show that the solution of intercept c in (4.9) is zero if $\mathbf{y}$ and each column of $\mathbf{X}$ contain centered scores.

4.6. Show that regression model (4.5) can be rewritten as

$$\mathbf{y} = \widetilde{\mathbf{X}}\boldsymbol{\beta} + \mathbf{e}, \tag{4.46}$$

with $\widetilde{\mathbf{X}} = \begin{bmatrix} x_{11} & \cdots & x_{1p} & 1 \\ \vdots & & \vdots & \vdots \\ x_{n1} & \cdots & x_{np} & 1 \end{bmatrix}$ an $n \times (p+1)$ matrix and

$\boldsymbol{\beta} = \begin{bmatrix} b_1 \\ \vdots \\ b_p \\ c \end{bmatrix}$ a $(p+1) \times 1$ vector.

4.7. Show that $\hat{\boldsymbol{\beta}} = \left(\widetilde{\mathbf{X}}'\widetilde{\mathbf{X}}\right)^{-1}\widetilde{\mathbf{X}}'\mathbf{y}$ is the least squares solution of $\boldsymbol{\beta}$ for (4.46), i.e., $\left\|\mathbf{y} - \widetilde{\mathbf{X}}\boldsymbol{\beta}\right\|^2$ is minimized for $\boldsymbol{\beta} = \hat{\boldsymbol{\beta}}$, using the facts in Appendix A.2.2.

4.8. Show that (4.12) can be rewritten as $\hat{\mathbf{b}} = \mathbf{V}_{XX}^{-1}\mathbf{v}_{Xy}$. Here, $\mathbf{V}_{XX} = n^{-1}\mathbf{X}'\mathbf{J}\mathbf{X}$ is the covariance matrix among explanatory variables, and $\mathbf{v}_{Xy} = n^{-1}\mathbf{X}'\mathbf{J}\mathbf{y}$ is the vector containing the covariances between explanatory and dependent variables, with n the number of individuals.

4.9. Show that (4.40) can be rewritten as $\hat{\mathbf{b}}_S = \mathbf{R}_{XX}^{-1}\mathbf{r}_{Xy}$. Here, $\mathbf{R}_{XX} = n^{-1}\mathbf{D}^{-1}\mathbf{X}'\mathbf{J}\mathbf{X}\mathbf{D}^{-1}$ is the correlation matrix among explanatory variables, and $\mathbf{r}_{Xy} = n^{-1}d^{-1}\mathbf{D}^{-1}\mathbf{X}'\mathbf{J}\mathbf{y}$ is the vector containing the correlation coefficients between explanatory and dependent variables, with n the number of individuals, $\mathbf{D}$ the diagonal matrix whose jth diagonal element is the standard deviation for the jth variable in $\mathbf{X}$ and d the standard deviation of the elements in $\mathbf{y}$.

4.10. Argue that $\mathbf{J}\mathbf{X}\hat{\mathbf{b}}$ in Fig. 4.2b is the image of a pencil reflected in a mirror, when the pencil and mirror stand for $\mathbf{J}\mathbf{y}$ and $\Xi(\mathbf{J}\mathbf{X})$, respectively, with $p = 2$.

4.11. In some procedures, a combination of function $f(\boldsymbol{\theta})$ and another one $g(\boldsymbol{\theta})$, i.e.,

$$f(\boldsymbol{\theta}) + \tau g(\boldsymbol{\theta}), \tag{4.47}$$

is minimized, where $\boldsymbol{\theta}$ is a parameter vector, τ is a given scalar value, and $g(\boldsymbol{\theta})$ is called a *penalty function* in that it penalizes $\boldsymbol{\theta}$ for increasing $g(\boldsymbol{\theta})$. In a special version of regression analysis (Hoerl and Kennard 1970), function $f(\boldsymbol{\theta})$

is defined as $f(\mathbf{b}) = \|\mathbf{y} - \mathbf{Xb}\|^2$ for a dependent variable vector $\mathbf{y}$ ($n \times 1$) and explanatory variable matrix $\mathbf{X}$ ($n \times p$), while a penalty function is defined as $g(\mathbf{b}) = \|\mathbf{b}\|^2$ which penalizes $\mathbf{b}$ for having a large squared norm. That is, $\|\mathbf{y} - \mathbf{Xb}\|^2 + \tau\|\mathbf{b}\|^2$ is minimized over $\mathbf{b}$ for a given τ. Show that the solution is given by $\mathbf{b} = (\mathbf{X}'\mathbf{X} + \tau\mathbf{I}_p)^{-1}\mathbf{X}'\mathbf{y}$.

Chapter 5
Principal Component Analysis (Part 1)

In regression analysis (Chap. 4), variables are classified as dependent and explanatory variables. Such a distinction does not exist in *principal component analysis* (*PCA*), which is introduced in this chapter. A single data matrix $\mathbf{X}$ is analyzed in PCA. This was originally formulated by Hotelling (1933) and is also rooted by Pearson (1901). PCA can be formulated apparently in different manners, as found in this chapter and the next. In some textbooks, PCA is firstly formulated as in Sect. 6.3 (in the next chapter), or the formulation found in this chapter is not described. However, the author believes that the latter formulation should precede the former one, in order to comprehend what PCA is. According to ten Berge and Kiers (1996), in which the formulations of PCA are classified into types based on Hotelling (1933), Pearson (1901), and Rao (1973), the formulation in this chapter is based on Pearson, while the next chapter is based on Hotelling.

5.1 Reduction of Variables into Components

PCA is usually used for an n-individuals $\times$ p-variables centered data matrix $\mathbf{X}$, with (3.20), i.e., $\mathbf{1}_n'\mathbf{X} = \mathbf{0}_p'$. Table 5.1(B) shows an example of $\mathbf{X}$ which is a 6 students $\times$ 5 courses matrix of the centered scores transformed from the test scores in Table 5.1(A).

For such a data matrix $\mathbf{X}$, PCA can be formulated with

$$\mathbf{X} = \mathbf{F}\mathbf{A}' + \mathbf{E}. \tag{5.1}$$

Here, $\mathbf{F}$ is an n-individuals $\times$ *m-components* matrix whose elements are called *principal component* (*PC*) *scores*, $\mathbf{A}$ is a p-variables $\times$ *m-components* matrix whose elements are called component *loadings*, and $\mathbf{E}$ contains errors, with

$$m \leq \text{rank}(\mathbf{X}) \leq \min(n, p). \tag{5.2}$$

© Springer Nature Singapore Pte Ltd. 2016
K. Adachi, *Matrix-Based Introduction to Multivariate Data Analysis*,
DOI 10.1007/978-981-10-2341-5_5

Table 5.1 Test scores for four courses, M (mathematics), P (physics), C (chemistry), and B (biology), with their averages and standard deviations (SD) (artificial example)

| | (A) Raw scores | | | | (B) Centered scores | | | | (C) Standard scores | | | |
	M	P	C	B	M	P	C	B	M	P	C	B
S1	69.0	66.4	77.0	74.1	−4.9	−10.6	0.3	5.3	−0.45	−0.70	0.02	0.38
S2	67.2	53.6	53.9	58.7	−6.7	−23.4	−22.8	−10.1	−0.61	−1.54	−1.75	−0.73
S3	78.6	96.9	97.3	96.2	4.7	19.9	20.6	27.4	0.43	1.31	1.58	1.97
S4	84.4	87.7	83.9	69.8	10.5	10.7	7.2	1.0	0.96	0.70	0.55	0.07
S5	56.3	68.7	72.1	56.8	−17.6	-8.3	−4.6	−12.0	−1.62	−0.55	−0.35	−0.86
S6	87.9	88.8	76.0	57.2	14.0	11.8	−0.7	−11.6	1.29	0.78	−0.05	−0.83
Average	73.9	77.0	76.7	68.8	0.0	0.0	0.0	0.0	0.00	0.00	0.00	0.00
SD	10.9	15.2	13.0	13.9	10.9	15.2	13.0	13.9	1.00	1.00	1.00	1.00

The term "*components*" roughly means those entities into which p-variables are *summarized* or *reduced*. The kth columns of $\mathbf{F}$ and $\mathbf{A}$ are called the kth components.

Inequality (5.2) implies that (5.1) takes the form

$$\mathbf{X} = \mathbf{F}\,\mathbf{A}' + \mathbf{E} \tag{5.3}$$

That is, $\mathbf{X}$ is assumed to be approximated by the product of unknown matrices $\mathbf{F}$ and transposed $\mathbf{A}$, with the number of columns (components) in $\mathbf{F}$ and $\mathbf{A}$ being smaller than that of $\mathbf{X}$, as illustrated by the rectangles in (5.3).

The matrices to be obtained in PCA are PC score matrix $\mathbf{F}$ and loading matrix $\mathbf{A}$. For obtaining them, a *least squares method* is used; the sum of the squares of the errors in $\mathbf{E} = \mathbf{X} - \mathbf{F}\mathbf{A}'$,

$$f(\mathbf{F}, \mathbf{A}) = ||\mathbf{E}||^2 = ||\mathbf{X} - \mathbf{F}\mathbf{A}'||^2, \tag{5.4}$$

is minimized over $\mathbf{F}$ and $\mathbf{A}$. When $\mathbf{X}$ is the 6×4 matrix in Table 5.1(B) and m is set to 2, the function (5.4) is minimized for the matrices $\mathbf{F}$ and $\mathbf{A}$ shown in Table 5.2

Table 5.2 Matrices $\mathbf{F}$, $\mathbf{A}$, and $\mathbf{W}$ obtained for the data set in Table 5.1(B)

| F (PC scores) | | | A (loadings) | | | W (weights) | | |
	C1	C2		C1	C2		C1	C2
S1	−0.23	−0.93	M	7.16	6.87	M	0.01	0.05
S2	−1.46	−0.18	P	14.28	4.34	P	0.03	0.03
S3	1.65	−1.04	C	12.52	−1.96	C	0.02	−0.01
S4	0.62	0.73	B	11.02	−7.86	B	0.02	−0.06
S5	−0.81	−0.41						
S6	0.25	1.82						

(whose **W** is introduced later). There, it should be noticed that **A** is of variables $\times$ components, i.e., not transposed as in (5.1) or (5.3). As found in the table, the students (individuals), which have been assessed by four kinds of scores (variables) in **X**, are described by the two kinds of PC scores in **F**, while the relationships of the PC scores to the original four variables are described in **A**. How **F** and **A** are interpreted is explained in Sect. 5.4. Currently, readers need only to keep in mind that the original four variables have been reduced to two components, which implies that variables are explained by the components whose number is smaller than that of variables. Such a reduction is called *reduced rank approximation*, which is detailed in Appendix A.4.3.

5.2 Singular Value Decomposition

PCA solutions are given through the *singular value decomposition (SVD)* introduced in the note below. As SVD is one of the most important properties of matrices, carefully memorizing the following note as *absolute truth* is strongly recommended.

Note 5.1. Singular Value Decomposition (SVD)
Any $n \times p$ matrix **X** with rank(**X**) $= r$ can be decomposed as

$$\mathbf{X} = \mathbf{K\Lambda L'}. \tag{5.5}$$

Here, **K** $(n \times r)$ and **L** $(p \times r)$ satisfy

$$\mathbf{K'K} = \mathbf{L'L} = \mathbf{I}_r \tag{5.6}$$

and

$$\mathbf{\Lambda} = \begin{bmatrix} \lambda_1 & & \\ & \ddots & \\ & & \lambda_r \end{bmatrix} \tag{5.7}$$

is an $r \times r$ diagonal matrix whose diagonal elements are positive and arranged in decreasing order:

$$\lambda_1 \geq \cdots \geq \lambda_r > 0. \tag{5.8}$$

This decomposition is called singular value decomposition (SVD) and λ_k (the kth diagonal element of $\mathbf{\Lambda}$) is called the kth largest *singular value* of **X**.

In (5.7), the blank elements of $\mathbf{\Lambda}$ represent their being zero, with this expression used in the remaining parts of this book. SVD is described in more detail in Appendix A3, where the theorem in the above note is presented as a compact form of SVD in Theorem A.3.2.

Another expression of the SVD explained in Note 5.1 is given next:

Note 5.2. Another Expression of SVD (1)

Let us express the matrices $\mathbf{K}$ and $\mathbf{L}$ in Note 5.1 as $\mathbf{K} = [\mathbf{k}_1, \ldots, \mathbf{k}_m, \mathbf{k}_{m+1}, \ldots, \mathbf{k}_r] = [\mathbf{K}_m, \mathbf{K}_{[m]}]$ and $\mathbf{L} = [\mathbf{l}_1, \ldots, \mathbf{l}_m, \mathbf{l}_{m+1}, \ldots, \mathbf{l}_r] = [\mathbf{L}_m, \mathbf{L}_{[m]}]$, with

$$\mathbf{K}_m = [\mathbf{k}_1, \ldots, \mathbf{k}_m] \text{ and } \mathbf{L}_m = [\mathbf{l}_1, \ldots, \mathbf{l}_m] \text{ (the first } m \text{ columns)}, \tag{5.9}$$

$$\mathbf{K}_{[m]} = [\mathbf{k}_{m+1}, \ldots, \mathbf{k}_r] \text{ and } \mathbf{L}_{[m]} = [\mathbf{l}_{m+1}, \ldots, \mathbf{l}_r] \text{ (the remaining columns)}. \tag{5.10}$$

Then, (5.6) can be rewritten as $\mathbf{k}'_u\mathbf{k}_u = \mathbf{l}'_u\mathbf{l}_u = 1$ and $\mathbf{k}'_u\mathbf{k}_v = \mathbf{l}'_u\mathbf{l}_v = 0$ for $u \neq v$ ($u = 1, \ldots, r$; $v = 1, \ldots, r$). Further, (5.5) can be rewritten as $\mathbf{X} = \lambda_1\mathbf{k}_1\mathbf{l}'_1 + \cdots + \lambda_m\mathbf{k}_m\mathbf{l}'_m + \lambda_{m+1}\mathbf{k}_{m+1}\mathbf{l}'_{m+1} + \cdots + \lambda_r\mathbf{k}_r\mathbf{l}'_r$, which is expressed in matrix form as

$$\mathbf{X} = \mathbf{K}\mathbf{\Lambda}\mathbf{L}' = \mathbf{K}_m\mathbf{\Lambda}_m\mathbf{L}'_m + \mathbf{K}_{[m]}\mathbf{\Lambda}_{[m]}\mathbf{L}'_{[m]}, \tag{5.11}$$

with

$$\mathbf{\Lambda}_m = \begin{bmatrix} \lambda_1 & & \\ & \ddots & \\ & & \lambda_m \end{bmatrix} \text{ and } \mathbf{\Lambda}_{[m]} = \begin{bmatrix} \lambda_{m+1} & & \\ & \ddots & \\ & & \lambda_r \end{bmatrix}; \text{ i.e.,} \tag{5.12}$$

$$\mathbf{\Lambda} = \begin{bmatrix} \mathbf{\Lambda}_m & \\ & \mathbf{\Lambda}_{[m]} \end{bmatrix}.$$

Further, SVD has the following important property, which is directly related to the PCA solution minimizing (5.4):

Note 5.3. SVD and Least Squares Solution

Let $\mathbf{X}$ be an $n \times p$ matrix whose SVD is defined as in Notes 5.1 and 5.2, $\mathbf{F}$ be an $n \times m$ matrix, and $\mathbf{A}$ be a $p \times m$ matrix, with $m \leq r = \text{rank}(\mathbf{X}) \leq \min(n, p)$. Then,

$$f(\mathbf{F}\mathbf{A}') = ||\mathbf{X} - \mathbf{F}\mathbf{A}'||^2 \tag{5.13}$$

is minimized for

$$\mathbf{F}\mathbf{A}' = \mathbf{K}_m\mathbf{\Lambda}_m\mathbf{L}'_m, \tag{5.14}$$

with $\mathbf{K}_m$, $\mathbf{L}_m$, and $\mathbf{\Lambda}_m$ defined as in (5.9) and (5.12).

The fact in the above note is proved by Theorem A.4.5 with (A.4.17) in Appendix A.4.3. The theorem is referred to as Eckart and Young's (1936) theorem in some of the literature.

/ in Note 5.1 and the solution in Note 5.3. The SVD (5.5) for the **X** in Table 5.1 (B) is given as

X				K				Λ				L′			
-4.9	-10.6	0.3	5.3	-0.10	-0.38	0.19	-0.71	56.57				0.31	0.62	0.54	0.48
-6.7	-23.4	-22.8	-10.1	-0.60	-0.07	0.52	0.42		28.10			0.60	0.38	-0.17	-0.68
4.7	19.9	20.6	27.4	0.67	-0.42	0.07	0.44			15.72		0.68	-0.37	-0.40	0.49
10.5	10.7	7.2	1.0	0.25	0.30	0.05	-0.33				5.16	-0.29	0.58	-0.72	0.25
-17.6	-8.3	-4.6	-12.0	-0.33	-0.17	-0.82	0.11								
14.0	11.8	-0.7	-11.6	0.10	0.74	-0.02	0.09								

(with $=$ between X and K)

$$(5.15)$$

Note 5.3 thus shows that the solution of **FA′** for minimizing (5.13) is given by

$\mathbf{K}_m$		$\mathbf{\Lambda}_m$		$\mathbf{L}_m'$			
-0.10	-0.38	56.57		0.31	0.62	0.54	0.48
-0.60	-0.07		28.10	0.60	0.38	-0.17	-0.68
0.67	-0.42						
0.25	0.30						
-0.33	-0.17						
0.10	0.74						

$$\mathbf{FA'} = \qquad\qquad\qquad\qquad\qquad\qquad\qquad\qquad\qquad (5.16)$$

We should note that SVD provides the solution of **FA′** in function (5.4) for PCA, but not each of **F** and **A** is given. Their solutions are generally expressed as

$$\mathbf{F} = \mathbf{K}_m \mathbf{\Lambda}_m^{\alpha} \mathbf{S}, \qquad\qquad (5.17)$$

$$\mathbf{A} = \mathbf{L}_m \mathbf{\Lambda}_m^{1-\alpha} \mathbf{S}'^{-1}, \qquad\qquad (5.18)$$

with α and **S** being arbitrary scalar and nonsingular matrices, respectively. We can easily verify that (5.17) and (5.18) meet (5.14). That is, there are multiple solutions for **F** and **A**. One of the solutions has been shown in Table 5.2, as explained in Sect. 5.4.

5.3 Formulation with a Weight Matrix

Notes 5.1 and 5.2 lead to the following facts:

Note 5.4. Another Expression of SVD (2)
Let $\mathbf{K}_m$, $\mathbf{L}_m$, and $\mathbf{\Lambda}_m$ be the matrices defined in Note 2. The post-multiplication of $\mathbf{K}_m$ and $\mathbf{L}_m$ by $\mathbf{\Lambda}_m$ can be expressed as

$$\mathbf{K}_m\mathbf{\Lambda}_m = \mathbf{XL}_m, \tag{5.19}$$

$$\mathbf{L}_m\mathbf{\Lambda}_m = \mathbf{X}'\mathbf{K}_m. \tag{5.20}$$

The facts in the above note are proved in Appendix A.3.3.

By comparing (5.19) with (5.17), we can rewrite it as $\mathbf{F} = \mathbf{K}_m\mathbf{\Lambda}_m\mathbf{\Lambda}_m^{\alpha-1}\mathbf{S} = \mathbf{XL}_m\mathbf{\Lambda}_m^{\alpha-1}\mathbf{S}$, i.e.,

$$\mathbf{F} = \mathbf{XW} \tag{5.21}$$

with

$$\mathbf{W} = \mathbf{L}_m\mathbf{\Lambda}_m^{\alpha-1}\mathbf{S} \tag{5.22}$$

a p-variables $\times$ m-components matrix that we refer to as a *weight matrix*. Equation (5.21) shows that the PC score matrix $\mathbf{F}$ is expressed as the data matrix post-multiplied by the weight matrix.

This fact shows that PCA may be formulated, by using (5.21) in (5.4), as minimizing

$$f(\mathbf{W}, \mathbf{A}) = ||\mathbf{X} - \mathbf{XWA}'||^2 \tag{5.23}$$

over $\mathbf{W}$ and $\mathbf{A}$. This minimization is equivalent to minimizing (5.4) over $\mathbf{F}$ and $\mathbf{A}$. Some authors have first presented (5.23) rather than (5.4) as the loss function for PCA, where the term loss function refers to the one to be minimized; its examples are (5.23), (5.4), and (4.7).

Equation (5.21) implies that the resulting PC scores are *centered* ones with

$$\mathbf{1}'_n\mathbf{F} = \mathbf{0}'_m, \text{ i.e., } \mathbf{JF} = \mathbf{F}, \tag{5.24}$$

when PCA is performed for a data matrix of centered scores with $\mathbf{1}'_n\mathbf{X} = \mathbf{0}'_p$, since $\mathbf{1}'_n\mathbf{F} = \mathbf{1}'_n\mathbf{XW} = \mathbf{0}'_p\mathbf{W} = \mathbf{0}'_m$, and it is equivalent to $\mathbf{JF} = \mathbf{F}$, as proved in Note 3.1.

5.4 Constraints for Components

For selecting a single set of $\mathbf{F}$ and $\mathbf{A}$ from the multiple solutions satisfying (5.17) and (5.18), we must impose constraints on $\mathbf{F}$ and $\mathbf{A}$. There are various types of constraints, and one of them is that

$$\frac{1}{n}\mathbf{F}'\mathbf{F} = \mathbf{I}_m, \tag{5.25}$$

$$\mathbf{A}'\mathbf{A} \text{ is a diagonal matrix whose}$$
$$\text{diagonal elements are arranged in decreasing order.} \quad (5.26)$$

The solution that satisfies this constraint is given as:

$$\mathbf{F} = \sqrt{n}\mathbf{K}_m, \quad (5.27)$$

$$\mathbf{A} = \frac{1}{\sqrt{n}}\mathbf{L}_m\mathbf{\Lambda}_m, \quad (5.28)$$

which are derived from (5.17) by setting $\alpha = 0$ and $\mathbf{S} = n^{1/2}\mathbf{I}_m$. We can verify that (5.27) and (5.28) satisfy (5.25) and (5.26) by noting that (5.6) and (5.9) imply $\mathbf{K}'_m\mathbf{K}_m = \mathbf{L}'_m\mathbf{L}_m = \mathbf{I}_m$. Under (5.25) and (5.26), the weight matrix is expressed as

$$\mathbf{W} = \sqrt{n}\mathbf{L}_m\mathbf{\Lambda}_m^{-1}, \quad (5.29)$$

which is derived from (5.22) by setting $\alpha = 0$ and $\mathbf{S} = n^{1/2}\mathbf{I}_m$. Table 5.2 shows the solutions of (5.27), (5.28), and (5.29) for the data in Table 5.1(B).

To consider the implications of constraints (5.25) and (5.26), we express the columns of $\mathbf{F}$ and $\mathbf{A}$ as $\mathbf{F} = [\mathbf{f}_1, \ldots, \mathbf{f}_m]$ and $\mathbf{A} = [\mathbf{a}_1, \ldots, \mathbf{a}_m]$, where the elements of $\mathbf{f}_k$ are called the kth PC scores and those of $\mathbf{a}_k$ are called the kth loadings ($k = 1, \ldots, m$). Let us note (5.24) and recall (3.22). They show that the left-hand side $n^{-1}\mathbf{F}'\mathbf{F}$ in (5.25) is the inter-component covariance matrix of PC scores, whose diagonal elements $n^{-1}\mathbf{f}'_k\mathbf{f}_k$ are variances, and whose off-diagonal elements $n^{-1}\mathbf{f}'_k\mathbf{f}_l$ ($k \neq l$) are covariances. The variances and covariances are constrained to be one and zero, respectively, in (5.25). This implies the following:

[1] PC scores are standardized.
[2] The kth PC scores are uncorrelated with the lth PC ones with $\mathbf{f}'_k\mathbf{f}_l = 0$ for $k \neq l$.

Similarly, the constraint of $\mathbf{A}'\mathbf{A}$ being a diagonal matrix in (5.26) is rewritten as $\mathbf{a}_k'\mathbf{a}_l = 0$ for $k \neq l$, which does not imply that $\mathbf{a}_k$ is uncorrelated with $\mathbf{a}_l$, since $\mathbf{1}_p'\mathbf{A} \neq \mathbf{0}_m$, in general, but allows the loadings to have the following property:

[3] The kth loading vector $\mathbf{a}_k$ is orthogonal to the lth one $\mathbf{a}_l$.

The properties are desirable in that [2] and [3] allow different components to be distinct and [1] makes it easier to compare PC scores between different components. Further, [1] leads to the following property:

[4] $\mathbf{A}(p \times m)$ is the covariance matrix between p-variables and m-components, in particular the correlation matrix when $\mathbf{X}$ is standardized.

It is proved as follows: we can use (5.20) and (5.27) to rewrite (5.28) as

$$\mathbf{A} = \frac{1}{\sqrt{n}}\mathbf{L}_m\mathbf{\Lambda}_m = \frac{1}{\sqrt{n}}\mathbf{X}'\mathbf{K}_m = \frac{1}{n}\mathbf{X}'\mathbf{F}, \quad (5.30)$$

Table 5.3 Matrices **F**, **A**, **W** obtained for the data set in Table 5.1(C)

F (PC scores)			A (loadings)			W (weights)		
	C1	C2		C1	C2		C1	C2
S1	−0.23	−0.78	M	0.70	0.66	M	0.24	0.79
S2	−1.43	0.13	P	0.94	0.20	P	0.33	0.24
S3	1.58	−1.10	C	0.94	−0.24	C	0.33	−0.28
S4	0.66	0.72	B	0.77	−0.56	B	0.27	−0.66
S5	−0.92	−0.74						
S6	0.33	1.76						

which equals (3.25) and is the covariance matrix for **X** and **F**, since of $1'_n\mathbf{X} = \mathbf{0}'_p$ and (5.24). Further, if **X** is standardized, (5.30) is the correlation matrix, because of property [1].

Note that the loading matrix **A** in Table 5.2 is the covariance matrix for **X** and **F**, but is not their correlation matrix, since it is the result for the data set which is not standardized. On the other hand, Table 5.3 shows the PCA solution for standardized scores in Table 5.1(C), where the constraints (5.25) and (5.26) are imposed. The **A** in Table 5.3 is the correlation matrix between variables and components. The solution has been given through SVD:

$$
\begin{array}{cccc}
\mathbf{X} & \mathbf{K} & \mathbf{\Lambda} & \mathbf{L'}
\end{array}
$$

X	K	Λ	L'
-0.45 -0.70 0.02 0.38	-0.09 -0.32 -0.26 0.72	4.15	0.41 0.56 0.56 0.46
-0.61 -1.54 -1.75 -0.73	-0.58 0.05 -0.55 -0.41	2.24	0.72 0.22 -0.26 -0.61
0.43 1.31 1.58 1.97 =	0.65 -0.45 -0.13 -0.44	1.23	-0.51 0.47 0.4 -0.6
0.96 0.70 0.55 0.07	0.27 0.29 0.02 0.33	0.38	0.23 -0.65 0.68 -0.25
-1.62 -0.55 -0.35 -0.86	-0.38 -0.30 0.77 -0.10		
1.29 0.78 -0.05 -0.83	0.14 0.72 0.16 -0.10		

$$ (5.31) $$

with **X** being the matrix in Table 5.1(C).

5.5 Interpretation of Loadings

Let us define the columns of matrices as $\mathbf{X} = [\mathbf{x}_1, \ldots, \mathbf{x}_p]$, $\mathbf{A'} = [\mathbf{a}_1, \ldots, \mathbf{a}_p]$, and $\mathbf{E} = [\mathbf{e}_1, \ldots, \mathbf{e}_p]$, with $\mathbf{x}_j$, $\mathbf{a}_j$, and $\mathbf{e}_j$ ($j = 1, \ldots, p$) corresponding to variable j (i.e., $\mathbf{a}_j'$ the jth row vector of **A**). Then, the PCA model (5.1) is rewritten as follows:

$$ \mathbf{x}_j = \mathbf{Fa}_j + \mathbf{e}_j. \tag{5.32} $$

It takes the same form as (4.5) except that (5.32) does not include an intercept. That is, PCA can be regarded as the *regression* of $\mathbf{x}_j$ onto **F**. When $m = 2$, as in Table 5.3, (5.32) is expressed as follows:

$$\mathbf{x}_j = a_{j1}\mathbf{f}_1 + a_{j2}\mathbf{f}_2 + \mathbf{e}_j \quad (j = 1, \ldots, p), \tag{5.33}$$

with $\mathbf{F} = [\mathbf{f}_1, \mathbf{f}_2]$ and $\mathbf{a}_j = [a_{j1}, a_{j2}]'$. That is, $\mathbf{f}_1$ and $\mathbf{f}_2$ can be viewed as the *explanatory* variables for a dependent variable $\mathbf{x}_j$, with loadings a_{j1} and a_{j2} as the *regression coefficients*. The equation is further rewritten as follows:

$$x_{ij} = a_{j1}f_{i1} + a_{j2}f_{i2} + e_{ij} \quad (j = 1, \ldots, p), \tag{5.34}$$

using $\mathbf{X} = (x_{ij})$, $\mathbf{F} = (f_{ik})$, and $\mathbf{A} = (a_{jk})$.

On the basis of (5.34), we can interpret the loadings in Table 5.3 as follows:

[A1]   All a_{j1} $(j = 1, \ldots, p)$ show fairly large positive values for all variables (courses), which implies that students with higher values of f_{i1} (the 1st PC score) tend to show higher scores x_{ij} for all courses $(j = 1, \ldots, p)$. The first component can thus be interpreted as standing for "a general ability" common to M, P, C, and B.

[A2] a_{j2} $(j = 1, 2)$ show positive loadings for M and P, but negative ones for C and B. Component 2 can be interpreted as standing for "a specific ability" advantageous for M and P, but disadvantageous for C and B.

As described in (5.30), the loadings in Table 5.3 can also be regarded as the correlation coefficients of variables to components. For example, courses P and C are very highly correlated with Component 1, since the corresponding coefficient 0.94 is close to the upper limit.

5.6   Interpretation of Weights

The role of weight matrix $\mathbf{W}$ is easily understood by rewriting (5.21) as follows:

$$\mathbf{f}_k = \mathbf{X}\mathbf{w}_k = w_{1k}\mathbf{x}_1 + \cdots + w_{pk}\mathbf{x}_p, \tag{5.35}$$

with $\mathbf{w}_k = [w_{1k}, \ldots, w_{pk}]'$ the kth column of $\mathbf{W} = (w_{jk})$. In (5.35), we find that the elements in $\mathbf{W}$ provide the *weights* by which variables are multiplied to form the PC scores in $\mathbf{F}$. We can further rewrite (5.35) as

$$f_{ik} = w_{1k}x_{i1} + \cdots + w_{pk}x_{ip}. \tag{5.36}$$

This allows us to interpret the $\mathbf{W}$ in Table 5.3 as follows:

[W1]   All w_{j1} $(j = 1, \ldots, p)$ show positive values for all variables (courses), which shows that the first PC score is the sum of all variables positively weighted. Thus, the score can be interpreted as standing for "a general ability" common to M, P, C, and B.

[W2] w_{j2} $(j = 1, 2)$ show positive values for M and P, but negative ones for C and
B; the second PC scores are higher for students who are superior in M and
P, while the scores are lower for those who are superior in C and B. The
scores can thus be interpreted as standing for "a specific ability" advantageous for M and P, but disadvantageous for C and B.

Those interpretations are congruous with [A1] and [A2] in the last section.

5.7 Percentage of Explained Variance

In this section, we consider assessing the amount of the errors for the resulting
solutions. Substituting SVD (5.5) and solution (5.14) into $\mathbf{X}$ and $\mathbf{FA}'$, respectively,
in the squared sum of errors (5.4), its resulting value can be expressed as follows:

$$\begin{aligned}
||\mathbf{E}||^2 &= ||\mathbf{K\Lambda L}' - \mathbf{K}_m\mathbf{\Lambda}_m\mathbf{L}'_m||^2 \\
&= \mathrm{tr}\mathbf{\Lambda K}'\mathbf{K\Lambda L}' - 2\mathrm{tr}\mathbf{L\Lambda K}'\mathbf{K}_m\mathbf{\Lambda}_m\mathbf{L}'_m + \mathit{tr}\mathbf{L}_m\mathbf{\Lambda}_m\mathbf{K}'_m\mathbf{K}_m\mathbf{\Lambda}_m\mathbf{L}'_m \quad (5.37) \\
&= \mathrm{tr}\mathbf{\Lambda}^2 - \mathrm{tr}\mathbf{\Lambda}_m^2 \geq 0.
\end{aligned}$$

Here, we have used $\mathbf{K}_m'\mathbf{K}_m = \mathbf{L}_m'\mathbf{L}_m = \mathbf{I}_m$ and

$$\mathbf{K}'\mathbf{K}_m = \mathbf{L}'\mathbf{L}_m = \begin{bmatrix} 1 & & & \\ & \ddots & & \\ & & 1 & \\ 0 & \cdots & 0 & \\ & & \vdots & \\ 0 & \cdots & 0 \end{bmatrix} :$$

$\mathbf{K}'\mathbf{K}_m = \mathbf{L}'\mathbf{L}_m$ equals the $r \times m$ matrix whose first m rows are those of $\mathbf{I}_m$ and the
remaining rows are filled with zeros, which follow from the fact that (5.6) can be
rewritten as $\mathbf{k}_u'\mathbf{k}_u = \mathbf{l}_u'\mathbf{l}_u = 1$ and $\mathbf{k}_u'\mathbf{k}_v = \mathbf{l}_u'\mathbf{l}_v = 0$ for $u \neq v$ $(u = 1, ..., m; v = 1, ...,$
$m)$, as described in Note 5.2. Dividing both sides of (5.37) by $\mathrm{tr}\mathbf{\Lambda}^2$ leads to

$$\frac{||\mathbf{E}||^2}{\mathrm{tr}\mathbf{\Lambda}^2} = 1 - \mathrm{PEV}_m \geq 0, \quad (5.38)$$

with

$$\mathrm{PEV}_m = \frac{\mathrm{tr}\mathbf{\Lambda}_m^2}{\mathrm{tr}\,\mathbf{\Lambda}^2} = \frac{\mathrm{tr}\mathbf{\Lambda}_m^2}{||\mathbf{X}||^2}. \quad (5.39)$$

Here, we have used

$$\|\mathbf{X}\|^2 = \text{tr}\mathbf{X}'\mathbf{X} = \text{tr}\mathbf{L}\mathbf{\Lambda}\mathbf{K}'\mathbf{K}\mathbf{\Lambda}\mathbf{L}' = \text{tr}\mathbf{\Lambda}^2. \tag{5.40}$$

Since (5.38) expresses the *largeness of errors* with taking a nonnegative value, (5.39) indicates the *smallness of errors*, i.e., how well $\mathbf{FA}'$ approximates $\mathbf{X}$, by taking a value within the range [0, 1]. Some different terms are used for proportion (5.39). One of them is the *proportion of explained variance* (PEV), since (5.39) can be rewritten as

$$\text{PEV}_m = \frac{\frac{1}{n}\text{tr}(\mathbf{K}_m\mathbf{\Lambda}_m\mathbf{L}'_m)'\mathbf{K}_m\mathbf{\Lambda}_m\mathbf{L}'_m}{\frac{1}{n}\text{tr}\mathbf{X}'\mathbf{X}} = \frac{\frac{1}{n}\text{tr}(\mathbf{FA}')'\mathbf{FA}'}{\text{tr}\,\mathbf{V}}, \tag{5.41}$$

with $\mathbf{V} = n^{-1}\mathbf{X}'\mathbf{X}$ the covariance matrix given in (3.22); the denominator of (5.41) is the sum of the variances of p-variables, while the numerator is the sum of the variances of the columns of $\mathbf{FA}'$, i.e., (5.14), since (5.24) implies that $\mathbf{FA}'$ is centered with $\mathbf{1}'_n\mathbf{FA}' = \mathbf{0}'_p$.

The PEV for the solution with $m = 2$ in Table 5.3 is obtained as

$$\text{PEV}_2 = \frac{4.15^2 + 2.24^2}{4.15^2 + 2.24^2 + 1.23^2 + 0.38^2} = \frac{22.24}{23.90} = 0.93, \tag{5.42}$$

using (5.31). This implies that 93 % of the data variances are accounted for by two components; in other words, 7 % (= 100 − 93) of the variances remain unexplained. If we adopt the $m = 3$ solution, then the PEV is given as follows:

$$\text{PEV}_3 = \frac{4.15^2 + 2.24^2 + 1.23^2}{4.15^2 + 2.24^2 + 1.23^2 + 0.38^2} = \frac{23.75}{23.90} = 0.99.$$

The PEV for the solution with $m = 2$ in Table 5.2 is obtained as

$$\text{PEV}_2 = \frac{56.57^2 + 28.10^2}{56.57^2 + 28.10^2 + 15.72^2 + 5.16^2} = \frac{3989.78}{4263.52} = 0.94, \tag{5.43}$$

using (5.15) and (5.16). This differs from (5.42); the difference is not due to a round-off error. This shows that the PCA solution for a centered data matrix without standardization differs from that for the standard scores for the same data matrix. The latter solution cannot be straightforwardly transformed from the former, which differs from the regression analysis in the last chapter.

5.8 High-Dimensional Data Analysis

Recently, we have often encountered data sets with many more variables than individuals, i.e., an $n \times p$ data matrix $\mathbf{X}$ with $p \gg n$. Such a data set is said to be *high-dimensional* (e.g., Kock 2014). In order to find a few components underlying a

number of variables, PCA is useful. In this section, we illustrate PCA for
high-dimensional data using Yeung and Ruzzo's (2001) gene expression data with
$n = 17$ time points and $p = 384$ genes. The data matrix is publicly available at
http://faculty.washington.edu/kayee/pca.

We performed PCA for the data set with $m = 4$. The solution shows
$PEV_4 = 0.81$, which implies that 81 % of the variances in 384 variables are
explained by only four components. For the resulting *loading* matrix, we performed
a *varimax rotation*, which is described in Chap. 13, for the following reason:

Note 5.5. Rotation of Components
If constraint (5.26) is removed and only (5.25) is considered, (5.1) can be
rewritten as follows:

$$\mathbf{X} = \mathbf{FA}' + \mathbf{E} = \mathbf{FTT}'\mathbf{A}' + \mathbf{E} = \mathbf{F_T}\mathbf{A}'_\mathbf{T} + \mathbf{E}. \tag{5.44}$$

Here,

$$\mathbf{F_T} = \mathbf{FT} \text{ and } \mathbf{A_T} = \mathbf{AT}, \tag{5.45}$$

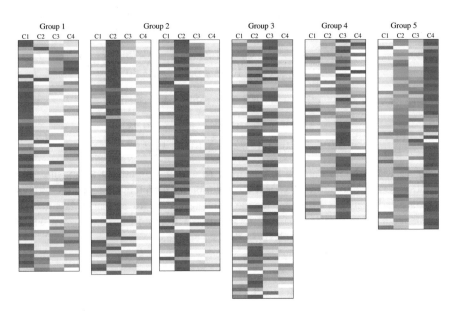

Fig. 5.1 Loadings for gene expression data with *red* and *blue* expressing positive and negative
values, respectively

with $\mathbf{T}$ a special matrix satisfying $\mathbf{T}'\mathbf{T} = \mathbf{T}\mathbf{T}' = \mathbf{I}_m$, which is detailed in Appendix A.1.2. If $\mathbf{F}$ meets (5.25), $\mathbf{F}_T$ also satisfies it:

$$\frac{1}{n}\mathbf{F}_T'\mathbf{F}_T = \frac{1}{n}\mathbf{T}'\mathbf{F}'\mathbf{F}\mathbf{T} = \mathbf{T}'\mathbf{T} = \mathbf{I}_m. \qquad (5.46)$$

Equations (5.44) and (5.46) imply that if $\mathbf{F}$ and $\mathbf{A}$ are the PCA solution that minimizes (5.4) subject to (5.25), so are $\mathbf{F}_T$ and $\mathbf{A}_T$.

The above $\mathbf{T}$ can be chosen by the *rotation* techniques in Chap. 13, so that the resulting $\mathbf{A}_T$ is easily interpreted.

The resulting loading matrix $\mathbf{A}_T$ is of 384×4, which is too big to capture the values of its elements. Such a matrix can be effectively presented by a *heat map*, in which the largeness of the absolute values of each element is represented as the *depth of color* in the cell corresponding to each element. Figure 4 shows a heat map for the resulting loadings, block-wise. There, the blocks correspond to the five groups into which the 384 genes are known to be categorized; each block is a matrix whose rows and columns are occupied by the genes in the corresponding group and the four components, respectively, though the genes in Group 2 are divided into two blocks. The solution is considered to be reasonable, as each phase has a unique feature of the loadings: The genes in Groups 1, 2, 4, and 5 positively load Components 1, 2, 3, and 4, respectively, while those in Group 3 positively load both Components 2 and 3.

5.9 Bibliographical Notes

Jolliffe (2002) exhaustively details various aspects of PCA. A subject that has not been treated in this book is the graphical *biplot* methodology for jointly representing the PC scores of individuals and the loadings of variables in a single configuration (Gower et al. 2011). The author of the present book has proposed a modified PCA procedure for easily capturing the biplot (Adachi 20).

A three-way data array is often observed whose element can be expressed as x_{ijk}, with i, j, and k standing for an individual, a variable, and an occasion, respectively, for example. The PCA formulation in this chapter is extended to the approximation of the three-way data array by the reduced components. Such extended PCA procedures are called *three-way PCA*, and some procedures have been developed, as found in Adachi (2016), Kroonenberg (2008), Smilde et al. (2004); and Tucker (1966).

Exercises

5.1. Write Eqs. (5.5–5.8) ten times, using different characters for matrices, in order to learn SVD by heart.

5.2. Show that (5.5) can be rewritten as $\mathbf{X} = \tilde{\mathbf{K}}\tilde{\mathbf{\Lambda}}\tilde{\mathbf{L}}'$ for $n \geq p$. Here, $\tilde{\mathbf{K}}$ and $\tilde{\mathbf{L}}$ are the $n \times p$ and $p \times p$ matrices, respectively, satisfying $\tilde{\mathbf{K}}'\tilde{\mathbf{K}} = \tilde{\mathbf{L}}'\tilde{\mathbf{L}} = \mathbf{I}_p$, while $\tilde{\mathbf{\Lambda}}$ is a $p \times p$ diagonal matrix with

$$\tilde{\mathbf{\Lambda}} = \begin{bmatrix} \lambda_1 & & & \\ & \ddots & & \\ & & \lambda_r & \\ & & & {}_{p-r}\mathbf{O}_{p-r} \end{bmatrix}.$$

This is the extended version of SVD in Appendix A.3.1.

5.3. Show that the error matrix $\mathbf{E} = \mathbf{X} - \mathbf{FA}'$ resulting in the minimization of (5.4) is expressed as $\mathbf{E} = \mathbf{K}_{[m]}\mathbf{\Lambda}_{[m]}\mathbf{L}_{[m]}'$, with its right-hand side defined as in (5.11) and the resulting PC scores are uncorrelated with the errors with $\mathbf{F}'\mathbf{E} = {}_m\mathbf{O}_p$.

5.4. Show that (5.14) can be rewritten as $\mathbf{FA}' = \lambda_1\mathbf{k}_1\mathbf{l}_1' + \cdots + \lambda_m\mathbf{k}_m\mathbf{l}_m'$ using (5.9).

5.5. Without condition $m \leq \text{rank}(\mathbf{X})$, show that the problem in Note 5.3 would be trivial, with its solution $\mathbf{FA}' = \mathbf{X}$.

5.6. If constraint (5.25) was replaced by $\mathbf{F}'\mathbf{F} = \mathbf{A}'\mathbf{A}$, show that (5.27) and (5.28) must be replaced by $\mathbf{F} = \mathbf{K}_m\mathbf{\Lambda}_m^{1/2}$ and $\mathbf{A} = \mathbf{L}_m\mathbf{\Lambda}_m^{1/2}$, respectively.

5.7. Show that the SVD in Notes 5.1 and 5.2 implies $\mathbf{K}_m = \mathbf{XL}_m\mathbf{\Lambda}_m^{-1}$.

5.8. Discuss the similarities and differences between the loading matrix $\mathbf{A}$ and the weight matrix $\mathbf{W}$.

5.9. Show $\text{PEV}_m \leq \text{PEV}_{m+1}$ for (5.39).

5.10. Let us define

$$\mathbf{X}^+ = \mathbf{L}\mathbf{\Lambda}^{-1}\mathbf{K}' \tag{5.47}$$

for the matrix $\mathbf{X}$ whose SVD is defined in Note 5.1. Show that $\mathbf{X}^+$ satisfies $\mathbf{XX}^+\mathbf{X} = \mathbf{X}$, $\mathbf{X}^+\mathbf{XX}^+ = \mathbf{X}^+$, $(\mathbf{XX}^+)' = \mathbf{XX}^+$, and $(\mathbf{X}^+\mathbf{X})' = \mathbf{X}^+\mathbf{X}$. Matrix (5.47) is called the *Moore–Penrose inverse* of $\mathbf{X}$, which plays various important roles in statistics.

5.11. If $\mathbf{X}$ is nonsingular, show that its inverse matrix $\mathbf{X}^{-1}$ is a special case of the Moore–Penrose inverse $\mathbf{X}^+$ (5.47).

5.12. Show that (5.47) can be rewritten as $\mathbf{X}^+ = (\mathbf{X}'\mathbf{X})^+\mathbf{X}' = \mathbf{X}'(\mathbf{XX}')^+$

5.13. Show that $\mathbf{X}(\mathbf{X}'\mathbf{X})^+\mathbf{X}' = \mathbf{KK}'$ using the SVD in Note 5.1, with $(\mathbf{X}'\mathbf{X})^+$ the Moore–Penrose inverse of $\mathbf{X}'\mathbf{X}$ defined as in Exercise 5.10.

5.14. Let us define an $n \times q$ matrix $\mathbf{Y} = (\mathbf{I}_n - \mathbf{X}'^+\mathbf{X}')\mathbf{Q}$, with $\mathbf{Q}$ an arbitrary $n \times q$ matrix and $\mathbf{X}^+$ the Moore–Penrose inverse of $\mathbf{X}$ ($n \times p$) expressed as (5.47). Show $\mathbf{Y}'\mathbf{X} = {}_q\mathbf{O}_p$.

5.15. Show that the Moore–Penrose inverse (5.47) is defined for every matrix.

5.16. As with SVD and the Moore–Penrose inverse, *QR decomposition* is also defined for every matrix. Here, the QR decomposition of $\mathbf{A}(p \times m)$ is expressed as $\mathbf{A} = \mathbf{QR}$, with $\mathbf{Q}'\mathbf{Q} = \mathbf{I}_m$ and the elements of $\mathbf{R} = (r_{jk})$ ($m \times m$) being zero for $j > k$. Verify that

A					Q					R			
-0.27	-1.74	1.24	1.58		-0.09	-0.32	-0.26	0.72		3	-2	4	6
-1.74	1.46	-2.57	-6.48		-0.58	0.05	-0.55	-0.41		0	6	-5	3
1.95	-4.00	4.85	0.09	=	0.65	-0.45	-0.13	-0.44		0	0	0	2
0.81	1.20	-0.37	4.18		0.27	0.29	0.02	0.33		0	0	0	5
-1.14	-1.04	-0.02	-2.14		-0.38	-0.30	0.77	-0.10					
0.42	4.04	-3.04	2.82		0.14	0.72	0.16	-0.10					

represents a QR decomposition.

Chapter 6
Principal Component Analysis (Part 2)

In this chapter, principal component analysis (PCA) is reformulated. The loss function to be minimized is the same as that in the previous chapter, but the constraints for the matrices are different. This reformulation gives two purposes of PCA that were not found in the previous chapter. They are [1] forming a weighted composite score with the maximum variance and [2] *visualizing* a *high-dimensional invisible distribution* of individuals. In Sects. 6.1 and 6.2, the reformulation of PCA is mathematically described, followed by illustrations of the two purposes in Sects. 6.3, 6.4, and 6.5. Finally, a subject parallel to that in Sect. 5.7 is treated in Sect. 6.6.

6.1 Reformulation with Different Constraints

Let $\mathbf{X}$ denotes an n-individuals $\times$ p-variables centered data matrix with $\mathbf{1}_n'\mathbf{X} = \mathbf{0}_p'$, as in the previous section. As described there, PCA is formulated as minimizing (5.4), which is equivalent to (5.23), i.e., minimizing

$$f(\mathbf{W}, \mathbf{A}) = ||\mathbf{X} - \mathbf{F}\mathbf{A}'||^2 = ||\mathbf{X} - \mathbf{X}\mathbf{W}\mathbf{A}'||^2 \qquad (6.1)$$

over weight matrix $\mathbf{W}$ and loading matrix $\mathbf{A}$ with $\mathbf{F} = \mathbf{X}\mathbf{W}$ containing PC scores. The solutions for $\mathbf{W}$ and $\mathbf{A}$ are expressed as (5.18) and (5.22), which are presented again here:

$$\mathbf{A} = \mathbf{L}_m\mathbf{\Lambda}_m^{1-a}\mathbf{S}'^{-1}, \qquad (6.2)$$

$$\mathbf{W} = \mathbf{L}_m\mathbf{\Lambda}_m^{a-1}\mathbf{S}. \qquad (6.3)$$

© Springer Nature Singapore Pte Ltd. 2016
K. Adachi, *Matrix-Based Introduction to Multivariate Data Analysis*,
DOI 10.1007/978-981-10-2341-5_6

Here, α and $\mathbf{S}$ are arbitrary scalar and nonsingular matrices, respectively, which show that multiple solutions exist. To select a single solution among them, we consider the following constraints:

$$\mathbf{F'F} = \mathbf{W'X'XW} = \text{a diagonal matrix whose}$$
$$\text{diagonal elements are arranged in descending order,} \tag{6.4}$$

$$\mathbf{W'W} = \mathbf{I}_m, \tag{6.5}$$

which differ from constraints (5.25) and (5.26) in the last chapter. Then, the solution for $\mathbf{W}$ and $\mathbf{A}$ is expressed as:

$$\mathbf{W} = \mathbf{A} = \mathbf{L}_m. \tag{6.6}$$

Both matrices are identical. Using this and (5.19) leads to $\mathbf{F} = \mathbf{XW} = \mathbf{XL}_m = \mathbf{K}_m\mathbf{\Lambda}_m.$ This fact and $\mathbf{K}'_m\mathbf{K}_m = \mathbf{I}_m$ imply

$$\mathbf{F'F} = \mathbf{W'X'XW} = \mathbf{\Lambda}^2_m. \tag{6.7}$$

We thus find that the resulting $\mathbf{F} = \mathbf{XW}$ satisfies (6.4), because of (5.7) and (5.12). The identity of $\mathbf{W}$ to $\mathbf{A}$ in (6.6) shows that we may rewrite (6.1) as:

$$f(\mathbf{W}, \mathbf{A}) = ||\mathbf{X} - \mathbf{XWW'}||^2 \tag{6.8}$$

without $\mathbf{A}$.

6.2   Maximizing the Sum of Variances

Minimization of (6.8) subject to (6.4) and (6.5) is equivalent to maximizing

$$g(\mathbf{W}) = \text{tr}\frac{1}{n}\mathbf{F'F} = \frac{1}{n}\text{tr}\mathbf{W'X'XW} \tag{6.9}$$

subject to the same constraints. The equivalence is shown by expanding (6.8) as:

$$f(\mathbf{W}) = \text{tr}\mathbf{X'X} - 2\text{tr}\mathbf{X'XWW'} + \text{tr}\mathbf{WW'X'XWW'}$$
$$= \text{tr}\mathbf{X'X} - 2\text{tr}\mathbf{W'X'XW} + \text{tr}\mathbf{W'X'XWW'W}. \tag{6.10}$$

Using (6.5), the function (6.10) can be further rewritten as:

$$f(\mathbf{W}) = \text{tr}\mathbf{X'X} - 2\text{tr}\mathbf{W'X'XW} + \text{tr}\mathbf{W'X'XW} = \text{tr}\mathbf{X'X} - \text{tr}\mathbf{W'X'XW}. \tag{6.11}$$

Here, we should note that only $-\mathrm{tr}\mathbf{W'X'XW}$ is a function of $\mathbf{W}$ in the right-hand side. It implies that the minimization of $f(\mathbf{W})$ over $\mathbf{W}$ is equivalent to minimizing $-\mathrm{tr}\mathbf{W'X'XW}$ or maximizing $\mathrm{tr}\mathbf{W'X'XW}$. Further, this maximization is equivalent to $\mathrm{tr}\mathbf{W'X'XW}$ divided by n, i.e., (6.9).

Thus, PCA can also be formulated as maximizing (6.9) subject to (6.4) and (6.5). Here, the matrix $n^{-1}\mathbf{F'F}$ in (6.9) is the covariance matrix of PC scores between components, since $\mathbf{F}$ is centered with (5.24). Thus, the diagonal elements of $n^{-1}\mathbf{F'F}$ are the variances of m PC scores, implying that (6.9) is the sum of the *variances* of the 1st, ..., *mth PC scores*:

$$g(\mathbf{W}) = \frac{1}{n}\mathbf{f}_1'\mathbf{f}_1 + \cdots + \frac{1}{n}\mathbf{f}_m'\mathbf{f}_m = \sum_{k=1}^{m}\left(\frac{1}{n}\mathbf{f}_k'\mathbf{f}_k\right) \tag{6.12}$$

We can also rewrite (6.9) as:

$$g(\mathbf{W}) = \mathrm{tr}\mathbf{W}'\left(\frac{1}{n}\mathbf{X'X}\right)\mathbf{W} = \mathrm{tr}\mathbf{W'VW}, \tag{6.13}$$

where

$$\mathbf{V} = \frac{1}{n}\mathbf{X'X} \tag{6.14}$$

is the covariance matrix for centered $\mathbf{X}$. In some books, PCA is introduced with the following decomposition:

Note 6.1. Eigenvalue Decomposition of a Covariance Matrix
The singular value decomposition $\mathbf{X} = \mathbf{K\Lambda L'}$ in (5.5) with (5.6) leads to $\mathbf{X'X} = \mathbf{L\Lambda^2 L'}$. Comparing it with $\mathbf{X'X} = n\mathbf{V}$ following from (6.14), we have $n\mathbf{V} = \mathbf{L\Lambda^2 L'}$. This equation can be rewritten as:

$$\mathbf{V} = \mathbf{L\Delta L'}. \tag{6.15}$$

Here,

$$\mathbf{\Delta} = \begin{bmatrix} \delta_1 & & \\ & \ddots & \\ & & \delta_r \end{bmatrix} = \frac{1}{n}\mathbf{\Lambda}^2, \tag{6.16}$$

with $r = \mathrm{rank}(\mathbf{X})$ and $\delta_1 \geq \cdots \geq \delta_r \geq 0$. Decomposition (6.15) is referred to as the *eigenvalue decomposition* (*EVD*) or *spectral decomposition* of $\mathbf{V}$, δ_k ($k = 1, ..., r$) is called the kth largest *eigenvalue* of $\mathbf{V}$, and the kth column of $\mathbf{L}$ is called the *eigenvector* of $\mathbf{V}$ corresponding to δ_k.

6.3 Weighted Composite Scores with Maximum Variance

Let us express the columns of a data matrix as $\mathbf{X} = [\mathbf{x}_1, \ldots, \mathbf{x}_p]$. An example of $\mathbf{X}$ with $n = 9$ (examinees) and $p = 3$ (tests) is given in Table 6.1(B), which contains the centered scores of the raw ones in (A). They are the scores of the entrance examinations for a company. The examinations consist of the following items:

ES: essay,

IN: interview,

PR: presentation,

which define the three variables in $\mathbf{X}$.

We perform PCA for this data set with the number of components m equaling one, i.e., $\mathbf{W} = \mathbf{w}_1$ ($p \times 1$) and $\mathbf{F} = \mathbf{f}_1 = \mathbf{X}\mathbf{w}_1$ ($n \times 1$) being vectors. By defining $\mathbf{w}_1 = [w_{11}, \ldots, w_{p1}]'$, the PC score vector $\mathbf{f}_1$ is written as:

$$\mathbf{f}_1 = \mathbf{X}\mathbf{w}_1 = w_{11}\mathbf{x}_1 + \cdots + w_{p1}\mathbf{x}_p = w_{11}\text{ES} + w_{21}\text{IN} + w_{31}\text{PR}, \qquad (6.17)$$

with the abbreviations for the variables in Table 6.1 used in the right-hand side. Here, $\mathbf{f}_1$ is found to contain the *weighted composite scores* for the examinees, i.e., the sum of the variables in $\mathbf{x}_j$ weighted by w_{j1}.

Using $\mathbf{W} = \mathbf{w}_1$ and $\mathbf{F} = \mathbf{f}_1 = \mathbf{X}\mathbf{w}_1$, function (6.9) or (6.12) is rewritten as:

$$g(\mathbf{w}_1) = \mathbf{w}_1'\left(\frac{1}{n}\mathbf{X}'\mathbf{X}\right)\mathbf{w}_1 = \mathbf{w}_1'\mathbf{V}\mathbf{w}_1 = \frac{1}{n}\mathbf{f}_1'\mathbf{f}_1. \qquad (6.18)$$

This stands for the variance of the weighted composite scores in (6.17); their variance is defined as $n^{-1}\mathbf{f}_1'\mathbf{J}\mathbf{f}_1 = \mathbf{w}_1'(n^{-1}\mathbf{X}'\mathbf{J}\mathbf{X})\mathbf{w}_1 = \mathbf{w}(n^{-1}\mathbf{X}'\mathbf{X})\mathbf{w}_1 = n^{-1}\mathbf{f}_1'\mathbf{f}_1$, since

Table 6.1 Scores for the entrance exam and its PCA scores, which are artificial examples found in Adachi (2016)

Examinee	(A) Raw scores			(B) Centered scores			(C) PC scores		
	ES	IN	PR	ES	IN	PR	1st	2nd	3rd
1	88	70	65	21.2	4.3	−3.0	10.8	−19.0	0.6
2	52	78	88	−14.8	12.3	20.0	13.3	24.3	−1.8
3	77	87	89	10.2	21.3	21.0	31.3	4.7	−1.4
4	35	40	43	−31.8	−25.7	−25.0	−46.5	10.9	−3.3
5	60	43	40	−6.8	−22.7	−28.0	−34.8	−11.4	−0.7
6	97	95	91	30.2	29.3	23.0	46.9	−10.3	−1.1
7	48	62	83	−18.8	−3.7	15.0	−1.8	23.4	6.5
8	66	66	65	−0.8	0.3	−3.0	−2.0	−0.9	−2.2
9	78	50	48	11.2	−15.7	−20.0	−17.1	−21.6	3.4
Average	66.8	65.7	68.0	0.0	0.0	0.0	0.0	0.0	0.0
Variance	358.0	324.2	380.2	358.0	324.2	380.2	793.3	260.4	8.5

$\mathbf{X}$ is centered with $\mathbf{X} = \mathbf{JX}$. This variance is to be maximized subject to (6.5), i.e., $\mathbf{w}_1'\mathbf{w}_1 = 1$ for $m = 1$ (where (6.4) may not be considered for $m = 1$, since $\mathbf{F}'\mathbf{F} = \mathbf{f}_1'\mathbf{f}_1$ is a single scalar). That is, the PC scores in $\mathbf{f}_1$ are the composite scores obtained by *weighting the variables* so that the *variance of the scores are maximized*, in other words, so that *individuals are best distinguished*.

PCA for the data set in Table 6.1(B) provides

$$\mathbf{w}_1 = [0.47, 0.63, 0.62]',   \tag{6.19}$$

which implies that

$$\text{PC score} = 0.47\,\text{ES} + 0.63\,\text{IN} + 0.62\,\text{PR}   \tag{6.20}$$

is to be obtained for each examinee. For example, the centered scores for the second examinee are -14.8, 12.3, and 20.0, thus, that examinee's first PC score is obtained as:

$$0.47 \times (-14.8) + 0.63 \times 12.3 + 0.62 \times 20.0 = 13.3.   \tag{6.21}$$

The PC scores computed for all examinees in this way are shown in the first column of Table 6.1(C).

In everyday life, we often use a composite score:

$$\text{Simple sum score} = \mathbf{x}_1 + \cdots + \mathbf{x}_p = \text{ES} + \text{IN} + \text{PR},   \tag{6.22}$$

i.e., the sum of the equally weighted variables. As compared to this score, the PC score (6.20) is more useful for distinguishing individuals.

6.4 Projecting Three-Dimensional Vectors Onto Two-Dimensional Ones

Though the maximization of (6.9) (for $m = 1$) was considered in the last section, the purpose of this section is to explain that the minimization of (6.8) implies *projecting* a three-dimensional (3D) space onto one that is two-dimensional (2D), for $p = 3$, as in Table 6.1(B), and $m = 2$. For that purpose, let us use $\mathbf{F} = \mathbf{XW}$ in (6.8) to rewrite it as

$$f(\mathbf{W}) = ||\mathbf{X} - \mathbf{XWW}'||^2 = ||\mathbf{X} - \mathbf{FW}'||^2.   \tag{6.8'}$$

Further, we use row vector $\tilde{\mathbf{x}}_i'$ $(1 \times p)$ for the data vector of individual i, and $\tilde{\mathbf{f}}_i'$ $(1 \times m)$ for the PC score vector of i:

$$
\mathbf{X} = \begin{bmatrix} \tilde{\mathbf{x}}_1' \\ \vdots \\ \tilde{\mathbf{x}}_i' \\ \vdots \\ \tilde{\mathbf{x}}_n' \end{bmatrix}, \mathbf{F} = \begin{bmatrix} \tilde{\mathbf{f}}_1' \\ \vdots \\ \tilde{\mathbf{f}}_i' \\ \vdots \\ \tilde{\mathbf{f}}_n' \end{bmatrix} = \mathbf{XW} = \begin{bmatrix} \tilde{\mathbf{x}}_1'\mathbf{W} \\ \vdots \\ \tilde{\mathbf{x}}_i'\mathbf{W} \\ \vdots \\ \tilde{\mathbf{x}}_n'\mathbf{W} \end{bmatrix}.
\tag{6.23}
$$

Then, the rows of $\mathbf{FW}' = \mathbf{XWW}'$ are expressed as:

$$
\mathbf{FW}' = \begin{bmatrix} \tilde{\mathbf{f}}_1'\mathbf{W}' \\ \vdots \\ \tilde{\mathbf{f}}_i'\mathbf{W}' \\ \vdots \\ \tilde{\mathbf{f}}_n'\mathbf{W}' \end{bmatrix} = \mathbf{XWW}' = \begin{bmatrix} \tilde{\mathbf{x}}_1'\mathbf{WW}' \\ \vdots \\ \tilde{\mathbf{x}}_i'\mathbf{WW}' \\ \vdots \\ \tilde{\mathbf{x}}_n'\mathbf{WW}' \end{bmatrix}.
\tag{6.24}
$$

Using (6.23) and (6.24) in (6.8′), it is rewritten as:

$$
f(\mathbf{W}) = \left\| \begin{bmatrix} \tilde{\mathbf{x}}_1' \\ \vdots \\ \tilde{\mathbf{x}}_i' \\ \vdots \\ \tilde{\mathbf{x}}_n' \end{bmatrix} - \begin{bmatrix} \tilde{\mathbf{x}}_1'\mathbf{WW}' \\ \vdots \\ \tilde{\mathbf{x}}_i'\mathbf{WW}' \\ \vdots \\ \tilde{\mathbf{x}}_n'\mathbf{WW}' \end{bmatrix} \right\|^2 = \left\| \begin{bmatrix} \tilde{\mathbf{x}}_1' \\ \vdots \\ \tilde{\mathbf{x}}_i' \\ \vdots \\ \tilde{\mathbf{x}}_n' \end{bmatrix} - \begin{bmatrix} \tilde{\mathbf{f}}_1'\mathbf{W}' \\ \vdots \\ \tilde{\mathbf{f}}_i'\mathbf{W}' \\ \vdots \\ \tilde{\mathbf{f}}_n'\mathbf{W}' \end{bmatrix} \right\|^2.
\tag{6.8″}
$$

When $p = 3$ and $m = 2$, the minimization of (6.8″) amounts to matching individuals' data vectors $\tilde{\mathbf{x}}_i' = [x_{i1}, x_{i2}, x_{i3}]$ to $\tilde{\mathbf{x}}_i'\mathbf{WW}' = \tilde{\mathbf{f}}_i'\mathbf{W}'(1 \times 3)$, which can be expressed as:

$$
\tilde{\mathbf{x}}_i'\mathbf{WW}' = \tilde{\mathbf{f}}_i'\,\mathbf{W}' = [f_{i1}, f_{i2}] \begin{bmatrix} \mathbf{w}_1' \\ \mathbf{w}_2' \end{bmatrix} = f_{i1}\mathbf{w}_1' + f_{i2}\mathbf{w}_2',
\tag{6.25}
$$

with $\mathbf{W} = [\mathbf{w}_1, \mathbf{w}_2]$ and $\tilde{\mathbf{f}}_i' = [f_{i1}, f_{i2}]$. A key point involves capturing what (6.25) geometrically stands for. This is explained in the following two paragraphs.

As the data vector $\tilde{\mathbf{x}}_i' = [x_{i1}, x_{i2}, x_{i3}]$ is 1×3, $\tilde{\mathbf{x}}_i'$ can be depicted in a 3D space, as in Fig. 6.1a; $\tilde{\mathbf{x}}_i'$ is the line extending to the coordinate $[x_{i1}, x_{i2}, x_{i3}]$. There, we can also depict a plane whose direction in the 3D space is defined by vectors $\mathbf{w}_1'$ and $\mathbf{w}_2'$. As found there, the *projection* of $\tilde{\mathbf{x}}_i'$ on the plane is expressed as (6.25), where the projection refers to the vector that extends to the intersection of the plane and the line drawn from $\tilde{\mathbf{x}}_i'$, vertical to the plane. Further, the PC scores in $\tilde{\mathbf{f}}_i' = [f_{i1}, f_{i2}] = \tilde{\mathbf{x}}_i'\,\mathbf{W}$ stand for the coordinates of the projection within the plane. Why this fact holds is explained in Appendix A1.4. The plane seen head-on is shown in Fig. 6.1b. There, the first and second PC scores in $[f_{i1}, f_{i2}]'$ are the coordinates on the

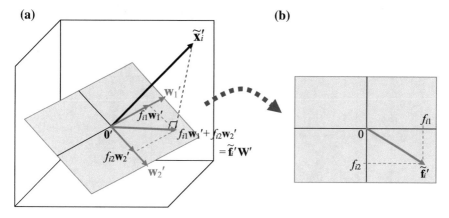

Fig. 6.1 The projection of a data vector onto a plane (**a**) with its front view (**b**)

horizontal and vertical axes of the plane. Below, we note the difference in this plane compared with that used in Chap. 4:

Note 6.2. Differences from Fig. 4.2

The plane in Fig. 4.2 differs from the one in Fig. 6.1a and the remaining ones in this chapter, in that *variable* vectors extend on the plane in Fig. 4.2, while *individuals'* vectors extend/are distributed on the planes in the figures appearing in this chapter.

We can freely define spaces (i.e., planes); which spaces are to be considered depends on one's research interests.

Now, let us recall function (6.8″), which is minimized over $\mathbf{W} = [\mathbf{w}_1', \mathbf{w}_2']'$ in PCA. This minimization implies bringing the projection (6.25) as close to $\tilde{\mathbf{x}}_i'$ as possible for all $i = 1, \ldots, n$. In other words, one purpose of PCA is to find the matrix $\mathbf{W} = [\mathbf{w}_1', \mathbf{w}_2']'$ that defines the direction of the plane so that the *projections* (6.25) are *closest* to the *original data vectors* $\tilde{\mathbf{x}}_i'$. The plane obtained by PCA is thus called a *best fitting plane*, since it is closest, i.e., the best fitted to the data vectors.

We illustrate the above case with the data in Table 6.1(B), whose data vectors $\tilde{\mathbf{x}}_i'$ ($i = 1, \ldots, 9$) can be depicted as in Fig. 6.2a. Here, the endpoints of the vectors $\tilde{\mathbf{x}}_i'$ have been indicated by circles (not by lines as in Fig. 6.1a) for the sake of ease in viewing. For the data set, PCA provides the solution of $\mathbf{W} = [\mathbf{w}_1', \mathbf{w}_2']'$ with $\mathbf{w}_1$ given by (6.19) and

$$\mathbf{w}_2 = [-0.84, 0.11, 0.53]'. \tag{6.26}$$

These vectors define the best fitting plane in Fig. 6.2b, on which the projections $\tilde{\mathbf{f}}_i'\mathbf{W}'$ ($i = 1, \ldots, 9$) for data vectors $\tilde{\mathbf{x}}_i'$ exist. A head-on view of the plane is shown in Fig. 6.2c. Here, the coordinates of the points are the first and second PC scores in

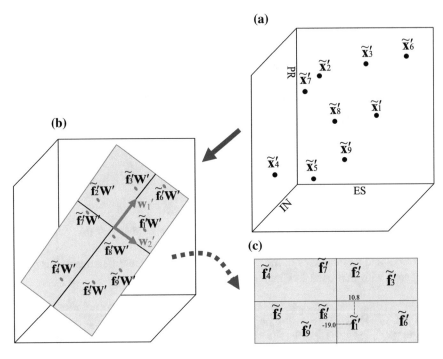

Fig. 6.2 Projections of data vectors (**a**) on a plane (**b**) with its front view (**c**)

$\tilde{\mathbf{f}}_i' = [f_{i1}, f_{i2}]$, whose values are presented in Table 6.1(C). For example, the PC score vector for examinee 1 is found to be $\tilde{\mathbf{f}}_i' = [10.8, -19.0]$ in the table, and it is located at the point with the coordinates $[10.8, -19.0]$ in Fig. 6.2c. Here, the second PC score, -19.0, has been obtained as $f_{12} = -0.84 \times 21.2 + 0.11 \times 4.3 + 0.53 \times (-3.0) = -19.0$ using his/her centered scores $[21.2, 4.3, -3.0]$ and the weights in (6.26).

This section deals with the logic in PCA by which the *original 3D* data distributions (as in Fig. 6.2a) are *projected* on a *2D* plane (as in (b)), whose front view is a scatter plot (as in (c)). This 2D plot is useful; in that it is easier to capture than the original 3D plot. However, this section is merely a preparation for the one that follows, where the distributions in the space of a higher dimension can be projected onto a lower dimensional space in the same way as in this section. It is one of the most important benefits gained by using PCA.

6.5 Visualization of Invisible Distributions

We consider a new data set in Table 6.2 (Committee for Guiding Psychological Experiments 1985). It contains the results of the rating by participants for to what extent 12 adjectives characterize 14 occupational categories. The table shows the

Table 6.2 Impressions of 14 occupational categories rated for 12 adjectives

Category	Noble	Useful	Good	Big	Powerful	Strong	Quick	Noisy	Young	Faithful	Strict	Busy
Monk	3.2	2.7	3.7	2.8	2.6	2.6	2.2	1.4	1.7	3.3	3.8	1.8
Bank clerk	3.4	3.5	3.4	2.5	2.2	2.6	3.2	2.1	3.6	4.1	4.7	4.2
Cartoonist	3.0	3.2	3.5	2.2	2.1	2.2	3.3	3.4	4.1	3.4	1.3	4.3
Designer	3.2	3.2	3.5	2.6	2.5	2.6	3.6	2.9	4.2	3.2	1.5	4.0
Nurse[a]	4.2	4.6	4.5	3.1	3.0	3.2	2.8	3.3	4.1	4.5	2.3	4.9
Professor	4.0	4.0	3.8	3.4	3.2	3.1	2.4	1.5	1.6	3.7	3.9	3.0
Doctor[b]	4.0	4.8	3.9	3.5	3.8	3.7	3.2	2.1	2.6	3.7	3.6	4.5
Policeman	3.7	4.6	4.1	3.4	4.0	4.1	4.3	3.4	3.5	4.2	4.4	4.0
Journalist	3.6	4.3	3.7	2.9	3.5	3.6	4.7	4.2	4.1	3.9	3.7	5.0
Sailor	3.6	3.6	3.5	3.5	4.2	4.2	3.5	3.5	3.7	3.5	2.5	3.5
Athlete	3.7	3.2	3.7	3.9	4.7	4.7	4.9	3.5	4.2	3.7	2.8	4.1
Novelist	3.4	3.7	3.5	3.1	2.7	2.4	2.3	1.8	2.3	3.3	2.9	3.3
Actor	3.2	3.2	3.6	2.9	2.2	2.5	3.3	3.3	3.4	2.8	1.8	4.3
Stewardess	3.2	3.8	3.8	2.8	2.3	2.4	3.9	2.5	4.7	3.9	2.3	4.3
Average[c]	3.5	3.7	3.7	3.0	3.1	3.1	3.4	2.8	3.4	3.7	3.0	3.9

[a]In nursing school
[b]Medical doctor
[c]Column average

average rating values for the categories on a scale of 1–5. For example, let us note the final column "busy": the busyness of "bank clerk" is rated at 4.2, while that of "professor" is 3.0, that is, people think that bank clerks are busier than professors.

Let $\mathbf{X} = (x_{ij})$ (14 × 12) contains the centered scores of the data in Table 6.2. For example, x_{32} is 3.2–3.7 = −0.5 (the usefulness of "cartoonist" minus the average of usefulness). Can we depict the distribution of the 14 categories' scores on the 12 variables? That would require a 12-dimensional (12D) space with its 12 coordinate axes orthogonally intersected. Unfortunately, a space of dimensionality $m > 3$ can *neither be drawn nor seen* by us, as *we live in a 3D world!* However, such a *high-dimensional space* can be considered in logic, i.e., mathematically, regardless of how high the dimensionality is.

Let us suppose that $\tilde{\mathbf{x}}_i' = [x_{i1}, \ldots, x_{i,12}]$ ($i = 1, \ldots, 14$; categories) are distributed in a 12D space as depicted in Fig. 6.3a. PCA for $\mathbf{X}$ yields the weight matrix $\mathbf{W} = [\mathbf{w}_1, \mathbf{w}_2]$ in Table 6.3(A). It defines the *best fitting plane* on which the projections $\tilde{\mathbf{f}}_i'\mathbf{W}'$ ($i = 1, \ldots, 14$) are located, as illustrated in Fig. 6.3b. This plane can be seen head-on, as shown in Fig. 6.4. There, the 14 categories are plotted, with their coordinates being the PC scores $\tilde{\mathbf{f}}_i' = [f_{i1}, f_{i2}]$, whose values are obtained as in Table 6.3(B) using the centered scores for Table 6.2 and the weights in Table 6.3 (A).

Although the original distribution of $\tilde{\mathbf{x}}_i'$ in Fig. 6.3a was *invisible*, the projection of $\tilde{\mathbf{x}}_i'$ on the best fitting 2D plane is *visible*, as found in Fig. 6.4. This shows that a benefit of PCA is the *visualization* of a *high-dimensional invisible space*. The resulting plot in Fig. 6.4 can be captured in the same manner as for a usual *map*; two objects close to each other can be viewed as similar, while those that are distant can be regarded as dissimilar. For example, Fig. 6.4 shows that "designer" and "cartoonist" are similar occupations, while "monk" and "journalist" are very different.

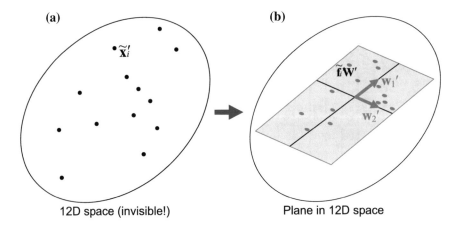

(a)   **(b)**

12D space (invisible!)   Plane in 12D space

Fig. 6.3 Projecting the distributions in a 12D space (**a**) on a 2D plane (**b**)

Table 6.3 Weights and PC scores obtained for the centered scores transformed from the data in Table 6.2

(A) W (weights)

	$\mathbf{w}_1$	$\mathbf{w}_2$
Noble	0.03	0.18
Useful	0.12	0.23
Good	0.04	0.07
Big	0.06	0.25
Powerful	0.22	0.46
Strong	0.26	0.42
Quick	0.44	0.09
Noisy	0.48	−0.07
Young	0.50	−0.27
Faithful	0.09	0.15
Strict	−0.19	0.59
Busy	0.39	−0.09

(B) F (PC scores)

	$\mathbf{f}_1$	$\mathbf{f}_2$
Monk	−3.46	0.27
Bank clerk	−0.91	0.19
Cartoonist	0.43	−2.57
Designer	0.43	−1.93
Nurse	1.06	−0.25
Professor	−2.41	1.40
Doctor	−0.24	1.60
Policeman	1.13	2.11
Journalist	2.16	0.72
Sailor	0.95	0.67
Athlete	2.24	1.30
Novelist	−2.06	−0.29
Actor	0.04	−1.78
Stewardess	0.65	−1.43

6.6 Goodness of Projection

It should be noticed that the original distribution in Fig. 6.3a is not perfectly reflected on the plane in Fig. 6.3b, which in turn gives Fig. 6.4; some information in the original distribution has been *lost* in Figs. 6.3b and 6.4. The *amount of the loss* can be assessed by the resulting value of loss function (6.8) or (6.8″), since it expresses the differences between the data vectors $\tilde{\mathbf{x}}_i'$ in Fig. 6.3a and their projections $\tilde{\mathbf{x}}_i' \mathbf{WW}' = \tilde{\mathbf{f}}_i'\mathbf{W}'(1 \times 3)$ in Fig. 6.3b.

The resulting value of (6.8), into which solution (6.6) is substituted, is expressed as

Fig. 6.4 Front view of the
plane in Fig. 6.3b

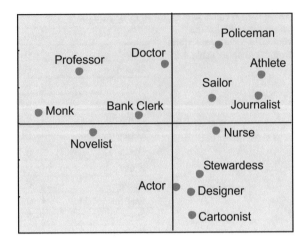

$$||\mathbf{X} - \mathbf{X}\mathbf{L}_m\mathbf{L}'_m||^2 = ||\mathbf{X} - \mathbf{K}_m\mathbf{L}_m\mathbf{L}'_m||^2 = ||\mathbf{K}\mathbf{\Lambda}\mathbf{L}' - \mathbf{K}_m\mathbf{\Lambda}_m\mathbf{L}'_m||^2, \qquad (6.27)$$

where we have used (5.5) and (5.19). It can be found that (6.27) is equivalent to
(5.37), which implies that the *proportion of explained variance* (5.39), i.e.,
$PEV_m = \mathrm{tr}\mathbf{\Lambda}_m^2/\mathrm{tr}\mathbf{\Lambda}^2 = \mathrm{tr}\mathbf{\Lambda}_m^2/||\mathbf{X}||^2$, is also an index for the *goodness of projection*.

For the centered scores of the data in Table 6.2, PCA gives $\mathrm{tr}\mathbf{\Lambda}_2^2 = 64.1$ and
$\mathrm{tr}\mathbf{\Lambda}^2 = ||\mathbf{X}||^2 = 86.6$, thus $PEV_2 = 64.1/86.6 = 0.74$. This implies that 74 % (= 0.74
× 100 %) of the information of the distribution in Fig. 6.3a is reflected in Fig. 6.4;
the former invisible distribution is visualized in the latter and, furthermore, we can
see 74 % of the former. This demonstrates the benefit of PCA.

6.7 Bibliographical Notes

As described in Sect. 5.9, various aspects of PCA are exhaustively detailed in
Jolliffe (2002). Visualization as a benefit of PCA in the natural sciences has been
illustrated in Izenman (2008) and Koch (2014), which are among the advanced
books recommended for a deeper understanding of multivariate analysis, though the
term visualization is not used in those books.

Here, we must mention *sparse PCA* for obtaining the sparse weight matrix
W (Jolliffe et al. 2003; Zou et al. 2006). Here, a sparse matrix refers to a matrix
including a number of zero elements. That is, sparse PCA refers to the modified
PCA procedures in which the elements of **W** to be zero is computationally chosen
jointly with estimating the values of the nonzero elements. The resulting **W** can be
easily interpreted, as only nonzero elements may be noted. The developments in
sparse PCA are summarized well in Hastie et al. (2015) and Trendafilov (2014).

This author and the author of the present book have proposed sparse PCA for obtaining sparse loading matrix $\mathbf{A}$ rather than $\mathbf{W}$ (Adachi and Trendafilov 2015b).

Exercises

6.1. Show $\mathrm{tr}\mathbf{V} = \mathrm{tr}\Lambda$ for $\mathbf{V}$ and Λ in Note 6.1.

6.2. Show that the eigenvalue decomposition (EVD) in Note 6.1 implies $\mathbf{V}\mathbf{l}_k = \delta_k \mathbf{l}_k$ ($k = 1, \ldots, r$) with $\mathbf{l}_k$ the kth column of $\mathbf{L}$. The equation is called the eigen equation for $\mathbf{V}$.

6.3. Show that the EVD in Note 6.1 can be rewritten as $\mathbf{X}'\mathbf{X} = \mathbf{L}\Lambda^2\mathbf{L}'$ and post-multiplying both of its sides by $\mathbf{L}\mathbf{L}'$ leads to $\mathbf{X}'\mathbf{X}\mathbf{L}\mathbf{L}' = \mathbf{X}'\mathbf{X}$, i.e.,

$$\mathbf{X}'\mathbf{X}(\mathbf{I}_p - \mathbf{L}\mathbf{L}') =_p \mathbf{O}_p. \tag{6.28}$$

6.4. Show that (6.28) leads to $(\mathbf{I}_p - \mathbf{L}\mathbf{L}')\mathbf{X}'\mathbf{X}(\mathbf{I}_p - \mathbf{L}\mathbf{L}') = {}_p\mathbf{O}_p$, which implies

$$\mathbf{X}(\mathbf{I}_p - \mathbf{L}\mathbf{L}') =_n \mathbf{O}_p, \tag{6.29}$$

using the fact that $\mathbf{M}'\mathbf{M} = {}_p\mathbf{O}_p$ implies $\mathbf{M} = {}_n\mathbf{O}_p$ for $\mathbf{M}$ being $n \times p$.

6.5. Show that the SVD in Note 5.1 can be derived from the EVD in Note 6.1, noting the fact that (6.29) implies $\mathbf{X} = \mathbf{X}\mathbf{L}\Lambda^{-1}\Lambda\mathbf{L}'$ and $\mathbf{X}\mathbf{L}\Lambda^{-1}$ can be regarded as $\mathbf{K}$ in Note 5.1.

6.6. A square matrix $\mathbf{N}$ is said to be *nonnegative definite* if $f(\mathbf{w}) = \mathbf{w}'\mathbf{N}\mathbf{w} \geq 0$ for any vector $\mathbf{w}$. It is known that $\mathbf{S}$ being nonnegative definite and symmetric is equivalent to the property of $\mathbf{S}$ that it can be rewritten as $\mathbf{S} = \mathbf{B}\mathbf{B}'$. Show that the covariance matrix $\mathbf{V} = n^{-1}\mathbf{X}'\mathbf{J}\mathbf{X}$ is nonnegative definite.

6.7. A square matrix $\mathbf{P}$ is said to be *positive definite*, if $f(\mathbf{w}) = \mathbf{w}'\mathbf{P}\mathbf{w} > 0$ for any vector $\mathbf{w}$. If $\mathbf{D}$ is a diagonal matrix, show that $\mathbf{D}$ being positive definite is equivalent to all diagonal elements of $\mathbf{D}$ being positive.

6.8. Let $v(\mathbf{f}_k)$ denote the variance of the kth PC scores, i.e., the elements in $\mathbf{f}_k = \mathbf{X}\mathbf{w}_k$. Show that $v(\mathbf{f}_k)$ equals δ_k, i.e., the kth eigenvalue of $\mathbf{V}$ defined in Note 6.1.

6.9. Show that the vectors $\tilde{\mathbf{f}}_i'\mathbf{W}'$ and $\tilde{\mathbf{x}}_i'-\tilde{\mathbf{f}}_i'\mathbf{W}'$ intersect orthogonally, as in Fig. 6.1, i.e., $\tilde{\mathbf{f}}_i'\mathbf{W}'(\tilde{\mathbf{x}}_i' - \mathbf{W}\tilde{\mathbf{f}}_i') = 0$.

6.10. If constraint (6.4) is removed and only (6.5) is imposed in PCA, show that (6.6) is replaced by $\mathbf{W} = \mathbf{A} = \mathbf{L}_m\mathbf{T}$, with $\mathbf{T}$ the orthonormal matrix satisfying (A.1.6) in Appendix A.1.2.

Chapter 7
Cluster Analysis

The term "*cluster*" is synonymous with both "*group*" as a noun and "*classify*" as a verb. *Cluster analysis*, which is also simply called *clustering*, generally refers to the procedures for computationally classifying (i.e., clustering) individuals into groups (i.e., clusters) so that similar individuals are classified into the same group and mutually dissimilar ones are allocated to different groups. There are various procedures for performing cluster analysis. One of the most popular of these, called *k-means clustering* (*KMC*), which was first presented by MacQueen (1967), is introduced here.

7.1 Membership Matrices

An example of a membership matrix is given here:

	Australia	UK	USA
Mick		1	
Kieth	1		
Ronny			1
Charly		1	
Bill			1

$\mathbf{G} = (g_{ik}) =$

It indicates the nationalities of individuals, and the blank cells stand for the elements taking zero. In general, a *membership matrix* $\mathbf{G} = (g_{ik})$ is defined as the matrix of *n-individuals* $\times$ *K-clusters* satisfying

$$g_{ik} = \begin{cases} 1 & \text{if individual } i \text{ belongs to cluster } k, \\ 0 & \text{otherwise} \end{cases} \tag{7.1}$$

© Springer Nature Singapore Pte Ltd. 2016
K. Adachi, *Matrix-Based Introduction to Multivariate Data Analysis*,
DOI 10.1007/978-981-10-2341-5_7

$$\mathbf{G1}_K = \mathbf{1}_n. \tag{7.2}$$

These equations imply that each row of $\mathbf{G}$ has only one element taking 1, i.e., each individual belongs to only one cluster. Such a matrix is also called an *indicator matrix* or a *design matrix*. One purpose of clustering procedures including KMC is to obtain $\mathbf{G}$ from an *n*-individuals × *p*-variables data matrix $\mathbf{X}$.

7.2 Example of Clustering Results

For a data matrix $\mathbf{X}$, KMC provides a membership matrix $\mathbf{G}$ together with a *K-clusters* × *p-variables cluster feature matrix* $\mathbf{C}$, which expresses how each cluster is characterized by variables.

Before explaining how to obtain $\mathbf{G}$ and $\mathbf{C}$, we show the KMC solution for the 14-occupations × 12-adjectives data matrix $\mathbf{X} = (x_{ij})$ in Table 6.2 in the last chapter. It describes to what extent the occupations are characterized by the adjectives. For the data matrix, KMC with K set at 4 provides the solutions of $\mathbf{G}$ and $\mathbf{C}$ shown in Tables 7.1 and 7.2. First, let us note the resulting *membership* matrix $\mathbf{G}$ in Table 7.1. The cluster numbers 1, 2, 3, 4 are merely for the purpose of distinguishing different clusters; $\mathbf{G}$ simply shows that the occupations having 1 in the same column belong to the same cluster. For example, monk, professor, and novelist are members of the same cluster. Next, let us note Table 7.2. There, the resulting *cluster feature* matrix $\mathbf{C}$ is shown, which describes the values of variables characterizing each cluster. For example, Cluster 2, whose members include bank clerk and doctor, are found to be very useful, strict, and busy.

Table 7.1 Membership matrix $\mathbf{G}$ obtained for the data in Table 6.2

Occupation	Cluster 1	Cluster 2	Cluster 3	Cluster 4
Monk	1			
Bank clerk		1		
Cartoonist			1	
Designer			1	
Nurse			1	
Professor	1			
Doctor		1		
Policeman				1
Journalist				1
Sailor				1
Athlete				1
Novelist	1			
Actor			1	
Stewardess			1	

Table 7.2 Cluster feature matrix **C** obtained for the data in Table 6.2

Cluster	Noble	Useful	Good	Big	Powerful	Strong	Quick	Noisy	Young	Faithful	Strict	Busy
1	3.5	3.5	3.7	3.1	2.8	2.7	2.3	1.6	1.9	3.4	3.5	2.7
2	3.7	4.2	3.7	3.0	3.0	3.2	3.2	2.1	3.1	3.9	4.2	4.4
3	3.4	3.6	3.8	2.7	2.4	2.6	3.4	3.1	4.1	3.6	1.8	4.4
4	3.7	3.9	3.8	3.4	4.1	4.2	4.4	3.7	3.9	3.8	3.4	4.2

7.3 Formulation

KMC is underlain by the model

$$\mathbf{X} = \mathbf{GC} + \mathbf{E},\qquad(7.3)$$

with $\mathbf{E}$ containing errors. To obtain $\mathbf{G}$ and $\mathbf{C}$, a *least squares method* is used; the sum of squared errors

$$f(\mathbf{G}, \mathbf{C}) = \|\mathbf{E}\|^2 = \|\mathbf{X} - \mathbf{GC}\|^2 \qquad(7.4)$$

is minimized over $\mathbf{G}$ and $\mathbf{C}$ subject to $\mathbf{G}$ satisfying (7.1) and (7.2).

For the sake of ease in understanding (7.3) and (7.4), we use the example of $\mathbf{X}$ in Fig. 7.1, which is more compact than the data set in Table 6.2. In Fig. 7.1, a 10×2 data matrix $\mathbf{X}$ is shown together with a scatter plot of the 10 row vectors in $\mathbf{X}$. For this data matrix, KMC with $K = 3$ gives the solution expressed as follows:

$$
\begin{array}{c}
\mathbf{X} \\
\begin{bmatrix}
1 & 4 \\
7 & 3 \\
6 & 1 \\
8 & 6 \\
3 & 5 \\
5 & 7 \\
9 & 2 \\
2 & 3 \\
8 & 4 \\
6 & 9
\end{bmatrix}
\end{array}
=
\begin{array}{c}
\mathbf{G} \\
\begin{bmatrix}
1 & & \\
& 1 & \\
& & 1 \\
& 1 & \\
1 & & \\
& 1 & \\
& & 1 \\
1 & & \\
& & 1 \\
& 1 &
\end{bmatrix}
\end{array}
\begin{array}{c}
\mathbf{C} \\
\begin{bmatrix}
2.0 & 4.0 \\
6.3 & 7.3 \\
7.5 & 2.5
\end{bmatrix}
\end{array}
+
\begin{array}{c}
\mathbf{E} \\
\begin{bmatrix}
-1.0 & 0.0 \\
-0.5 & 0.5 \\
-1.5 & -1.5 \\
1.7 & -1.3 \\
1.0 & 1.0 \\
-1.3 & -0.3 \\
1.5 & -0.5 \\
0.0 & -1.0 \\
0.5 & 1.5 \\
-0.3 & 1.7
\end{bmatrix}
\end{array}
\qquad(7.5)
$$

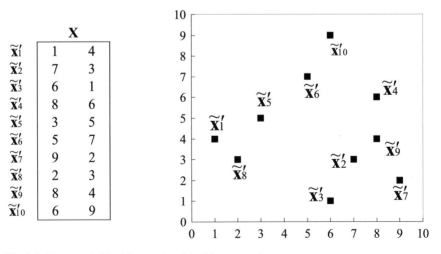

	$\mathbf{X}$	
$\widetilde{\mathbf{x}}_1'$	1	4
$\widetilde{\mathbf{x}}_2'$	7	3
$\widetilde{\mathbf{x}}_3'$	6	1
$\widetilde{\mathbf{x}}_4'$	8	6
$\widetilde{\mathbf{x}}_5'$	3	5
$\widetilde{\mathbf{x}}_6'$	5	7
$\widetilde{\mathbf{x}}_7'$	9	2
$\widetilde{\mathbf{x}}_8'$	2	3
$\widetilde{\mathbf{x}}_9'$	8	4
$\widetilde{\mathbf{x}}_{10}'$	6	9

Fig. 7.1 Data matrix $\mathbf{X}$ with a scatter plot of its row vectors

Here, model (7.3) is shown into which the data and the resulting solution were substituted. Further, we can obtain the *product* of **G** and **C** to rewrite (7.5) as:

$$
\begin{array}{cccc}
\mathbf{X} & \mathbf{GC} & \mathbf{E} & \\
\begin{array}{cc} 1 & 4 \\ 7 & 3 \\ 6 & 1 \\ 8 & 6 \\ 3 & 5 \\ 5 & 7 \\ 9 & 2 \\ 2 & 3 \\ 8 & 4 \\ 6 & 9 \end{array}
=
\begin{array}{cc} 2.0 & 4.0 \\ 7.5 & 2.5 \\ 7.5 & 2.5 \\ 6.3 & 7.3 \\ 2.0 & 4.0 \\ 6.3 & 7.3 \\ 7.5 & 2.5 \\ 2.0 & 4.0 \\ 7.5 & 2.5 \\ 6.3 & 7.3 \end{array}
+
\begin{array}{cc} -1.0 & 0.0 \\ -0.5 & 0.5 \\ -1.5 & -1.5 \\ 1.7 & -1.3 \\ 1.0 & 1.0 \\ -1.3 & -0.3 \\ 1.5 & -0.5 \\ 0.0 & -1.0 \\ 0.5 & 1.5 \\ -0.3 & 1.7 \end{array}
& (7.6)
\end{array}
$$

where white, light gray, and dark gray have been used for the background colors of the rows corresponding to Clusters 1, 2, and 3, respectively. In (7.6), we find that the ith row of **X** is matched to the row of $\mathbf{C} = [\tilde{\mathbf{c}}_1, \tilde{\mathbf{c}}_2, \tilde{\mathbf{c}}_3]'$ associated with the cluster into which individual i is classified; for example, $\tilde{\mathbf{x}}_4' = [8, 6]$ is matched to $\tilde{\mathbf{c}}_2' = [6.3, 7.3]$.

Solution (7.6) can be illustrated graphically, as in Fig. 7.2, in which the rows of **X** and **C** are plotted. There, we can find that $\tilde{\mathbf{c}}_k'$ (the k th row of **C**) expresses the *representative point* of cluster k that is located at the *center* of the individuals ($\tilde{\mathbf{x}}_i'$s) belonging to that cluster. For this reason, **C** is also called a *cluster center matrix*. In Fig. 7.2, each of the *lines* connects $\tilde{\mathbf{x}}_i'$ for individual i and $\tilde{\mathbf{c}}_k'$ for the cluster including i. The lines in the figure indicate the row vectors of error matrix

$$
\mathbf{E} = \begin{bmatrix} \tilde{\mathbf{e}}_1' \\ \vdots \\ \tilde{\mathbf{e}}_n' \end{bmatrix}.
$$

For example, the line extending from center $\tilde{\mathbf{c}}_3'$ to $\tilde{\mathbf{x}}_9'$ indicates $\tilde{\mathbf{e}}_9' = \tilde{\mathbf{x}}_9' - \tilde{\mathbf{c}}_3'$, with $\tilde{\mathbf{e}}_i'$ the ith row of **E**. Here, we should note that the function (7.4) to

Fig. 7.2 Joint plot of the rows of X in Fig. 7.1 with those of C

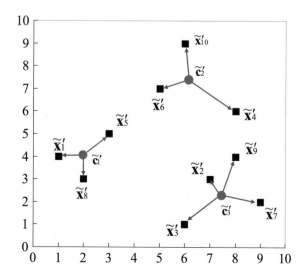

be minimized is rewritten as $\|\mathbf{E}\|^2 = \|\tilde{\mathbf{e}}_1\|^2 + \cdots + \|\tilde{\mathbf{e}}_n\|^2$. Its minimization is restated as minimizing the *sum of the squared lengths of the lines* in Fig. 7.2, which implies making each individual vector $(\tilde{\mathbf{x}}_i')$ close to the center of the cluster $(\tilde{\mathbf{c}}_k')$ including the individual.

7.4   Iterative Algorithm

Let us remember that the PCA solution is obtained through (5.14) and the solution for regression analysis is given by (4.9) and (4.12); those solutions are expressed explicitly as formulas. On the other hand, the KMC solution minimizing (7.4) *cannot* be given explicitly by a formula. In general, statistical analysis procedures can be classified into the following two types:

[1]   those *with explicit solutions* (as in regression analysis and PCA)
[2]   those *without explicit solutions* (as in KMC)

How are solutions for [2] obtained? They can be attained with *iterative algorithms*, where steps are iterated for finding the solution. There are some types of iterative algorithms, as described in Appendix A.6.1.

The algorithm for KMC is formed using the following fact: although the $\mathbf{G}$ and $\mathbf{C}$ minimizing (7.4), i.e., $f(\mathbf{G}, \mathbf{C})$, cannot be expressed as formulas, the matrices

$$\mathbf{C} \text{ that minimizes } f(\mathbf{G}, \mathbf{C}) \text{ while } \mathbf{G} \text{ is fixed at a specified matrix} \qquad (7.7)$$

and

$$\mathbf{G} \text{ that minimizes } f(\mathbf{G}, \mathbf{C}) \text{ while } \mathbf{C} \text{ is fixed at a specified matrix} \qquad (7.8)$$

can be explicitly given, as shown in the next sections. This fact allows us to form the iterative *algorithm for KMC*, described by the following steps:

Step 1. Set $\mathbf{G}$ and $\mathbf{C}$ to specified matrices $\mathbf{G}_{[t]}$ and $\mathbf{C}_{[t]}$, respectively, with $t = 0$.
Step 2. Obtain $\mathbf{C}$ defined as (7.7) with $\mathbf{G}$ being fixed at $\mathbf{G}_{[t]}$, and express the resulting $\mathbf{C}$ as $\mathbf{C}_{[t+1]}$.
Step 3. Obtain $\mathbf{G}$ defined as (7.8) with $\mathbf{C}$ fixed at $\mathbf{C}_{[t+1]}$, and express the resulting $\mathbf{G}$ as $\mathbf{G}_{[t+1]}$.
Step 4. Finish and regard $\mathbf{C}_{[t+1]}$ and $\mathbf{G}_{[t+1]}$ as the solution, if convergence is reached; otherwise, go back to Step 2 and increase t by one.

Here, t stands for the number of iterations, and the convergence in Step 4 is explained later. The central part of the algorithm is the *alternate iteration of Steps 2 and 3*. *With this iteration*, the value of function (7.4) *decreases monotonically* (or remains unchanged), regardless of what is used for the specified matrices in Step 1, as described in the following paragraphs.

Let us consider the value of (7.4) at Step 1, i.e., $f(\mathbf{G}_{[0]}, \mathbf{C}_{[0]}) = f(\mathbf{G}_{[t]}, \mathbf{C}_{[t]})$ for $t = 0$, which is followed by Steps 2 and 3, providing $f(\mathbf{G}_{[0]}, \mathbf{C}_{[1]})$ and $f(\mathbf{G}_{[1]}, \mathbf{C}_{[1]})$, respectively. They are found to satisfy

$$f\left(\mathbf{G}_{[0]}, \mathbf{C}_{[0]}\right) \geq f\left(\mathbf{G}_{[0]}, \mathbf{C}_{[1]}\right) \geq f\left(\mathbf{G}_{[1]}, \mathbf{C}_{[1]}\right). \tag{7.9}$$

Here, the first inequality $f(\mathbf{G}_{[0]}, \mathbf{C}_{[0]}) \geq f(\mathbf{G}_{[0]}, \mathbf{C}_{[1]})$ follows from the fact that $\mathbf{C}_{[1]}$ is the matrix $\mathbf{C}$ that minimizes $f(\mathbf{G}, \mathbf{C})$ with $\mathbf{G}$ fixed to $\mathbf{G}_{[0]}$ as found in (7.7), and the second inequality $f(\mathbf{G}_{[0]}, \mathbf{C}_{[1]}) \geq f(\mathbf{G}_{[1]}, \mathbf{C}_{[1]})$ follows from (7.8), i.e., $\mathbf{G}_{[1]}$ being the matrix $\mathbf{G}$ that minimizes $f(\mathbf{G}, \mathbf{C}_{[1]})$.

As described in Step 4, unless convergence is reached, the algorithm must go back to Step 2, with an increase in t from one to two. Then, Steps 2 and 3 are performed again to have $\mathbf{C}_{[2]}$ and $\mathbf{G}_{[2]}$, which allows (7.9) to be followed by two inequalities as:

$$f\left(\mathbf{G}_{[0]}, \mathbf{C}_{[0]}\right) \geq f\left(\mathbf{G}_{[0]}, \mathbf{C}_{[1]}\right) \geq f\left(\mathbf{G}_{[1]}, \mathbf{C}_{[1]}\right) \geq f\left(\mathbf{G}_{[1]}, \mathbf{C}_{[2]}\right) \geq f\left(\mathbf{G}_{[2]}, \mathbf{C}_{[2]}\right) \tag{7.10}$$

We can generalize (7.9) and (7.10) as:

$$f\left(\mathbf{G}_{[t]}, \mathbf{C}_{[t]}\right) \geq f\left(\mathbf{G}_{[t]}, \mathbf{C}_{[t+1]}\right) \geq f\left(\mathbf{G}_{[t+1]}, \mathbf{C}_{[t+1]}\right) \tag{7.11}$$

for $t = 0, 1, 2 \ldots$, where $\mathbf{C}_{[t+1]}$ denotes the matrix $\mathbf{C}$ obtained in Step 2 and $\mathbf{G}_{[t+1]}$ denotes the matrix $\mathbf{G}$ obtained in Step 3 at the tth iteration. That is, the value of $f(\mathbf{G}, \mathbf{C})$ *decreases monotonically* with an *increase* in t so that the value is expected to *converge* to the minimum. Convergence can be defined as having a difference in the value of (7.4) from the previous round of iteration that is small enough to be ignored, i.e.,

$$f\left(\mathbf{G}_{[t]}, \mathbf{C}_{[t]}\right) - f\left(\mathbf{G}_{[t+1]}, \mathbf{C}_{[t+1]}\right) \leq \varepsilon, \tag{7.12}$$

with ε being a small value, such as 0.1^6 or 0.1^5.

Figure 7.3 shows the change in $f(\mathbf{G}, \mathbf{C})$ with the iterative KMC algorithm for the data in Fig. 7.1, where the elements of the specified matrices in Step 1 were

Fig. 7.3 Values of $f(\mathbf{G}, \mathbf{C})$ at steps in the t-iteration ($t = 0$, 1, 2, 3) with subscripts 2 and 3 indicating Steps 2 and 3

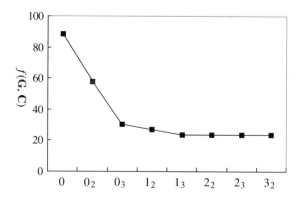

randomly chosen. In Fig. 7.3, we find the monotonic decrease in the $f(\mathbf{G}, \mathbf{C})$ value with t, and the value is unchanged from $t = 2$ to 3, i.e., convergence is reached at $t = 3$. The matrices $\mathbf{C}$ and $\mathbf{G}$ at this time are their solution in (7.5). The computations that were used in Steps 2 and 3 are described in the following two sections.

7.5   Obtaining Cluster Features

In this section, we consider Step 2 from the previous section, i.e., obtaining the *cluster feature matrix* $\mathbf{C}$ defined as (7.7). The matrix $\mathbf{C}$ can be found that minimizes $f(\mathbf{C}) = \|\mathbf{X} - \mathbf{G}_{[t]}\mathbf{C}\|^2$, i.e., the function (7.4) with $\mathbf{G}$ fixed at $\mathbf{G}_{[t]}$. This $\mathbf{C}$ is given by

$$\mathbf{C}_{[t+1]} = (\mathbf{G}'_{[t]}\mathbf{G}_{[t]})^{-1}\mathbf{G}'_{[t]}\mathbf{X} = \mathbf{D}^{-1}\mathbf{G}'_{[t]}\mathbf{X}, \tag{7.13}$$

with $\mathbf{D} = \mathbf{G}_{[t]}'\mathbf{G}_{[t]}$, as explained in Appendix A.2.2. There, we can compare (7.4) with (A.2.11) to find that (A.2.12) leads to (7.13).

Let us consider what matrix $\mathbf{D} = \mathbf{G}_{[t]}'\mathbf{G}_{[t]}$ is, with a simple example of $\mathbf{G}_{[t]}$:

$$\text{If } \mathbf{G}_{[t]} = \begin{bmatrix} & 1 \\ 1 & \\ & & 1 \\ 1 & \\ & 1 \\ & 1 \end{bmatrix}, \quad \text{then}$$

$$\mathbf{D} = \begin{bmatrix} & 1 & & 1 \\ 1 & & & & 1 & 1 \\ & & 1 \end{bmatrix} \begin{bmatrix} & 1 \\ 1 & \\ & & 1 \\ 1 & \\ & 1 \\ & 1 \end{bmatrix} = \begin{bmatrix} 2 & \\ & 3 \\ & & 1 \end{bmatrix}.$$

In general, $\mathbf{D} = \mathbf{G}_{[t]}'\mathbf{G}_{[t]}$ is a $K \times K$ diagonal matrix whose kth diagonal element d_k is the number of individuals belonging to cluster k. Thus, the inverse matrix $\mathbf{D}^{-1}$ is found to be the diagonal matrix whose kth diagonal element is $1/d_k$. Further, in the above   example, $\mathbf{D}^{-1}\mathbf{G}_{[t]}' = \begin{bmatrix} & 1/2 & & 1/2 \\ 1/3 & & & & 1/3 & 1/3 \\ & & 1 \end{bmatrix}$.   This   is post-multiplied by 6×2 $\mathbf{X}$ to give an example of (7.13):

$$\mathbf{C}_{[t+1]} = \mathbf{D}^{-1}\mathbf{G}'_{[t]}\mathbf{X} = \begin{bmatrix} & 1/2 & & 1/2 & & \\ 1/3 & & & & 1/3 & 1/3 \\ & & 1 & & & \end{bmatrix} \begin{bmatrix} x_{11} & x_{12} \\ x_{21} & x_{22} \\ x_{31} & x_{32} \\ x_{41} & x_{42} \\ x_{51} & x_{52} \\ x_{61} & x_{62} \end{bmatrix}$$

$$= \begin{bmatrix} \bar{x}_{11} & \bar{x}_{12} \\ \bar{x}_{21} & \bar{x}_{22} \\ \bar{x}_{31} & \bar{x}_{32} \end{bmatrix}. \tag{7.14}$$

Here,

$$\bar{x}_{kj} = \frac{1}{n_k} \sum_{i \in cluster\ k} x_{ij}, \tag{7.15}$$

with n_k the number of members in cluster k and $\sum_{i \in cluster\ k} x_{ij}$ denoting the summation of x_{ij} over the individuals belonging to cluster k. That is, (7.15) is the *average* of the data within *cluster k* for variable j, and such cluster averages are the elements of (7.13) and its example (7.14). The term "*k-means*" originates in the fact that the averages (i.e., means) of clusters play an important role in the algorithm.

7.6   Obtaining Memberships

In this section, Step 3 from Sect. 7.4 is considered; it is shown that the *membership* matrix $\mathbf{G}$ can be obtained that minimizes $f(\mathbf{G}) = \|\mathbf{X} - \mathbf{G}\mathbf{C}_{[t+1]}\|^2$, i.e., the function (7.4) with $\mathbf{C}$ fixed at $\mathbf{C}_{[t+1]}$.

Using $\tilde{\mathbf{g}}'_i$ for the ith row of $\mathbf{G}$, the function $f(\mathbf{G})$ can be rewritten as:

$$\|\mathbf{X} - \mathbf{G}\mathbf{C}_{[t+1]}\|^2 = \left\| \begin{bmatrix} \tilde{\mathbf{x}}'_1 \\ \vdots \\ \tilde{\mathbf{x}}'_n \end{bmatrix} - \begin{bmatrix} \tilde{\mathbf{g}}'_1 \\ \vdots \\ \tilde{\mathbf{g}}'_n \end{bmatrix} \mathbf{C}_{[t+1]} \right\|^2 = \sum_{i=1}^{n} \|\tilde{\mathbf{x}}'_i - \tilde{\mathbf{g}}'_i \mathbf{C}_{[t+1]}\|^2, \tag{7.16}$$

which is the sum of the least squares function of $\tilde{\mathbf{g}}'_i$,

$$f_i(\tilde{\mathbf{g}}'_i) = \|\tilde{\mathbf{x}}'_i - \tilde{\mathbf{g}}'_i \mathbf{C}_{[t+1]}\|^2, \tag{7.17}$$

over $i = 1, \ldots, n$. Here, it should be noted that $\tilde{\mathbf{g}}'_i$ appears only in $f_i(\tilde{\mathbf{g}}'_i)$, i.e., not in the other functions $f_u(\tilde{\mathbf{g}}'_u)$ with $u \neq i$. This implies that the optimal $\tilde{\mathbf{g}}'_i$, which minimizes (7.17), can be obtained independently of $\tilde{\mathbf{g}}'_u$ with $u \neq i$; the repetition of obtaining the optimal $\tilde{\mathbf{g}}_i'$ over $i = 1, \ldots, n$ provides the rows of the membership matrix $\mathbf{G}$ that minimizes (7.16)

Let us recall (7.1) and (7.2), i.e., that $\tilde{\mathbf{g}}_i'$ is filled with zeros except for one element taking 1. For example, if $K = 3$ and individual i belongs to cluster 2, then $\tilde{\mathbf{g}}_i' = [0, 1, 0]$, thus, $\tilde{\mathbf{g}}_i' \mathbf{C}_{[t+1]} = \tilde{\mathbf{c}}_2'$ and (7.17) can be rewritten as $\| \tilde{\mathbf{x}}_i' - \tilde{\mathbf{c}}_2' \|^2$, with $\tilde{\mathbf{c}}_k'$ the kth row of $\mathbf{C}_{[t+1]}$. This example allows us to find that (7.17) takes one of K distinct values as

$$f_i(\tilde{\mathbf{g}}_i') = \|\tilde{\mathbf{x}}_i' - \tilde{\mathbf{g}}_i'\mathbf{C}\|^2 = \begin{cases} \|\tilde{\mathbf{x}}_i' - \tilde{\mathbf{c}}_1'\|^2 & \text{if} \quad \tilde{\mathbf{g}}_i' = [1, 0, 0, \cdots, 0] \\ \|\tilde{\mathbf{x}}_i' - \tilde{\mathbf{c}}_2'\|^2 & \text{if} \quad \tilde{\mathbf{g}}_i' = [0, 1, 0, \cdots, 0] \\ \quad \vdots \\ \|\tilde{\mathbf{x}}_i' - \tilde{\mathbf{c}}_k'\|^2 & \text{if} \quad \tilde{\mathbf{g}}_i' = [0, 0, 0, \cdots, 1] \end{cases}, \qquad (7.18)$$

Therefore, we can compare the largeness of $\|\tilde{\mathbf{x}}_i' - \tilde{\mathbf{c}}_k'\|^2$ across $k = 1, \ldots, K$ to select the vector $\tilde{\mathbf{g}}_i'$ corresponding to the minimal one among $\|\tilde{\mathbf{x}}_i' - \tilde{\mathbf{c}}_k'\|^2$, $k = 1, \ldots, K$. This selection is formally expressed as:

$$g_{ik} = \begin{cases} 1 & \text{if} \|\tilde{\mathbf{x}}_i - \tilde{\mathbf{c}}_k'\|^2 = \min_{1 \le l \le K} \|\tilde{\mathbf{x}}_i - \tilde{\mathbf{c}}_l'\|^2 \\ 0 & \text{otherwise} \end{cases}. \qquad (7.19)$$

The selected vector is the optimal $\tilde{\mathbf{g}}_i' = [g_{i1}, \ldots, g_{iK}]$ minimizing (7.17). Repeating the selection (7.19) over $i = 1, \ldots, n$ provides the vectors $\tilde{\mathbf{g}}_1', \ldots, \tilde{\mathbf{g}}_n'$ that form the rows of $\mathbf{G}_{[t+1]}$ to be obtained.

7.7 Brief Description of Algorithm

The steps of the KMC algorithm in Sect. 7.4 can be rewritten in a simpler manner (without using the subscript t indicating the number of iteration) as follows:

Step 1. Initialize $\mathbf{G}$.
Step 2. Obtain $\mathbf{C} = \mathbf{D}^{-1}\mathbf{G}'\mathbf{X}$
Step 3. Update $\mathbf{G}$ with (7.19)
Step 4. Finish if convergence is reached; otherwise, back to Step 2.

Here, the facts in Sects. 7.5 and 7.6 have been used in Steps 2 and 3. The phrase "*initialize* $\mathbf{G}$" in Step 1 refers to "set $\mathbf{G}$ to a matrix", as the elements of the latter matrix are called initial values. It should be noted that $\mathbf{C}$ may not be initialized in Step 1, since $\mathbf{C}$ is obtained in the next step.

Another version of the KMC algorithm can be formed in which rather $\mathbf{C}$ is initialized with Steps 2 and 3 interchanged. The version can be listed as follows:

Step 1. Initialize $\mathbf{C}$.
Step 2. Update $\mathbf{G}$ with (7.19).
Step 3. Obtain $\mathbf{C} = \mathbf{D}^{-1}\mathbf{G}'\mathbf{X}$.
Step 4. Finish if convergence is reached; otherwise, back to Step 2.

7.8 Bibliographical Notes

Everitt (1993) has intelligibly treated the cluster analysis procedures, including *hierarchal clustering*, which was not introduced in the present book. The recent developments in clustering have been exhaustively detailed in Gan, Ma, and Wu (2007). Hartigan and Wang (1979) have proposed a modified version of the algorithm described in this chapter.

Exercises

7.1. Show that (7.2) could not be satisfied if two or more elements took one with the other elements being zero in a row of $\mathbf{G}$.

7.2. Let $\mathbf{D} = \mathbf{G}'\mathbf{G}$. Show that $\mathbf{D}$ is a $K \times K$ diagonal matrix and $\mathbf{G}'\mathbf{1}_n = \mathbf{D}\mathbf{1}_K$ with the kth element of $\mathbf{D}\mathbf{1}_K$ and the kth diagonal element of $\mathbf{D}$ being the number of the individuals in group k.

7.4. Show that (7.4) can be rewritten as $\sum_{i=1}^{n} \left\| \tilde{\mathbf{x}}_i - \tilde{\mathbf{c}}_{y_i} \right\|^2$, with y_i the index representing the cluster to which individual i belongs and $\tilde{\mathbf{c}}'_{y_i}$ being the y_ith row of $\mathbf{C}$.

7.5. Show that (7.4) can be rewritten as $\sum_{i=1}^{n} \sum_{k=1}^{K} g_{ik} \left\| \tilde{\mathbf{x}}_i - \tilde{\mathbf{c}}_k \right\|^2$.

7.6. One drawback of the k-means clustering (KMC) is that it tends to give *local* minima, but not the *global* minimum. Here, the *global minimum* is defined as the minimum of $f(\theta)$ for all possible θ, using $f(\theta)$ for the loss function of θ (parameter vector or a parameter) to be minimized. On the other hand, a *local minimum* is defined as the minimum of $f(\theta)$ for the θ value within a restricted range. Those minima are illustrated in Fig. 7.4. To avoid the selection of θ for a local minimum as the solution, the algorithm is run multiple times by starting with different initial values in the procedures sensitive to local minima. Let us use $f(\theta_l)$ for the loss function value resulting in the lth run of the algorithm with $l = 1, \ldots, L$. Then, θ_{l*} is selected as the solution with $f(\theta_{l*}) = \min_{1 \leq l \leq L} f(\theta_l)$. Describe why this multi-run procedure decreases the possibility of selecting θ for a local minimum as the solution.

7.7. The iterative algorithm in Sects. 7.4 to 7.7 is included in a family of algorithms generally called *alternating least squares* (*ALS*) *algorithms*. In this exercise, let us consider an ALS algorithm for a problem different from

Fig. 7.4 Illustration of local minima and the global minimum

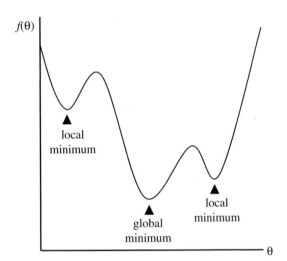

KMC. The problem is the minimization of $f(a, b) = \|\mathbf{y} - a\mathbf{x}_1 - ab\mathbf{x}_2\|^2$ over a and b for $n \times 1$ data vectors $\mathbf{y}$, $\mathbf{x}_1$, and $\mathbf{x}_2$. Here, it should be noted that the coefficient of $\mathbf{x}_2$ is the product of a and b. Show that $f(a, b)$ can be rewritten as $\|\mathbf{y} - a\mathbf{x}\|^2$ with $\mathbf{x} = \mathbf{x}_1 + b\mathbf{x}_2$ and also as $\|\mathbf{y}^* - b\mathbf{x}^*\|^2$ with $\mathbf{y}^* = \mathbf{y} - a\mathbf{x}_1$ and $\mathbf{x}^* = a\mathbf{x}_2$, leading to an ALS algorithm in which the minimization can be attained by the following steps:

Step 1. Initialize b.
Step 2. Obtain $\mathbf{x} = \mathbf{x}_1 + b\mathbf{x}_2$ to update a with $a = (\mathbf{x}'\mathbf{x})^{-1}\mathbf{x}'\mathbf{y}$.
Step 3. Obtain $\mathbf{y}^* = \mathbf{y} - a\mathbf{x}_1$ and $\mathbf{x}^* = a\mathbf{x}_2$ to update b with $b = (\mathbf{x}^{*\prime}\mathbf{x}^*)^{-1}\mathbf{x}^{*\prime}\mathbf{y}^*$.
Step 4. Finish if convergence is reached; otherwise, go back to Step 2.

7.8. Show that (7.4) can be decomposed as:

$$\|\mathbf{X} - \mathbf{GC}\|^2 = \|\mathbf{X} - \mathbf{GD}^{-1}\mathbf{G}'\mathbf{X}\|^2 + \|\mathbf{GD}^{-1}\mathbf{G}'\mathbf{X} - \mathbf{GC}\|^2, \qquad (7.20)$$

with $\mathbf{D} = \mathbf{G}'\mathbf{G}$, by noting
$\|\mathbf{X} - \mathbf{GC}\|^2 = \|\mathbf{X} - \mathbf{GD}^{-1}\mathbf{G}'\mathbf{X} + \mathbf{GD}^{-1}\mathbf{G}'\mathbf{X} - \mathbf{GC}\|^2 = \|\mathbf{X} - \mathbf{GD}^{-1}\mathbf{G}'\mathbf{X}\|^2 + \|\mathbf{GD}^{-1}\mathbf{G}'\mathbf{X} - \mathbf{GC}\|^2 + 2\mathrm{tr}(\mathbf{X} - \mathbf{GD}^{-1}\mathbf{G}'\mathbf{X})'(\mathbf{GD}^{-1}\mathbf{G}'\mathbf{X} - \mathbf{GC})$.
Hints are found in Appendix A.2.2.

7.9. Show that $\|\mathbf{GD}^{-1}\mathbf{G}'\mathbf{X} - \mathbf{GC}\|^2$ in (7.20) can be rewritten as $\|\mathbf{D}^{-1/2}\mathbf{G}'\mathbf{X} - \mathbf{D}^{1/2}\mathbf{C}\|^2$, i.e., (7.4) can be decomposed as:

$$\|\mathbf{X} - \mathbf{GC}\|^2 = \|\mathbf{X} - \mathbf{GD}^{-1}\mathbf{G}'\mathbf{X}\|^2 + \|\mathbf{D}^{-1/2}\mathbf{G}'\mathbf{X} - \mathbf{D}^{1/2}\mathbf{C}\|^2. \qquad (7.21)$$

7.10. De Soete and Carroll (1994) have proposed *reduced k-means analysis* (*RKM*) in which clustering is performed simultaneously with principal component analysis. In RKM, the matrix $\mathbf{C}$ ($K \times p$) in (7.4) is constrained as $\mathbf{C} = \mathbf{FA}'$. Here, $\mathbf{F}$ is $K \times m$, $\mathbf{A}$ is $p \times m$, $\mathbf{A}'\mathbf{A} = \mathbf{I}_m$, and $\mathbf{F}'\mathbf{DF}$ being a diagonal matrix whose diagonal elements are arranged in descending order, with $m \leq \min(K, p)$ and $\mathbf{D} = \mathbf{G}'\mathbf{G}$. That is, RKM is formulated as minimizing $\| \mathbf{X} - \mathbf{GFA}' \|^2$ over $\mathbf{G}$, $\mathbf{F}$, and $\mathbf{A}$ subject to the above constraints, (7.1) and (7.2). Show that an ALS algorithm for RKM can be formed by the following steps:

Step 1. Initialize $\mathbf{C} = \mathbf{FA}'$.
Step 2. Perform (7.19) to obtain $\mathbf{G}$.
Step 3. Perform SVD of $\mathbf{D}^{-1/2}\mathbf{G}'\mathbf{X}$, defined as $\mathbf{D}^{-1/2}\mathbf{G}'\mathbf{X} = \mathbf{K}\mathbf{\Lambda}\mathbf{L}'$.
Step 4. Obtain $\mathbf{C} = \mathbf{D}^{-1/2} \mathbf{k}_m\mathbf{\Lambda}_m\mathbf{L}_m'$ with $\mathbf{K}_m$ ($K \times m$) and $\mathbf{L}_m$ ($p \times m$) containing the first m columns of $\mathbf{K}$ and $\mathbf{L}$, respectively, and $\mathbf{\Lambda}_m$ ($m \times m$), the diagonal matrix whose lth diagonal element is that of $\mathbf{\Lambda}$.
Step 5. Set $\mathbf{F} = \mathbf{D}^{-1/2} \mathbf{k}_m\mathbf{\Lambda}_m$ and $\mathbf{A} = \mathbf{L}_m$ to finish if convergence is reached; otherwise, go back to Step 2.
Here, (7.21) has been used in Steps 3 and 4 with the hints for those steps found in Note 5.3.

7.11. Show that the algorithm in Sects. 7.4 to 7.7 can give a $\mathbf{G}$ whose columns include $\mathbf{0}_n$ during iteration, which belies the supposition that $\mathbf{D} = \mathbf{G}'\mathbf{G}$ is nonsingular and stops the algorithm, i.e., makes KMC fail.

Part III
Maximum Likelihood Procedures

This part starts with the introduction of the principle underlying the maximum likelihood method. This is followed by introductions to path analysis, factor analysis, and structural equation modeling, whose solutions are estimated by the maximum likelihood method. Their solutions can also be obtained by least squares methods, and the procedures in Part II can also be formulated with the maximum likelihood method. However, the latter are often introduced with the least squares methods, while the maximum likelihood method is often used for the procedures discussed in this part.

Chapter 8
Maximum Likelihood and Multivariate Normal Distribution

In the analysis procedures introduced in the last four chapters, parameters are estimated by the *least squares (LS) method*, as reviewed in Sect. 8.1. The remaining sections in this chapter serve to prepare readers for the following chapters, in which a *maximum likelihood (ML) method*, which differs from LS, is used for estimating parameters. That is, the ML method is introduced in Sect. 8.2, which is followed by describing the notion of *probability density function* and the ML method with *multivariate normal distribution*. Finally, ML-based *model selection* with *information criteria* is introduced.

8.1 Model, Parameter, Objective Function, and Optimization

This section deals with a very big subject: We discuss a general framework in which almost all statistical analysis procedures can be formulated; namely, a procedure is underlain by a *model* that can be expressed as:

$$\text{Data} \cong \phi(\Theta) \quad \text{or} \quad \text{Data} = \phi(\Theta) + \text{Errors}, \tag{8.1}$$

with Θ standing for the *parameters* to be obtained. For example, in k-means clustering (Chap. 6), Θ is $\{\mathbf{G}, \mathbf{C}\}$ and $\phi(\Theta) = \phi(\mathbf{G}, \mathbf{C}) = \mathbf{GC}$, as found in (7.3). Another example is regression analysis (Chap. 4). In its model (4.5), the "Data" in (8.1) is denoted as dependent variable vector $\mathbf{y}$, while $\Theta = \{\mathbf{b}, c\}$ and $\phi(\Theta) = \phi(\mathbf{b}, c) = \mathbf{Xb} + c\mathbf{1}$, with $\mathbf{X}$ containing explanatory variables.

An analysis procedure modeled as (8.1) obtains or estimates parameter Θ values. This is formulated as "Obtaining Θ that *optimizes* an *objective function* $obj(\Theta)$ subject to a *constraint* on Θ." This phrase is rewritten as:

© Springer Nature Singapore Pte Ltd. 2016
K. Adachi, *Matrix-Based Introduction to Multivariate Data Analysis*,
DOI 10.1007/978-981-10-2341-5_8

$$\text{Optimizing } obj(\Theta) \text{ over } \Theta \text{ subject to a constraint on } \Theta. \tag{8.2}$$

Here, the term "optimizes" refers to either "minimizes" or "maximizes," and some function can be used as $obj(\Theta)$. In Chaps. 4–7, *least squares* functions are used as $obj(\Theta)$, which are generally expressed as $\|\text{Data} - \phi(\Theta)\|^2$, i.e., the sum of the squared Errors $= \text{Data} - \phi(\Theta)$, with "optimizes" referring to "minimizes." The phrase "subject to a *constraint* on Θ" in (8.2) is not indispensable; whether the phrase is necessary or not depends on analysis procedures. For example, it is necessary in the k-means clustering in which $\mathbf{G}$ in $\Theta = \{\mathbf{G}, \mathbf{C}\}$ is constrained to satisfy (7.1) and (7.2), while the phrase is unnecessary in the regression analysis, in which $\Theta = \{\mathbf{b}, c\}$ is unconstrained.

A methodology formulated by rephrasing "Optimizing $obj(\Theta)$ over Θ" in (8.2) as "minimizing a least squares function" is generally called a *least squares (LS) method*, which is used for the procedures in Part 2. Another methodology, which is as important as the LS method, is introduced next.

8.2 Maximum Likelihood Method

A *maximum likelihood (ML) method* can be formulated by rephrasing "optimizing" and "an objective function" in (8.2) as "maximizing" and "probability," respectively. One feature of the ML method is that it uses the notion of probabilities, which are not used in the LS method. In this section, we introduce the ML method using a simple example.

We suppose that a black box contains *black* and *white* balls, with the total number of the balls known to be 100, but the number of *black/white* balls is unknown. We use θ for the number of black ones. Let us consider a case illustrated in Fig. 8.1: In order to estimate θ, a ball was chosen from the box and returned five times, which gave the data set

$$\mathbf{d} = [1, \quad 0, \quad 0, \quad 1, \quad 0]'. \tag{8.3}$$

Here, $d_i = 1$ and $d_i = 0$ indicate black and white balls chosen, respectively, with d_i the ith element of $\mathbf{d}$.

Let us consider the probability of the data set observed in (8.3). On the supposition of a ball randomly chosen, $P(d_i = 1|\theta)$ and $P(d_i = 0|\theta)$, which denote the probability of $d_i = 1$ observed (i.e., a *black* ball chosen) and that of $d_i = 0$ (i.e., a *white* one chosen), respectively, are expressed as:

$$P(d_i = 1|\theta) = \frac{\theta}{100} \quad \text{and} \quad P(d_i = 0|\theta) = 1 - \frac{\theta}{100}. \tag{8.4}$$

Further, we suppose the balls were chosen mutually independently. Then, the probability of the data set $\mathbf{d} = [1, 0, 0, 1, 0]$ observed in (8.3), i.e., $d_1 = 1$, $d_2 = 0$,

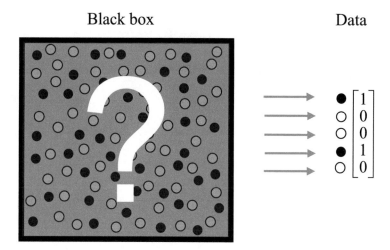

Fig. 8.1 Data of balls chosen from a *black box* that contains *white* and *black balls* with their numbers unknown

$d_3 = 0$, $d_4 = 1$, and $d_5 = 0$, is given by the product $P(d_1 = 1|\theta) \times P(d_2 = 0|\theta) \times P(d_3 = 0|\theta) \times P(d_4 = 1|\theta) \times P(d_5 = 1|\theta)$:

$$
\begin{aligned}
P(\mathbf{d}|\theta) &= \frac{\theta}{100} \times \left(1 - \frac{\theta}{100}\right) \times \left(1 - \frac{\theta}{100}\right) \times \frac{\theta}{100} \times \left(1 - \frac{\theta}{100}\right) \\
&= \left(\frac{\theta}{100}\right)^2 \left(1 - \frac{\theta}{100}\right)^3.
\end{aligned}
\tag{8.5}
$$

For estimating the value of θ, the ML method can be used. Without using mathematics, the idea of the method can be stated as:

> Obtaining the *parameter* value that
> the *occurrence of an event* is the *most likely*. (8.6)

Here, the *occurrence* of an *event* refers to the *observation* of a *data* set, i.e., observing **d** in (8.3), and *how likely it is that the event will occur* is measured by its *probability*. That is, we can use statistical terms to rephrase (8.6) as:

> Obtaining the *parameter* value that
> *maximizes* the *probability* of the *data* being observed. (8.7)

Therefore, the ML method for the data set in (8.3) is to obtain the value of θ that maximizes (8.5). Figure 8.2a shows the values of (8.5) for $\theta = 0, 1, \ldots, 100$. There, we can find that the solution of θ that maximizes the probability is 40. The ML method is similar to a human *psychological* process; *most people seem to think in a*

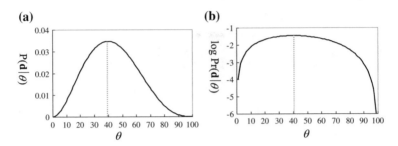

Fig. 8.2 Probability values (**a**) and their logarithms (**b**) against θ

manner similar to that in the ML method. For example, in order to determine who caused an event as "James caused it? Jim? Or Miller did?", one would consider the person most likely to cause the event is the person to be found!

Let us note that $P(\mathbf{d}|\theta)$ is treated as a *function of parameter* θ for a fixed $\mathbf{d}$ in the ML method (Fig. 8.2a), in contrast to cases where $P(\mathbf{d}|\theta)$ is regarded as expressing how probable it is that data set $\mathbf{d}$ occurs for a fixed value of θ. As in Fig. 8.2a, the probability, if it is treated as a function of parameters, is rephrased as *likelihood*, from which the name "maximum likelihood method" originates.

For the sake of ease in mathematical operation, the parameter value that maximizes the log of probability (*log likelihood*) rather than the probability (likelihood) is often obtained in the ML method, since a function, $f(y)$, and its logarithm, $\log f(y)$, take their maximums at the same value of y. The log of (8.5) is given by:

$$\log P(\mathbf{d}|\theta) = 2\log \frac{\theta}{100} + 3\log\left(1 - \frac{\theta}{100}\right). \tag{8.8}$$

Figure 8.2b shows the change in (8.8), where it is also found to attain its maximum for $\theta = 40$.

A solution in the ML method is called a *maximum likelihood estimate (MLE)*. The MLE $\theta = 40$ divided by 100 gives 0.4, which equals the proportion of black balls in (8.3). Thus, one may only glance at (8.3) to intuitively conjecture that θ is about 40, without using the ML method. However, when solutions cannot be intuitively conjectured, the ML method serves as an effective parameter estimation methodology.

8.3 Probability Density Function

In the last section, we used an example of cases where a variable can only take *discrete values* as 1 and 0. In the remaining sections of this chapter, we do not treat such discrete variables, but rather those variables taking continuous or almost continuous values.

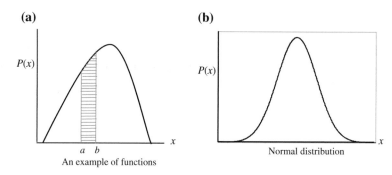

Fig. 8.3 Probability density functions

The probability of a genuinely *continuous variable* being a specific value cannot reasonably be defined. For example, "the probability of a person's stature being exactly 170.0 cm" stands for "the probability of it completely equaling 170.00000000… cm," which can be found to be zero. However, the probability can reasonably be defined for the *intervals* of a continuous variable by letting $P(x \pm \delta)$ be the probability of variable x taking the values within the interval of $x - \delta$ to $x + \delta$ with $\delta > 0$. The density of the probability is given by dividing $P(x \pm \delta)$ by the width of interval $\delta - (-\delta) = 2\delta$ as $P(x \pm \delta)/2\delta$. The density $P(x) = P(x \pm \delta)/2\delta$, in which the width 2δ is reduced to be small enough to be ignored, can be used to express how likely x is to take a specific value, and $P(x)$ is called a *probability density* or the *probability density function* (*PDF*) of variable x. An example of PDF is given in Fig. 8.3a. Its horizontal axis shows the values that x can take and its vertical axis indicates the value of PDF $P(x)$. The following two points should be known about PDF:

Note 8.1. Probability Density

A probability density stands for how likely it is that a value will be observed; an x value with a *greater* value of probability density $P(x)$ is *more likely* to be observed. For example, $P(a) < P(b)$ in Fig. 8.3a implies that $x = b$ is more likely to occur than $x = a$.

The probability density also has the following property (though it is not used, thus, it is unimportant in this book): The *area* below PDF $P(x)$ expresses a probability. In Fig. 7.3a, the probability of x taking the values within the interval $[a, b]$ is indicated by the area with the horizontal lines.

For a variable taking almost continuous values, its probability of being a specific value can be reasonably considered. For example, it makes sense to consider "the probability of a test score being 8" for a test whose scores take the integers from 0 to 10. However, such a variable is also usually treated as a continuous variable for which a probability density is defined, as it is more efficiently analyzed than in cases where it is treated as a discrete variable.

Among a variety of PDFs, the symmetric bell-shaped function shown in Fig. 8.3b is used in a number of univariate statistical procedures. The distribution of *x* with this PDF is called the *normal distribution* or *Gaussian distribution*, the latter name originating from the German mathematician Gauss, who derived the function. Its generalization is introduced next.

8.4 Multivariate Normal Distribution

For multivariate analysis, a PDF for multiple variables is needed, for example, in order to express how likely a person's stature, weight, and waist measurement are to show the values 170.6 cm, 65.3 kg, and 80.7 cm, respectively. As such a PDF,

$$P(\mathbf{x}|\boldsymbol{\mu}, \boldsymbol{\Sigma}) = \frac{1}{(2\pi)^{p/2}|\boldsymbol{\Sigma}|^{1/2}} \exp\left\{ -\frac{1}{2}(\mathbf{x} - \boldsymbol{\mu})'\boldsymbol{\Sigma}^{-1}(\mathbf{x} - \boldsymbol{\mu}) \right\} \qquad (8.9)$$

is very often used, where $\mathbf{x} = [x_1, \ldots, x_p]'$ is the $p \times 1$ vectors of *p*-variables, $\boldsymbol{\mu}$ is a $p \times 1$ vector containing fixed values, π ($\cong 3.14$) denotes the circle ratio, $\exp\{\bullet\} = e^{\{\bullet\}}$ with e ($\cong 2.72$) the base of the natural logarithm, $\boldsymbol{\Sigma}$ is not the symbol of summation but a $p \times p$ *positive-definite* matrix containing fixed values, and $|\boldsymbol{\Sigma}|$ denotes the *determinant* of $\boldsymbol{\Sigma}$. The positive definiteness and determinant are explained in the next notes.

> **Note 8.2. Nonnegative and Positive-Definite Matrices**
> A $p \times p$ square matrix $\mathbf{S}$ is said to be *nonnegative-definite* if $f(\mathbf{w}) = \mathbf{w}'\mathbf{Sw} \geq 0$ for any vector $\mathbf{w}$. It is known that $\mathbf{S}$ being nonnegative-definite and symmetric is equivalent to the property of $\mathbf{S}$ that it can be rewritten as $\mathbf{S} = \mathbf{BB}'$.
>   Nonnegative matrix $\mathbf{S}$ is particularly said to be *positive-definite* if $f(\mathbf{w}) = \mathbf{w}'\mathbf{Sw} \neq 0$, i.e., $f(\mathbf{w}) > \mathbf{w}'\mathbf{Sw}$ for any vector $\mathbf{w} \neq \mathbf{0}_p$. It is known that any positive-definite matrix is *nonsingular*, i.e., its inverse matrix exists, and this matrix is also positive-definite.

A determinant is defined for any square matrix and its function yielding a scalar. However, only the determinants of positive-definite matrices are treated in this book, which can be obtained as follows:

> **Note 8.3. Determinants**
> Let $\mathbf{S}$ be a $p \times p$ positive-definite matrix whose singular values are $\lambda_1, \ldots, \lambda_p$, and the determinant of $\mathbf{S}$ is given as:

$$|\mathbf{S}| = \lambda_1 \times \lambda_2 \times \cdots \times \lambda_p. \tag{8.10}$$

In general, a determinant has the following properties:

$$|\mathbf{SU}| = |\mathbf{S}| \times |\mathbf{U}|, \tag{8.11}$$

$$\left|\mathbf{S}^{-1}\right| = |\mathbf{S}|^{-1}. \tag{8.12}$$

Those two equations hold true, even if $\mathbf{S}$ or $\mathbf{U}$ is not positive-definite.

The distribution of $\mathbf{x}$ whose PDF is (8.9) is called a *multivariate normal* (*MVN*) *distribution*. The value of (8.9) for a specified $\mathbf{x}$ can be obtained, with $\boldsymbol{\mu}$ and $\boldsymbol{\Sigma}$ given. We next describe cautions for notations:

Note 8.4. Three Types of Vector Expressions for Data
Until the last chapter,

$$\mathbf{X} = \begin{bmatrix} \mathbf{x}_1, \ldots, \mathbf{x}_j, \ldots, \mathbf{x}_p \end{bmatrix} = \begin{bmatrix} \tilde{\mathbf{x}}_1' \\ \vdots \\ \tilde{\mathbf{x}}_i' \\ \vdots \\ \tilde{\mathbf{x}}_p' \end{bmatrix}$$

had been used for an n-individuals $\times$ p-variables data matrix; we had expressed the $p \times 1$ vector for individual i as $\tilde{\mathbf{x}}_i$ with the *tilde* symbol ($\sim$) attached to $\mathbf{x}_i$, in order to distinguish $\tilde{\mathbf{x}}_i$ from the vector $\mathbf{x}_j$ ($n \times 1$) standing for variable j.

The $p \times 1$ vector $\mathbf{x}$ in (8.9) is associated with $\tilde{\mathbf{x}}_i$. However, a *tilde* is not used in (8.9) for the sake of simplicity. We do not attach the *tilde* to the vectors standing for individuals from this chapter; they are expressed as

$$\mathbf{X} = \begin{bmatrix} \mathbf{x}_1' \\ \vdots \\ \mathbf{x}_i' \\ \vdots \\ \mathbf{x}_p \end{bmatrix}$$. This is the same for the other vectors. Thus, readers should be

careful about whether vectors stand for the rows of matrices or their columns.

The reason for vector $\mathbf{x}$ in (8.9) not having a subscript is that $\mathbf{x}$ is a *random vector*. This term means that the elements of that vector can take arbitrary values. Thus, $\mathbf{x}_i$ for any i can be substituted into $\mathbf{x}$, with the probability density of $\mathbf{x} = \mathbf{x}_i$ expressed as $P(\mathbf{x}_i)$.

Let us suppose that an infinite number of $\mathbf{x}$ with its PDF (8.9) is observed. The average vector ($p \times 1$) and inter-variable covariance matrix ($p \times p$) of those $\mathbf{x}$ are known to equal $\boldsymbol{\mu}$ and $\boldsymbol{\Sigma}$. Thus, vector $\mathbf{x}$ with its PDF (8.9) is said to *have* (or *follow*) *the MVN distribution with its mean vector* $\boldsymbol{\mu}$ *and covariance matrix* $\boldsymbol{\Sigma}$. This statement is denoted as:

$$\mathbf{x} \sim N_p(\boldsymbol{\mu}, \boldsymbol{\Sigma}), \tag{8.13}$$

where N and its subscript p stand for "normal distribution" and the number of variables, respectively.

The PDF (8.9) with $p = 2$, $\boldsymbol{\mu} = [165, 70]$, $\boldsymbol{\Sigma} = \begin{bmatrix} 150 & 136 \\ 136 & 159 \end{bmatrix}$ is drawn in

Fig. 8.4a. It resembles a bell. The vector $\mathbf{x}$ closer to the place corresponding to the top of $P(\mathbf{x}|\boldsymbol{\mu}, \boldsymbol{\Sigma})$ is more likely to be observed. A bird's-eye view of the distribution in (a) is shown in Fig. 8.4b. There, we can find that the *center* corresponding to the *top* of $P(\mathbf{x}|\boldsymbol{\mu}, \boldsymbol{\Sigma})$ is the *mean vector* $\boldsymbol{\mu}$. It is surrounded by *ellipses* which express the *contours* of $P(\mathbf{x}|\boldsymbol{\mu}, \boldsymbol{\Sigma})$; that is, each of the ellipses stands for the terminus of the vector $\mathbf{x}$ providing an equivalent value of $P(\mathbf{x}|\boldsymbol{\mu}, \boldsymbol{\Sigma})$. The shapes of those ellipses are known to be determined by the *covariance matrix* $\boldsymbol{\Sigma}$. If p is reduced to one, the shape of $P(\mathbf{x}|\boldsymbol{\mu}, \boldsymbol{\Sigma})$ is equal to that drawn in Fig. 8.3b. If $p \geq 3$, then we need graphs of more than three dimensions, which *cannot be drawn or seen*. But, we can imagine "a bell in a multidimensional space."

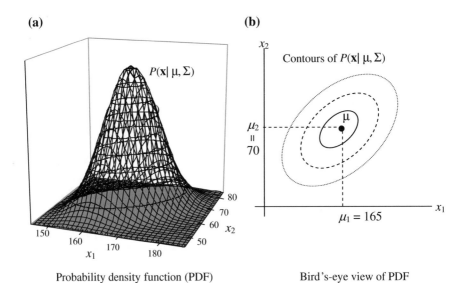

(a)   **(b)**

$P(\mathbf{x}|\boldsymbol{\mu}, \boldsymbol{\Sigma})$

Contours of $P(\mathbf{x}|\boldsymbol{\mu}, \boldsymbol{\Sigma})$

Probability density function (PDF)   Bird's-eye view of PDF

Fig. 8.4 Illustration of a multivariate normal distribution for $p = 2$, i.e., $\mathbf{x} = [x_1, x_2]'$

8.5 Maximum Likelihood Method for Normal Variables

In Fig. 8.4, the PDF $P(\mathbf{x}|\boldsymbol{\mu}, \boldsymbol{\Sigma})$ for MVN distribution is illustrated on the assumption that $\boldsymbol{\mu}$ and $\boldsymbol{\Sigma}$ are known. But, in practical situations, $\boldsymbol{\mu}$ and $\boldsymbol{\Sigma}$ are often unknown and $\mathbf{x}$ is observed as specific vectors $\mathbf{x}_i$ ($i = 1, \ldots, n$), for example, as the rows of $\mathbf{X}$ in Table 8.1. In this section, we consider *estimating parameters* $\boldsymbol{\mu}$ and $\boldsymbol{\Sigma}$ from an $n \times p$ *data* matrix $\mathbf{X} = [\mathbf{x}_1, \ldots, \mathbf{x}_n]'$ on the assumption that their row vectors follow the MVN distribution with its average vector $\boldsymbol{\mu}$ and covariance matrix $\boldsymbol{\Sigma}$:

$$\mathbf{x}_i \sim N_p(\boldsymbol{\mu}, \boldsymbol{\Sigma}) \quad (i = 1, \ldots, n). \tag{8.14}$$

For this estimation, we can use the *ML method* introduced in Sect. 7.2. The ML method for continuous variables can be expressed simply by attaching "density" to "probability" in (8.7), as:

Obtaining the *parameter* value

that *maximizes* the *probability density* of *data* observed. $\tag{8.15}$

By substituting $\mathbf{x}_i$ for $\mathbf{x}$ in (8.9), the probability density of $\mathbf{x} = \mathbf{x}_i$ is expressed as:

$$P(\mathbf{x}_i|\boldsymbol{\mu}, \boldsymbol{\Sigma}) = \frac{1}{(2\pi)^{p/2}|\boldsymbol{\Sigma}|^{1/2}} \exp\left\{-\frac{1}{2}(\mathbf{x}_i - \boldsymbol{\mu})'\boldsymbol{\Sigma}^{-1}(\mathbf{x}_i - \boldsymbol{\mu})\right\}. \tag{8.16}$$

For example, the probability density of $\mathbf{x} = \mathbf{x}_1$ in Table 8.1 is:

$$P(\mathbf{x}_1|\boldsymbol{\mu}, \boldsymbol{\Sigma}) = \frac{1}{(2\pi)^{p/2}|\boldsymbol{\Sigma}|^{1/2}} \exp\left\{-\frac{1}{2}([80, 77, 68]' - \boldsymbol{\mu})'\boldsymbol{\Sigma}^{-1}([80, 77, 68]' - \boldsymbol{\mu})\right\}. \tag{8.17}$$

Table 8.1 Data matrix showing scores of 11 students × 3 subject tests, with the first *five* and the remaining *six* students belonging to two different classes (artificial example)

		Physics	Chemistry	Biology
$\mathbf{X} =$	$\mathbf{x}_1'$	80	77	68
	$\mathbf{x}_2'$	65	46	70
	$\mathbf{x}_3'$	82	57	76
	$\mathbf{x}_4'$	66	61	60
	$\mathbf{x}_5'$	73	72	76
	$\mathbf{x}_6'$	79	84	89
	$\mathbf{x}_7'$	89	74	78
	$\mathbf{x}_8'$	67	60	61
	$\mathbf{x}_9'$	91	87	85
	$\mathbf{x}_{10}'$	81	64	72
	$\mathbf{x}_{11}'$	71	73	75

Mathematical operations for *probabilities* also hold for *probability densities* (Hogg et al. 2005). On the supposition that the rows of $\mathbf{X} = [\mathbf{x}_1, \ldots, \mathbf{x}_n]'$ are observed mutually independently, the *probability density* of the n rows in $\mathbf{X}$ being jointly observed is given by the *product* of (8.16) over $i = 1, \ldots, n$:

$$
\begin{aligned}
P(\mathbf{X}|\boldsymbol{\mu}, \boldsymbol{\Sigma}) &= \prod_{i=1}^{n} \left\{ \frac{1}{(2\pi)^{p/2}|\boldsymbol{\Sigma}|^{1/2}} \exp\left\{ -\frac{1}{2}(\mathbf{x}_i - \boldsymbol{\mu})'\boldsymbol{\Sigma}^{-1}(\mathbf{x}_i - \boldsymbol{\mu}) \right\} \right\} \\
&= \frac{1}{(2\pi)^{np/2}|\boldsymbol{\Sigma}|^{n/2}} \prod_{i=1}^{n} \exp\left\{ -\frac{1}{2}(\mathbf{x}_i - \boldsymbol{\mu})'\boldsymbol{\Sigma}^{-1}(\mathbf{x}_i - \boldsymbol{\mu}) \right\} \quad , \quad (8.18) \\
&= \frac{1}{(2\pi)^{np/2}|\boldsymbol{\Sigma}|^{n/2}} \exp\left\{ -\frac{1}{2}\sum_{i=1}^{n}(\mathbf{x}_i - \boldsymbol{\mu})'\boldsymbol{\Sigma}^{-1}(\mathbf{x}_i - \boldsymbol{\mu}) \right\}
\end{aligned}
$$

with the operator $\prod$ defined as follows:

Note 8.5. Repetition of Products

$$
\prod_{i=1}^{m} a_i = a_1 \times a_2 \times \cdots \times a_m
$$

The probability density, if it is treated as a function of parameters, is also rephrased as the *likelihood*. That is, (8.18) can be called the likelihood of $\boldsymbol{\mu}$ and $\boldsymbol{\Sigma}$ for the data matrix $\mathbf{X}$.

8.6 Maximum Likelihood Estimates of Means and Covariances

The $\boldsymbol{\mu}$ and $\boldsymbol{\Sigma}$ values are to be obtained in the ML method, such that the data matrix $\mathbf{X}$ is the *most likely* to be observed, i.e., the *MLE* of $\boldsymbol{\mu}$ and $\boldsymbol{\Sigma}$ is estimated that *maximizes* (8.18) or its logarithm. This is given by

$$
\log P(\mathbf{X}|\boldsymbol{\mu}, \boldsymbol{\Sigma}) = -\frac{np}{2}\log 2\pi - \frac{n}{2}\log|\boldsymbol{\Sigma}| - \frac{1}{2}\sum_{i=1}^{n}(\mathbf{x}_i - \boldsymbol{\mu})'\boldsymbol{\Sigma}^{-1}(\mathbf{x}_i - \boldsymbol{\mu}). \quad (8.19)
$$

Here, $-(np/2)\log 2\pi$ is a *constant irrelevant* to $\boldsymbol{\mu}$ and $\boldsymbol{\Sigma}$. Thus, the maximization of (8.19) is equivalent to maximizing the function:

$$
l(\boldsymbol{\mu}, \boldsymbol{\Sigma}) = \frac{n}{2}\log|\boldsymbol{\Sigma}| - \frac{1}{2}\sum_{i=1}^{n}(\mathbf{x}_i - \boldsymbol{\mu})'\boldsymbol{\Sigma}^{-1}(\mathbf{x}_i - \boldsymbol{\mu}), \quad (8.20)
$$

with the constant term deleted from (8.19). We refer to (8.20) as the *log likelihood* below. As shown in Appendix A.5.1, the MLE of $\boldsymbol{\mu}$ and $\boldsymbol{\Sigma}$ is given by:

$$\hat{\boldsymbol{\mu}} = \bar{\mathbf{x}} = \frac{1}{n}\sum_{i=1}^{n}\mathbf{x}_i, \tag{8.21}$$

$$\hat{\boldsymbol{\Sigma}} = \frac{1}{n}\sum_{i=1}^{n}(\mathbf{x}_i - \bar{\mathbf{x}})(\mathbf{x}_i - \bar{\mathbf{x}})' = \mathbf{V}. \tag{8.22}$$

Here, $\mathbf{V}$ is the matrix defined in (3.13), which equals $n^{-1}\sum_{i=1}^{n}(\mathbf{x}_i - \bar{\mathbf{x}})(\mathbf{x}_i - \bar{\mathbf{x}})'$, as shown next:

Note 8.6. Another Expression of V

Let us recall (3.13). It can be rewritten as $\mathbf{V} = n^{-1}\mathbf{X}'\mathbf{J}\mathbf{X} = n^{-1}(\mathbf{J}\mathbf{X})'\mathbf{J}\mathbf{X}$, where $\mathbf{J}\mathbf{X}$ contains the centered scores: $\mathbf{J}\mathbf{X} = \begin{bmatrix} (\mathbf{x}_1 - \bar{\mathbf{x}})' \\ \vdots \\ (\mathbf{x}_n - \bar{\mathbf{x}})' \end{bmatrix}$. Thus, $n^{-1}(\mathbf{J}\mathbf{X})'\mathbf{J}\mathbf{X}$ is found to equal $n^{-1}\sum_{i=1}^{n}(\mathbf{x}_i - \bar{\mathbf{x}})(\mathbf{x}_i - \bar{\mathbf{x}})'$.

In (8.21) and (8.22), we find that the MLE of $\boldsymbol{\mu}$ and $\boldsymbol{\Sigma}$ is found to equal the *average vector* and *covariance matrix* obtained from the *data* set, respectively.

Though both $\boldsymbol{\mu}$ and $\bar{\mathbf{x}}$ are referred to as average/mean vectors, and both $\boldsymbol{\Sigma}$ and $\mathbf{V}$ are called covariance matrices, $\boldsymbol{\mu}$ and $\boldsymbol{\Sigma}$ differ from $\bar{\mathbf{x}}$ and $\mathbf{V}$, in that the former are the parameters determining $N_p(\boldsymbol{\mu}, \boldsymbol{\Sigma})$, while $\bar{\mathbf{x}}$ and $\mathbf{V}$ are the statistics obtained from $\mathbf{X}$. However, the MLE of $\boldsymbol{\mu}$ and $\boldsymbol{\Sigma}$ equals $\bar{\mathbf{x}}$ and $\mathbf{V}$, respectively, as shown in (8.21) and (8.22) on the assumption of the rows of $\mathbf{X}$ following $N_p(\boldsymbol{\mu}, \boldsymbol{\Sigma})$ mutually independently. For distinguishing $\boldsymbol{\mu}$ and $\boldsymbol{\Sigma}$ from $\bar{\mathbf{x}}$ and $\mathbf{V}$, the latter statistics are called a *sample average vector* and a *sample covariance matrix*, respectively.

By substituting MLE (8.21) and (8.22) into the log likelihood (8.20), its maximum is expressed as:

$$\begin{aligned} l(\hat{\boldsymbol{\mu}}, \hat{\boldsymbol{\Sigma}}) &= -\frac{n}{2}\log|\mathbf{V}| - \frac{1}{2}\sum_{i=1}^{n}(\mathbf{x}_i - \bar{\mathbf{x}})'\mathbf{V}^{-1}(\mathbf{x}_i - \bar{\mathbf{x}}) \\ &= -\frac{n}{2}\log|\mathbf{V}| - \frac{1}{2}\operatorname{tr}\sum_{i=1}^{n}(\mathbf{x}_i - \bar{\mathbf{x}})'\mathbf{V}^{-1}(\mathbf{x}_i - \bar{\mathbf{x}}) \\ &= -\frac{n}{2}\log|\mathbf{V}| - \frac{1}{2}\operatorname{tr}\sum_{i=1}^{n}(\mathbf{x}_i - \bar{\mathbf{x}})(\mathbf{x}_i - \bar{\mathbf{x}})'\mathbf{V}^{-1} \\ &= -\frac{n}{2}\log|\mathbf{V}| - \frac{n}{2}\operatorname{tr}\mathbf{V}\mathbf{V}^{-1} = -\frac{n}{2}\log|\mathbf{V}| - \frac{np}{2}, \end{aligned} \tag{8.23}$$

which is used in the next section.

8.7   Model Selection

Cases exist for which several models are considered to explain a single data set, as illustrated in Fig. 8.5. *Model selection* refers to comparing models and selecting the *model best fitted* to a data set. An advantage of the ML method is that its MLE can be used for model selection with statistics generally called *information criteria*.

One statistic included in such information criteria was first derived by the Japanese statistician Hirotsugu Akaike (1927–2009). The statistic is known as *Akaike's* (1974) *information criterion* (*AIC*), which is defined as:

$$\text{AIC} = -2\,l(\hat{\Theta}) + 2\eta \qquad (8.24)$$

for a model in which η is the *number of parameters* to be estimated in the model, $\hat{\Theta}$ stands for a set of MLEs of parameters, and $l(\hat{\Theta})$ expresses the value of the log likelihood $l(\Theta)$ for $\Theta = \hat{\Theta}$. AIC is defined for each of the models considered for a data set, and the model with a *smaller* AIC value is regarded as the *better* model.

Following AIC, similar statistics have been proposed. Among them, a popular one is Schwarz's (1978) *Bayesian information criterion* (*BIC*), defined as:

$$\text{BIC} = -2\,l(\hat{\Theta}) + \eta \log n, \qquad (8.25)$$

with n the number of individuals in a data set. As with AIC, (8.25) is defined for each model, and a smaller value implies that the model is better. It should be noted that both (8.24) and (8.25) *penalize* a model for having *more parameters*, which can be related to the philosophy of science, as in the following note:

> **Note 8.7. Information Criteria and Philosophy**
> How information criteria such as (8.24) and (8.25) are derived is beyond the scope of this book. However, the following arguments are to be born in mind:
>   Let us view Fig. 8.5 by replacing *"model"* with *"theory"* and *"data set"* with *"phenomenon."* In the *philosophy of science*, it had been argued that the *goodness of a theory* should be evaluated by:

Fig. 8.5 Several models for a data set

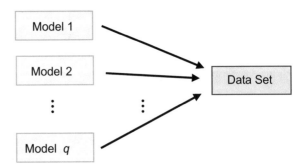

[1] how *well* it *explains* a phenomenon;

[2] how *simple* (*parsimonious*, in philosophical terms) it is. (e.g., Hempel 1966).

We can reasonably consider that [1] corresponds to the attained value of the *log likelihood* $\log l(\hat{\Theta})$ and [2] is associated with the *smallness* of η (*the number of parameters*). Thus, [1] and [2] are found to correspond to smaller values of (8.24) and (8.25). In this sense, *information criteria* can be viewed as a *mathematical validation* of the *philosophical argument*.

Sometimes, the model chosen by AIC is different from that by BIC. For such a case, the model must be chosen by users' subjective consideration. This shows that no absolute index exists for model selection, which should be kept in mind.

8.8 Assessment of Between-Group Heterogeneity

In order to illustrate model selection by information criteria, we consider two models for data matrix $\mathbf{X}$ in Table 8.1. *Model 1* is expressed as (8.14); all row vectors of $\mathbf{X}$ are assumed to follow an *identical MVN* distribution, $N_p(\boldsymbol{\mu}, \boldsymbol{\Sigma})$, in Model 1. On the other hand, let us consider *Model 2* expressed as:

$$\mathbf{x}_i \sim N_p(\boldsymbol{\mu}_1, \boldsymbol{\Sigma}_1) \quad \text{for } i = 1, \ldots, 5 \text{ and } \mathbf{x}_i \sim N_p(\boldsymbol{\mu}_2, \boldsymbol{\Sigma}_2) \text{ for } i = 6, \ldots, 11. \quad (8.26)$$

The first five row vectors and the remaining six are assumed to follow *different* MVN distributions in Model 2.

The MLEs for *Model 1* are given by (8.21) and (8.22) with $n = 11$, and their values are obtained as in Table 8.2(A). As found there, η (*the number of parameters*) is $3 + 6 = 9$, where 3 is the number of elements in $\boldsymbol{\mu}$ and 6 is the number of different covariances in $\boldsymbol{\Sigma}$; it has 3×3 elements, but the number of different ones is 6, since $\boldsymbol{\Sigma}$ is symmetric. Substituting those MLEs into (8.23), we have: $l(\hat{\boldsymbol{\mu}}, \hat{\boldsymbol{\Sigma}}) = -\frac{11}{2}\log(202139.9) - \frac{33}{2} = -285.268$. Further, this is substituted into $l(\hat{\Theta})$ in (8.24) and (8.25) to give AIC $= -2 \times (-285.268) + 2 \times 9 = 588.54$ and BIC $= -2 \times (-285.268) + 9\log 11 = 592.12$, as shown in Table 8.2(A).

Note 8.8. Base of Logarithm
In this book, "$\log x$" stands for "$\log_e x$", with $e \cong 2.72$ the base of the natural logarithm.

Next, let us obtain the MLE, AIC, and BIC for *Model 2*. On the supposition that the rows of $\mathbf{X} = [\mathbf{x}_1, \ldots, \mathbf{x}_n]'$ are observed mutually independently, the *probability density* of the n rows in $\mathbf{X}$ being observed is jointly expressed as:

Table 8.2 MLE, η (the number of parameters), AIC, and BIC for the data in Table 8.1

Model	(A) Model 1				(B) Model 2							
Parameter	$\hat{\boldsymbol{\mu}}$	$\hat{\boldsymbol{\Sigma}}$			$\hat{\boldsymbol{\mu}}_1$	$\hat{\boldsymbol{\Sigma}}_1$			$\hat{\boldsymbol{\mu}}_2$	$\hat{\boldsymbol{\Sigma}}_2$		
Physics	76.7	73.7	63.9	48.4	73.2	48.6	38.9	22.0	79.7	75.6	52.2	50.9
Chemistry	68.6		136.8	64.6	62.6		121.0	2.0	73.7		94.2	83.2
Biology	73.6			71.9	70.0			35.2	76.7			82.2
η	9				18							
AIC	588.54				189.45							
BIC	592.12				196.61							

$$P(\mathbf{X}|\boldsymbol{\mu}_1,\boldsymbol{\Sigma}_1,\boldsymbol{\mu}_2,\boldsymbol{\Sigma}_1) = \frac{1}{(2\pi)^{5p/2}|\boldsymbol{\Sigma}_1|^{5/2}}\exp\left\{-\frac{1}{2}\sum_{i=1}^{5}(\mathbf{x}_i-\boldsymbol{\mu}_1)'\boldsymbol{\Sigma}_1^{-1}(\mathbf{x}_i-\boldsymbol{\mu}_1)\right\}$$

$$\times \frac{1}{(2\pi)^{6p/2}|\boldsymbol{\Sigma}_2|^{6/2}}\exp\left\{-\frac{1}{2}\sum_{i=6}^{11}(\mathbf{x}_i-\boldsymbol{\mu}_2)'\boldsymbol{\Sigma}_2^{-1}(\mathbf{x}_i-\boldsymbol{\mu}_2)\right\},$$

$$(8.27)$$

because of (8.26), where

$$\frac{1}{(2\pi)^{5p/2}|\boldsymbol{\Sigma}_1|^{5/2}}\exp\left\{-\frac{1}{2}\sum_{i=1}^{5}(\mathbf{x}_i-\boldsymbol{\mu}_1)'\boldsymbol{\Sigma}_1^{-1}(\mathbf{x}_i-\boldsymbol{\mu}_1)\right\}$$

stands for the probability density of $\mathbf{x}_1, \ldots, \mathbf{x}_5$ being jointly observed, while

$$\frac{1}{(2\pi)^{6p/2}|\boldsymbol{\Sigma}_2|^{6/2}}\exp\left\{-\frac{1}{2}\sum_{i=6}^{11}(\mathbf{x}_i-\boldsymbol{\mu}_2)'\boldsymbol{\Sigma}_2^{-1}(\mathbf{x}_i-\boldsymbol{\mu}_2)\right\}$$

is the probability density for $\mathbf{x}_6, \ldots, \mathbf{x}_{11}$.

The *log likelihood* corresponding to (8.27) can be expressed as:

$$l(\boldsymbol{\mu}_1,\boldsymbol{\Sigma}_1,\boldsymbol{\mu}_2,\boldsymbol{\Sigma}_2) = l_1(\boldsymbol{\mu}_1,\boldsymbol{\Sigma}_1) + l_2(\boldsymbol{\mu}_2,\boldsymbol{\Sigma}_2).$$

$$(8.28)$$

where

$$l_1(\boldsymbol{\mu}_1,\boldsymbol{\Sigma}_1) = -\frac{5}{2}|\boldsymbol{\Sigma}_1| - \frac{1}{2}\sum_{i=1}^{5}(\mathbf{x}_i-\boldsymbol{\mu}_1)'\boldsymbol{\Sigma}_1^{-1}(\mathbf{x}_i-\boldsymbol{\mu}_2),$$

$$(8.29)$$

$$l_2(\boldsymbol{\mu}_2,\boldsymbol{\Sigma}_2) = -\frac{6}{2}|\boldsymbol{\Sigma}_2| - \frac{1}{2}\sum_{i=6}^{11}(\mathbf{x}_i-\boldsymbol{\mu}_2)'\boldsymbol{\Sigma}_2^{-1}(\mathbf{x}_i-\boldsymbol{\mu}_2),$$

$$(8.30)$$

with the constants irrelevant to parameters being deleted.

As found in (8.28), the log likelihood is decomposed into $l_1(\mu_1, \Sigma_1)$ and $l_2(\mu_2, \Sigma_2)$. Since they are functions of different sets of parameters, $\{\mu_1, \Sigma_1\}$ maximizing $l_1(\mu_1, \Sigma_1)$ and $\{\mu_2, \Sigma_2\}$ maximizing $l_2(\mu_2, \Sigma_2)$ are found to be the MLEs that maximize (8.28). By comparing (8.29) and (8.30) with (8.20), we can find that (8.29) or (8.30) is equivalent to log likelihood (8.20), in which μ and Σ have the subscript 1 or 2, and the series $i = 1, \ldots, n$ is replaced by $i = 1, \ldots, 5$ or $i = 6, \ldots, 11$. This fact, along with (8.21) and (8.22), shows that μ_1 and Σ_1 maximizing $l_1(\mu_1, \Sigma_1)$ are given by:

$$\hat{\mu}_1 = \frac{1}{5}\sum_{i=1}^{5} x_i, \quad \hat{\Sigma}_1 = \frac{1}{5}\sum_{i=1}^{5} (x_i - \hat{\mu}_1)(x_i - \hat{\mu}_1)',$$

while μ_2 and Σ_2 maximizing $l_2(\mu_2, \Sigma_2)$ are given by:

$$\hat{\mu}_2 = \frac{1}{6}\sum_{i=6}^{11} x_i, \quad \hat{\Sigma}_2 = \frac{1}{6}\sum_{i=6}^{11} (x_i - \hat{\mu}_2)(x_i - \hat{\mu}_2)'.$$

Those values are shown in Table 8.2(B), whose substitution into (8.28) gives the value of the *maximum log likelihood*:

$$\log l(\hat{\mu}_1, \hat{\Sigma}_1, \hat{\mu}_2, \hat{\Sigma}_2) = -\frac{5}{2}\log(98328.73) - \frac{15}{2} - \frac{6}{2}\log(36140.64) - \frac{18}{2} = -76.73,$$

with $\eta = 18$ for Model 2 and $n = 5 + 6 = 11$. Using them in (8.24) and (8.25), we get the AIC and BIC values in Table 8.2(B).

In Table 8.2, both the AIC and BIC are found to show that *Model 2* is better; the 11 students are classified into the two groups characterized by different MVN distributions. It should be kept in mind that comparing the AIC and BIC values is senseless; the comparison is to be made *within the same index*. Comparing AIC values for different data sets as well as BIC values for different data sets is also senseless. A model comparison must be made *for a single data set* (Fig. 8.5).

8.9   Bibliographical Notes

This chapter can serve as a preliminary stage before learning statistical inferences, which are not treated in the present book. *Statistical inferences* refer to the theories in which the relationships of the estimates of parameters to their true values are discussed on the basis of probabilities. One of the established books on elementary statistical inferences was written by Hogg et al. (2005). Books on multivariate statistical inferences include Anderson (2003), Rao (2001), Rencher and Christensen (2012), and Seber (1984). Detailed treatments of information criteria are found in Konishi (2014) and Konishi and Kitagawa (2008). In those books, properties of maximum likelihood estimates are detailed.

Exercises

8.1. Let **d** be an $n \times 1$ data vector whose m elements take one and whose remaining elements are zero, with ω the probability of an element in **d** taking one. The likelihood of parameter ω for the data set **d** is expressed as:

$$P(\omega) = \omega^m (1 - \omega)^{n-m}, \tag{8.31}$$

on the supposition that the elements in **d** are mutually independently observed. Show that the MLE of ω is given by m/n, using the fact that $d\log P(\omega)/d\omega$, i.e., the differentiation of the logarithm of (8.31) with respect to ω, is zero for the ω value being MLE, with $d\log\omega/d\omega = 1/\omega$ and $d\log(1 - \omega)/d\omega = -1/(1 - \omega)$.

8.2. The function

$$\phi(x|b, c) = \frac{1}{1 + \exp(-bx + c)} \tag{8.32}$$

is called a *logistic function* and is used for relating a continuous variable x to probability $\phi(x|b, c)$. Verify that the function $\phi(x|b, c)$ takes the forms in Fig. 8.6 with $0 \leq \phi(x|b, c) \leq 1$, by substituting some values into x with b and c fixed at specific values.

8.3. Let us suppose that the probability of engine i (= 1, ..., n) having trouble is expressed as $1/\{1 + \exp(-bx_i + c)\}$ with x_i the value of the variable for i explaining the trouble probability. Show that the likelihood of b and c can be expressed as:

$$\prod_{i=1}^{n} \left(\frac{1}{1 + \exp(-bx_i + c)} \right)^{d_i} \left(\frac{\exp(-bx_i + c)}{1 + \exp(-bx_i + c)} \right)^{1-d_i} \tag{8.33}$$

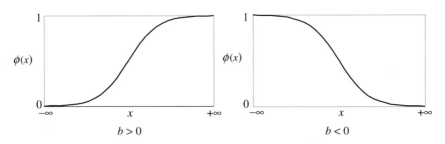

Fig. 8.6 Illustrations of logistic functions

for observed data x_i and d_i, $i = 1, \ldots, n$, with $d_i = 1$ if i has trouble; $d_i = 0$ otherwise. Here, $d_1, \ldots, d_n$ are assumed to be mutually independently observed.

8.4. Show that the logarithm of (8.33) can be written as:

$$\sum_{i=1}^{n} \{(1 - d_i)(-bx_i + c) - \log[1 + \exp(-bx_i + c)]\}. \qquad (8.34)$$

8.5. Let us consider another model for engine trouble in which the probability of engine i $(= 1, \ldots, n)$ having trouble is expressed as $1/\{1 + \exp(-\alpha z_i - \beta x_i + \gamma)\}$, with x_i the one in (8.33) and z_i the value of another explanatory variable for i. The likelihood for this model is expressed as:

$$\prod_{i=1}^{n} \left(\frac{1}{1 + \exp(-\alpha z_i - \beta x_i + \gamma)} \right)^{d_i} \left(\frac{\exp(-\alpha z_i - \beta x_i + \gamma)}{1 + \exp(-\alpha z_i - \beta x_i + \gamma)} \right)^{1 - d_i}. \qquad (8.35)$$

Let $\hat{b}$ and $\hat{c}$ denote the MLE of b and c for maximizing (8.33) or (8.34). On the other hand, let $\hat{\alpha}$, $\hat{\beta}$, and $\hat{\gamma}$ be the MLE of α, β, and γ for maximizing (8.35). Show that BIC is expressed as:

$$-2 \sum_{i=1}^{n} \{(1 - d_i)(-\hat{b}x_i + \hat{c}) - \log[1 + \exp(-\hat{b}x_i + \hat{c})]\} + 2\log n$$

for (8.33), while it is expressed as:

$$-2 \sum_{i=1}^{n} \{(1 - d_i)(-\hat{\alpha}z_i - \hat{\beta}x_i + \hat{\gamma}) - \log[1 + \exp(-\hat{\alpha}z_i - \hat{\beta}x_i + \hat{\gamma})]\} + 3\log n$$

for (8.35).

8.6. The similarity of the ML method to a human psychological process was mentioned with an example in Sect. 8.2. Present another example for illustrating the similarity.

8.7. If $x \sim N_p(\mu, \Sigma)$, it is known that $x + a \sim N_p(\mu + a, \Sigma)$ for fixed a. Use this fact to show the equivalence between $x \sim N_p(\mu, \Sigma)$ and $x = \mu + e$ with $e \sim N_p(0_p, \Sigma)$.

8.8. Use the fact that $|\sigma^2 I_p| = \sigma^{2p}$ to show that the probability density function (PDF) of $x \sim N_p(\mu, \sigma^2 I_p)$ is the product of $P(x_j|\mu_j, \sigma^2) = \frac{1}{\sqrt{2\pi}\sigma} \exp\left\{ -\frac{1}{2} \frac{(x_j - \mu_j)^2}{\sigma^2} \right\}$ over $j = 1, \ldots, p$, with x_j the jth element of x, μ_j that of μ, and $P(x_j|\mu_j, \sigma)$ being the PDF of the (univariate) normal distribution with its mean μ_j and variance σ^2.

8.9. If $\mathbf{x} \sim N_p(\boldsymbol{\mu}, \sigma^2\mathbf{I}_p)$, show that the MLE of σ^2 is given by $\hat{\sigma}^2 = \frac{1}{np}\sum_{i=1}^{n}\|\mathbf{x}_i - \bar{\mathbf{x}}\|^2$, with $\mathbf{x}_i$, $i = 1, \ldots, n$, the $p \times 1$ observed vectors for $\mathbf{x}$, and $\bar{\mathbf{x}} = n^{-1}\sum_{i=1}^{n}\mathbf{x}_i$.

8.10. Let us consider the model $\mathbf{x}_i = \mathbf{f}_i(\boldsymbol{\Theta}) + \mathbf{e}_i$ for $p \times 1$ data vectors $\mathbf{x}_i$, $i = 1, \ldots, n$, observed mutually independently, with $\mathbf{e}_i \sim N(\mathbf{0}_p, \sigma^2\mathbf{I}_p)$ and $\mathbf{f}_i(\boldsymbol{\Theta})$ a function of parameter $\boldsymbol{\Theta}$ yielding a $p \times 1$ data vector. Show the equivalence between the MLE of $\boldsymbol{\Theta}$ and the least squares estimate of $\boldsymbol{\Theta}$ minimizing $\sum_{i=1}^{n}\|\mathbf{x}_i - \mathbf{f}_i(\boldsymbol{\Theta})\|^2$, using the facts in Exercises 8.7 to 8.9.

8.10. For n_k vectors $\mathbf{x}_{ki}$ ($p \times 1$), $i = 1, \ldots, n_k$, observed mutually independently, for group $k = 1, 2, 3$, let us consider the following models:

Model 1. $\mathbf{x}_{ki} \sim N_p(\boldsymbol{\mu}, \boldsymbol{\Sigma})$: All observations follow an identical distribution.
Model 2. $\mathbf{x}_{ki} \sim N_p(\boldsymbol{\mu}_k, \boldsymbol{\Sigma}_k)$: Each group has a specific distribution.
Model 3. $\mathbf{x}_{1i} \sim N_p(\boldsymbol{\mu}_1, \boldsymbol{\Sigma}_1)$ and $\mathbf{x}_{ki} \sim N_p(\boldsymbol{\mu}_2, \boldsymbol{\Sigma}_2)$ for $k = 2, 3$: Group 1 differs from 2 and 3.

Express AIC for the models as functions of $\mathbf{x}_{ki}$ and the number of parameters.

8.11. For n_k vectors $\mathbf{x}_{ki}$ ($p \times 1$), $i = 1, \ldots, n_k$ observed mutually independently, for group $k = 1, \ldots, K$, let us consider the following models:

Model 1. $\mathbf{x}_{ki} \sim N_p(\boldsymbol{\mu}_k, \boldsymbol{\Sigma})$: The covariances are homogeneous among groups.
Model 2. $\mathbf{x}_{ki} \sim N_p(\boldsymbol{\mu}_k, \boldsymbol{\Sigma}_k)$: The covariances are heterogeneous.

Express BIC for the models as functions of $\mathbf{x}_{ki}$ and the number of parameters, using the facts described in Appendix A.5.2.

8.12. For n vectors $\mathbf{x}_i$ ($p \times 1$), $i = 1, \ldots, n$, observed mutually independently, let us consider the models:

Model 1. $\mathbf{x}_i \sim N_p(\boldsymbol{\mu}, \boldsymbol{\Sigma})$: The covariances are unconstrained.
Model 2. $\mathbf{x}_i \sim N_p(\boldsymbol{\mu}, \sigma^2\mathbf{I}_p)$: The covariances are constrained.

Express AIC for the models as functions of $\mathbf{x}_{ki}$ and the number of parameters.

Chapter 9
Path Analysis

Let us assume three variables, A, B, and C, to be analyzed. The regression analysis for predicting C from A and B is based on the causal model, with A and B causes and C the result. However, this model is not guaranteed to indicate the true relationships among A, B, and C. The true causal model might be "A causes B which causes C" or "A causes B and C." *Path analysis* is a procedure in which *users form causal models* by themselves and *select the model* fitted well to a data set. The origins of path analysis can be found in Wright's (1918, 1960) biometric studies and Haavelmo's (1943) econometric ones (Kaplan 2000).

9.1 From Multiple Regression Analysis to Path Analysis

In this chapter, we use a 60 students × 5 variables data matrix, which is divided into the two blocks in Table 9.1(A). The five variables concern a lecture:

IN: to what extent students were **in**terested in the lecture;
KN: the amount of prior **kn**owledge of the lecture subjects;
AB: how often students were **ab**sent from the lecture;
SH: study **h**ours that students took at home for the lecture; and
RE: **re**cords that students were finally given.

For this data set, the *regression analysis* for predicting RE is modeled as:

$$RE = b_1 \times IN + b_2 \times KN + b_3 \times AB + b_4 \times SH + c + error. \qquad (9.1)$$

This model can be expressed as the path diagram in Fig. 9.1a. There, *double-headed arrows* indicate linked variables being merely *correlated*, and *single-headed arrows* indicate the *causal* relationships; they extend from causes (*explanatory* variables) to a result (*dependent* variable). That is, regression analysis is based on the causal model with *multiple causes* and *a single result*.

© Springer Nature Singapore Pte Ltd. 2016
K. Adachi, *Matrix-Based Introduction to Multivariate Data Analysis*,
DOI 10.1007/978-981-10-2341-5_9

Table 9.1 A data set for five
variables for a lecture
(artificial example)

	IN	KN	AB	SH	RE
(A) *Raw data*					
1	4	54	13.8	120	82
2	7	68	0	150	96
3	4	66	19.6	90	82
4	4	68	17.5	90	80
5	4	68	35.1	60	70
6	4	66	24	90	58
7	3	76	26.1	30	82
8	3	66	32.2	60	66
9	2	58	41.2	0	40
10	6	70	1.1	150	90
11	6	98	10.6	60	90
12	2	48	48	60	44
13	4	70	11.9	150	98
14	6	76	13.7	120	90
15	3	50	39.7	90	70
16	5	62	11.8	120	96
17	3	52	25.2	60	60
18	2	74	34	0	54
19	3	52	33.1	90	64
20	5	70	13	150	86
21	5	80	9.5	150	88
22	1	56	39.7	0	48
23	7	74	11.5	180	84
24	4	60	15.5	90	80
25	1	64	53.6	0	52
26	5	60	23.4	150	80
27	5	50	16.7	180	74
28	5	66	13.9	90	74
29	5	76	26.2	120	80
30	5	62	10.4	120	88
31	3	64	25.5	60	78
32	3	62	27.4	60	68
33	3	72	37	30	64
34	3	74	22.8	90	90
35	6	68	24.2	180	94
36	3	64	35.8	60	76
37	5	70	16.8	90	94
38	4	58	17.5	90	90
39	2	56	25.2	0	58
40	5	64	9.4	120	90
41	5	66	6.2	120	86
42	3	52	38	30	48

(continued)

Table 9.1 (continued)

	IN	KN	AB	SH	RE
43	6	66	5.8	150	86
44	5	62	19.4	90	86
45	5	82	9.9	30	92
46	3	60	36.4	60	62
47	4	58	24	120	82
48	2	56	32.1	60	60
49	4	58	38.8	60	56
50	2	40	30.7	90	64
51	4	50	31.9	90	72
52	4	64	10.5	120	78
53	3	44	19.8	60	66
54	5	70	9.4	150	82
55	4	50	24.5	120	74
56	4	66	25.6	120	76
57	5	62	26	120	86
58	5	74	15.8	60	90
59	5	64	4.8	90	94
60	4	52	43.3	90	58
Av	4.03	63.47	22.78	90.5	75.77
SD	1.35	9.99	12.08	46.63	14.61
(B) *Centered data* $= X$					
1	−0.03	−9.47	−8.97	29.50	6.23
2	2.97	4.53	−22.78	59.50	20.23
3	−0.03	2.53	−3.17	−0.50	6.23
4	−0.03	4.53	−5.28	−0.50	4.23
5	−0.03	4.53	12.33	−30.50	−5.77
6	−0.03	2.53	1.23	−0.50	−17.77
7	−1.03	12.53	3.33	−60.50	6.23
8	−1.03	2.53	9.43	−30.50	−9.77
9	−2.03	−5.47	18.43	−90.50	−35.77
10	1.97	6.53	−21.68	59.50	14.23
11	1.97	34.53	−12.18	−30.50	14.23
12	−2.03	−15.47	25.23	−30.50	−31.77
13	−0.03	6.53	−10.88	59.50	22.23
14	1.97	12.53	−9.08	29.50	14.23
15	−1.03	−13.47	16.93	−0.50	−5.77
16	0.97	−1.47	−10.98	29.50	20.23
17	−1.03	−11.47	2.43	−30.50	−15.77
18	−2.03	10.53	11.23	−90.50	−21.77
19	−1.03	−11.47	10.33	−0.50	−11.77
20	0.97	6.53	−9.78	59.50	10.23

(continued)

Table 9.1 (continued)

	IN	KN	AB	SH	RE
21	0.97	16.53	−13.28	59.50	12.23
22	−3.03	−7.47	16.93	−90.50	−27.77
23	2.97	10.53	−11.28	89.50	8.23
24	−0.03	−3.47	−7.28	−0.50	4.23
25	−3.03	0.53	30.83	−90.50	−23.77
26	0.97	−3.47	0.63	59.50	4.23
27	0.97	−13.47	−6.08	89.50	−1.77
28	0.97	2.53	−8.87	−0.50	−1.77
29	0.97	12.53	3.43	29.50	4.23
30	0.97	−1.47	−12.38	29.50	12.23
31	−1.03	0.53	2.73	−30.50	2.23
32	−1.03	−1.47	4.63	−30.50	−7.77
33	−1.03	8.53	14.23	−60.50	−11.77
34	−1.03	10.53	0.03	−0.50	14.23
35	1.97	4.53	1.43	89.50	18.23
36	−1.03	0.53	13.03	−30.50	0.23
37	0.97	6.53	−5.97	−0.50	18.23
38	−0.03	−5.47	−5.28	−0.50	14.23
39	−2.03	−7.47	2.43	−90.50	−17.77
40	0.97	0.53	−13.38	29.50	14.23
41	0.97	2.53	−16.58	29.50	10.23
42	−1.03	−11.47	15.23	−60.50	−27.77
43	1.97	2.53	−16.98	59.50	10.23
44	0.97	−1.47	−3.38	−0.50	10.23
45	0.97	18.53	−12.88	−60.50	16.23
46	−1.03	−3.47	13.63	−30.50	−13.77
47	−0.03	−5.47	1.23	29.50	6.23
48	−2.03	−7.47	9.33	−30.50	−15.77
49	−0.03	−5.47	16.03	−30.50	−19.77
50	−2.03	−23.47	7.93	−0.50	−11.77
51	−0.03	−13.47	9.13	−0.50	−3.77
52	−0.03	0.53	−12.28	29.50	2.23
53	−1.03	−19.47	−2.97	−30.50	−9.77
54	0.97	6.53	−13.38	59.50	6.23
55	−0.03	−13.47	1.73	29.50	−1.77
56	−0.03	2.53	2.83	29.50	0.23
57	0.97	−1.47	3.23	29.50	10.23
58	0.97	10.53	−6.97	−30.50	14.23
59	0.97	0.53	−17.98	−0.50	18.23
60	−0.03	−11.47	20.53	−0.50	−17.77
Av	0.00	0.00	0.00	0.00	0.00
SD	1.35	9.99	12.08	46.63	14.61

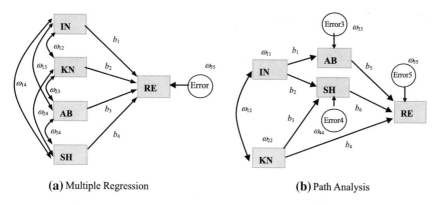

(a) Multiple Regression   **(b)** Path Analysis

Fig. 9.1 Multiple regression model and an example of a path analysis model for the data in Table 9.1

But, *other causal models* may better describe the relationships of variables. An example of other models is shown by the path diagram in Fig. 9.1b, in which it is considered that IN influences RE by way of AB and SH, while KN influences RE directly and by way of SH. In other words, [1] AB is influenced by IN; [2] SH is influenced by IN and KN; and [3] RE is influenced by KN, AB, and SH. These causal relationships are expressed as a set of regression analysis models:

$$AB = b_1 \times IN + c_3 + e_3,$$
$$SH = b_2 \times IN + b_3 \times KN + c_4 + e_4, \qquad (9.2)$$
$$RE = b_4 \times KN + b_5 \times AB + b_6 \times SH + c_5 + e_5,$$

with c_j and e_j ($j = 3, 4, 5$) intercepts and errors, respectively. Here, the subscripts 3, 4, and 5 attached to c and e merely correspond to AB, SH, and RE being the third, fourth, and fifth variables. The set of the three equations is equivalent to the path diagram in Fig. 9.1b, where the intercepts are omitted. Parameters $b_1, \ldots, b_6$ in (9.2) are called *path coefficients*.

In path analysis, the variables are classified into explanatory and dependent variables as follows:

[1] An *explanatory* variable is one to which no single-headed arrow extends in a path diagram; in Fig. 9.1b, IN and KN are explanatory variables. The *errors* e_3, e_4, and e_5 are also included in explanatory variables.
[2] A *dependent* variable is one to which at least a single-headed arrow extends; AB, SH, and RE in Fig. 9.1b are dependent variables.

Explanatory and dependent variables are also called *exogenous* and *endogenous* variables, respectively.

9.2 Matrix Expression

Table 9.1(B) contains the centered scores transformed from the raw scores in (A). It
is known that the path analysis for (A) and that for (B) give an identical solution,
except for the resulting *intercepts* (c_3, c_4, c_5) being *zero* in the *latter* analysis. We
thus *omit* the *intercepts* in the models for path analysis, for the sake of simplicity,
supposing that a data set to be analyzed contains *centered scores*. Thus, (9.2) is
simplified without c_3, c_4, and c_5 as:

$$
\begin{aligned}
\text{AB} &= b_1 \times \text{IN} + e_3, \\
\text{SH} &= b_2 \times \text{IN} + b_3 \times \text{KN} + e_4, \\
\text{RE} &= b_4 \times \text{KN} + b_5 \times \text{AB} + b_6 \times \text{SH} + e_3.
\end{aligned}
\tag{9.3}
$$

Using matrices and vectors, the three equations in (9.3) are summarized in the
following single equation:

$$
\begin{array}{ccccc}
\mathbf{x} & \mathbf{B} & \mathbf{x} & \mathbf{u} \\
\begin{bmatrix} \text{IN} \\ \text{KN} \\ \text{AB} \\ \text{SH} \\ \text{RE} \end{bmatrix}
=
\begin{bmatrix} & & & & \\ & & & & \\ b_1 & & & & \\ b_2 & b_3 & & & \\ & & b_4 & b_5 & b_6 \end{bmatrix}
\begin{bmatrix} \text{IN} \\ \text{KN} \\ \text{AB} \\ \text{SH} \\ \text{RE} \end{bmatrix}
+
\begin{bmatrix} \text{IN} \\ \text{KN} \\ e_1 \\ e_2 \\ e_3 \end{bmatrix}
\end{array}
\tag{9.4}
$$

with the blank cells in **B** occupied by zeros. Here, the first and second rows in the
left- and right-hand sides of (9.4) stand for "IN = IN" and "KN = KN," which
obviously hold true, and the remaining rows are found to equal (9.3). Any model
for path analysis can be expressed in the form of (9.4), i.e.,

$$
\mathbf{x} = \mathbf{Bx} + \mathbf{u},
\tag{9.5}
$$

where $\mathbf{x}$ is a $p \times 1$ random vector for p *observed* variables, $\mathbf{u}$ is a random vector
containing p *explanatory* variables, and $\mathbf{B}$ is the $p \times p$ *path coefficient* matrix, in
which the (i, j) element being nonzero implies that variable i is influenced by
variable j.

Equation (9.5) is rewritten as $\mathbf{x} - \mathbf{Bx} = \mathbf{u}$; thus, $(\mathbf{I}_p - \mathbf{B})\mathbf{x} = \mathbf{u}$. We can further
rewrite it as:

$$
\mathbf{x} = \left(\mathbf{I}_p - \mathbf{B} \right)^{-1} \mathbf{u},
\tag{9.6}
$$

supposing the existence of $(\mathbf{I}_p - \mathbf{B})^{-1}$, i.e., that the inverse matrix of $\mathbf{I}_p - \mathbf{B}$ can be
obtained. For the model in Fig. 9.1b, (9.6) is expressed in the concrete form:

$$
\begin{array}{c}
\mathbf{x} \\
\begin{array}{|c|}
\hline
\text{IN} \\
\text{KN} \\
\text{AB} \\
\text{SH} \\
\text{RE} \\
\hline
\end{array}
\end{array}
=
\begin{array}{c}
(\mathbf{I}_p - \mathbf{B})^{-1} \\
\begin{array}{|ccccc|}
\hline
1 & & & & \\
& 1 & & & \\
-b_1 & & 1 & & \\
-b_2 & -b_3 & & 1 & \\
& -b_4 & -b_5 & -b_6 & 1 \\
\hline
\end{array}
\end{array}^{-1}
\begin{array}{c}
\mathbf{u} \\
\begin{array}{|c|}
\hline
\text{IN} \\
\text{KN} \\
e_3 \\
e_4 \\
e_5 \\
\hline
\end{array}
\end{array}
\tag{9.7}
$$

9.3 Distributional Assumptions

Let us assume that *explanatory variable* vector $\mathbf{u}$ follows the multivariate normal (MVN) distribution with its mean vector $\mathbf{0}_p$ and covariance matrix $\mathbf{\Omega}$:

$$
\mathbf{u} \sim N_p(\mathbf{0}_p, \mathbf{\Omega}). \tag{9.8}
$$

The elements of the covariance matrix are described as:

$$
\mathbf{\Omega} =
\begin{array}{c}
\begin{array}{ccccc}
\text{IN} & \text{KN} & e_3 & e_4 & e_5
\end{array} \\
\begin{array}{c}
\text{IN} \\ \text{KN} \\ e_3 \\ e_4 \\ e_5
\end{array}
\begin{array}{|ccccc|}
\hline
\omega_{11} & \omega_{12} & & & \\
\omega_{12} & \omega_{22} & & & \\
& & \omega_{33} & & \\
& & & \omega_{44} & \\
& & & & \omega_{55} \\
\hline
\end{array}
\end{array}
\tag{9.9}
$$

for the model in Fig. 9.1b. Here, the blank cells indicate zero elements, which implies that *errors* are assumed to be *uncorrelated* with *explanatory variables* and that *errors* are assumed to be mutually *uncorrelated*. Those assumptions are found in Fig. 9.1b; they are not linked by paths there.

Here, we introduce a property of MVN variables without its proof:

Note 9.1. A Property of MVN Distribution
If $\mathbf{u}$ is a *random* vector with $\mathbf{u} \sim N_p(\mathbf{\mu}, \mathbf{\Omega})$, then

$$
\mathbf{A}\mathbf{u} + \mathbf{c} \sim N_p(\mathbf{A}\mathbf{\mu} + \mathbf{c}, \mathbf{A}\mathbf{\Omega}\mathbf{A}') \tag{9.10}
$$

for fixed $\mathbf{A}$ ($p \times p$) and $\mathbf{c}$ ($p \times 1$).

Here, the difference of random $\mathbf{u}$ to fixed $\mathbf{A}$ and $\mathbf{c}$ should be noted; the elements of $\mathbf{u}$ take a variety of values as $\mathbf{x}$ in Note 8.4, while the elements in $\mathbf{A}$ and $\mathbf{c}$ are constant.

Because of (9.6), (9.8), and (9.10), observed variable vector $\mathbf{x}$ is found to follow an MVN distribution as follows:

$$\mathbf{x} \sim N_p(\mathbf{0}_p, \Sigma), \tag{9.11}$$

with its covariance matrix

$$\Sigma = (\mathbf{I}_p - \mathbf{B})^{-1} \Omega (\mathbf{I}_p - \mathbf{B})^{-1'}. \tag{9.12}$$

9.4 Likelihood for Covariance Structure Analysis

Let $\mathbf{X}$ denote the centered data matrix in Table 9.1(B) and $\mathbf{x}_i'$ denote the ith row of $\mathbf{X}$. If $\mathbf{x}_i \sim N_p(\mu, \Sigma)$, the log likelihood for $\mathbf{X}$ is expressed as (8.20) in Chap. 8. However, in the path analysis model, μ is restricted to $\mathbf{0}_p$, as in (9.11), with Σ constrained as (9.12).

The substitution of $\mathbf{0}_p$ into μ in (8.20) leads to the *log likelihood* of Σ for path analysis:

$$
\begin{aligned}
l(\Sigma) &= -\frac{n}{2}\log|\Sigma| - \frac{1}{2}\sum_{i=1}^{n}\mathbf{x}_i'\Sigma^{-1}\mathbf{x}_i \\
&= -\frac{n}{2}\log|\Sigma| - \frac{1}{2}\mathrm{tr}\Sigma^{-1}\sum_{i=1}^{n}\mathbf{x}_i\mathbf{x}_i' \\
&= -\frac{n}{2}\log|\Sigma| - \frac{n}{2}\mathrm{tr}\Sigma^{-1}\left(\frac{1}{n}\sum_{i=1}^{n}\mathbf{x}_i\mathbf{x}_i'\right) = -\frac{n}{2}\log|\Sigma| - \frac{n}{2}\mathrm{tr}\Sigma^{-1}\mathbf{V},
\end{aligned}
\tag{9.13}
$$

where Σ is constrained as (9.12) and

$$\mathbf{V} = \frac{1}{n}\sum_{i=1}^{n}\mathbf{x}_i\mathbf{x}_i' = \frac{1}{n}\mathbf{X}'\mathbf{X} \tag{9.14}$$

is the inter-variable covariance matrix for the centered score matrix $\mathbf{X}$, as shown in (3.22).

Let us note that matrix Σ maximizing (9.13) is equivalent to the one maximizing

$$l^*(\Sigma) = \frac{n}{2}\log|\Sigma^{-1}\mathbf{V}| - \frac{n}{2}\mathrm{tr}\Sigma^{-1}\mathbf{V}, \tag{9.15}$$

since we can use (8.11) and (8.12) in (9.15) to rewrite it as:

$$l^*(\Sigma) = \frac{n}{2}\log(|\Sigma^{-1}| \times |\mathbf{V}|) - \frac{n}{2}\mathrm{tr}\Sigma^{-1}\mathbf{V} = -\frac{n}{2}\log|\Sigma| + \frac{n}{2}\log|\mathbf{V}| - \frac{n}{2}\mathrm{tr}\Sigma^{-1}\mathbf{V}. \tag{9.15'}$$

Its parts relevant to Σ are the same as in (9.13); that is, (9.15) can be regarded as the log likelihood equivalent to (9.13). The former is easier to treat than (9.13) in that the same matrix $\Sigma^{-1}V$ appears in the determinant and trace. We thus use (9.15) for the log likelihood of Σ from here. The log likelihoods for the procedures in Chaps. 10, 11, and 12 are also written in the form of (9.15).

> **Note 9.2. Covariance Structure Analysis**
> Likelihood (9.15) is a function of the covariance matrices V and Σ that are obtained from data and derived from a model as in (9.12), respectively. To distinguish the two matrices from one another, the data-based V is called a sample covariance matrix, while the model-based Σ is called a covariance structure. Further, the path analysis and the procedures in Chaps. 10, 11, and 12 are generally called covariance structure analysis, as those procedures share in common log likelihoods that are written in the form of (9.15) and differ only in the covariance structure; it is constrained as (9.12) in the path analysis, while constraints different from (9.12) are imposed upon Σ in the other procedures.

9.5 Maximum Likelihood Estimation

Substituting (9.12) into Σ in (9.15) leads to the *log likelihood* of *parameter* B and Ω for the data matrix X:

$$\log l^*(\mathbf{B}, \Omega) = \frac{n}{2}\log\left|(\mathbf{I}_p - \mathbf{B})'\Omega^{-1}(\mathbf{I}_p - \mathbf{B})\mathbf{V}\right| - \frac{n}{2}\mathrm{tr}(\mathbf{I}_p - \mathbf{B})'\Omega^{-1}(\mathbf{I}_p - \mathbf{B})\mathbf{V}.$$

$$(9.16)$$

Here, we have used the fact that the inverse matrix of (9.12) is expressed as:

$$\Sigma^{-1} = \left\{(\mathbf{I}_p - \mathbf{B})^{-1}\Omega(\mathbf{I}_p - \mathbf{B})'^{-1}\right\}^{-1} = (\mathbf{I}_p - \mathbf{B})'\Omega^{-1}(\mathbf{I}_p - \mathbf{B}).$$

because of (4.15) and (4.16).

In path analysis, log likelihood (9.16) is maximized (in other words, its negative $-\log l^*(\mathbf{B}, \Omega)$ is minimized) over $\mathbf{B}$ and Ω. In model (9.7), the number of parameters to be obtained is 12, since the nonzero elements in $\mathbf{B}$ and Ω are $b_1, \ldots, b_6$ and ω_{11}, $\omega_{22}, \omega_{33}, \omega_{44}, \omega_{55}$, and ω_{12}, respectively. Since the solution is not explicitly given, the minimization of $-\log l^*(\mathbf{B}, \Omega)$ is attained by iterative algorithms. A popular algorithm is the one using a *gradient method*, which is illustrated in Appendix A.6.3. Setting the vector θ in A.6.3 to $[b_1, \ldots, b_6, \omega_{11}, \omega_{22}, \omega_{33}, \omega_{44}, \omega_{55}, \omega_{12}]'$, the solution can be obtained. We express the solution of $\mathbf{B}$, Ω, and (9.12) as $\hat{\mathbf{B}}$, $\hat{\Omega}$, and $\hat{\Sigma}$, respectively.

9.6   Estimated Covariance Structure

For the data set in Table 9.1(B), the *solution* of the path analysis with its model (9.7) is given by

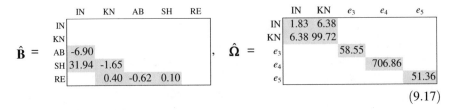

	IN	KN	AB	SH	RE
IN					
KN					
AB	-6.90				
SH	31.94	-1.65			
RE		0.40	-0.62	0.10	

$\hat{\mathbf{B}} = $ (above)

	IN	KN	e_3	e_4	e_5
IN	1.83	6.38			
KN	6.38	99.72			
e_3			58.55		
e_4				706.86	
e_5					51.36

$\hat{\Omega} = $ (above)

$$(9.17)$$

Figure 9.2a presents a path diagram with the values in the above solution shown in the corresponding parts. The *GFI statistic*, defined as:

$$\mathbf{GFI} = 1 - \frac{\mathrm{tr}\left(\hat{\Sigma}^{-1}\mathbf{V} - \mathbf{I}_p\right)^2}{\mathrm{tr}\left(\hat{\Sigma}^{-1}\mathbf{V}\right)^2}, \qquad (9.18)$$

is convenient for assessing whether a solution is satisfactory or not. Index (9.18) indicates the closeness of the sample covariance matrix $\mathbf{V}$ and the estimated co-variance structure

$$\hat{\Sigma} = \left(\mathbf{I}_p - \hat{\mathbf{B}}\right)^{-1}\hat{\Omega}\left(\mathbf{I}_p - \hat{\mathbf{B}}\right)^{-1\prime}, \qquad (9.19)$$

i.e., (9.12) in which the solutions of $\hat{\mathbf{B}}$ and $\hat{\Omega}$ have been substituted. If $\hat{\Sigma} = \mathbf{V}$, which implies that a model is fitted completely to a data set with $\hat{\Sigma}^{-1}\mathbf{V} = \mathbf{I}_p$, then (9.18) attains the one at the upper limit; the largeness of the GFI stands for how well solution-based covariances approximate sample covariances. A value of 0.9 is sometimes used as a benchmark with a GFI ≥ 0.9 showing a satisfactory model, though selecting 0.9 does not have any theoretical rationale.

The sample covariance matrix for the data in Table 9.1 and the estimated co-variance structure for the solution in Fig. 9.2a are given as:

	IN	KN	AB	SH	RE
IN	1.83			*Sym*	
KN	6.38	99.72			
AB	-12.65	-51.10	145.87		
SH	47.98	39.27	-350.54	2174.75	
RE	15.34	75.14	-144.42	443.12	213.48

$\mathbf{V} = $ (above)

	IN	KN	AB	SH	RE
IN	1.83			*Sym*	
KN	6.38	99.72			
AB	-12.65	-44.08	145.87		
SH	47.98	39.27	-331.26	2174.75	
RE	15.01	70.81	-139.75	431.23	207.71

$\hat{\Sigma} = $ (above)

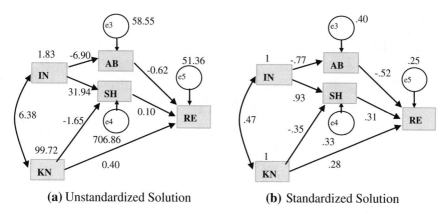

Fig. 9.2 Solution of the model in Fig. 8.1b for the data in Table 8.1

respectively, where $\hat{\Sigma}$ has been obtained by substituting solution (9.17) in (9.12), and the upper triangular elements in **V** and $\hat{\Sigma}$ have been omitted by writing "*Sym*," since they are *symmetric*. The above **V** and $\hat{\Sigma}$ are substituted in (9.18) to give a GFI of 0.984, which is higher than 0.90, suggesting that the solution in Fig. 9.2a is satisfactory.

9.7   Unstandardized and Standardized Solutions

The solution in Fig. 9.2a is called *unstandardized*, as it is obtained from unstandardized data with variables having different variances. Thus, it is senseless to compare the largeness of the resulting parameter values. For the comparison to make sense, we must obtain the *standardized* solution obtained for the standard scores transformed from the raw data. That solution is shown in Fig. 9.2b. There, it makes sense to compare the parameter values. For example, we can find AB to be the most influential for RE among the three explanatory variables AB, SH, and KN that extend paths to RE, since the absolute value of the coefficient (−0.52) attached to the path from AB to RE is the largest. Further, the sign of that coefficient is negative, implying that AB tends to considerably decrease RE. The covariance $\omega_{12} = 0.47$ in the standardized solution is viewed as the correlation coefficient, since all variables are standard scores.

Unstandardized and standardized solutions can be considered two *different expressions of the same solution*, as the maximum value of log likelihood (9.16) is equivalent between unstandardized and standardized solutions. The GFI is also so, and the standardized solution can be transformed easily from the unstandardized one, which is explained next.

Note 9.3. Scale Invariance

When data are standardized, the covariance matrix $\mathbf{V}$ in (9.14) is replaced by the corresponding correlation matrix $\mathbf{R} = \mathbf{D}^{-1}\mathbf{V}\mathbf{D}^{-1}$, i.e., (3.17), which is rewritten as:

$$\mathbf{V} = \mathbf{DRD}, \tag{9.20}$$

with $\mathbf{D}$ the diagonal matrix whose diagonal elements are the standard deviations of variables. Using (9.20) in log likelihood (9.15) for $\mathbf{V}$, it can be rewritten into the log likelihood for $\mathbf{R}$ as:

$$
\begin{aligned}
l(\Sigma) &= \frac{n}{2}\log\left|\Sigma^{-1}\mathbf{DRD}\right| - \frac{n}{2}\text{tr}\Sigma^{-1}\mathbf{DRD} \\
&= \frac{n}{2}\log\left|\mathbf{D}\Sigma^{-1}\mathbf{DR}\right| - \frac{n}{2}\text{tr}\left(\mathbf{D}\Sigma^{-1}\mathbf{D}\right)\mathbf{R} = \frac{n}{2}\log\left|\Sigma_{\mathrm{S}}^{-1}\mathbf{R}\right| - \frac{n}{2}\text{tr}\Sigma_{\mathrm{S}}^{-1}\mathbf{R}.
\end{aligned}
\tag{9.21}
$$

Here, we have used (4.16) and (8.11) with

$$\Sigma_{\mathrm{S}} = \mathbf{D}^{-1}\Sigma\mathbf{D}^{-1}. \tag{9.22}$$

It can be rewritten using (9.12) as:

$$
\begin{aligned}
\Sigma_{\mathrm{S}} &= \mathbf{D}^{-1}(\mathbf{I}_p - \mathbf{B})^{-1}\Omega(\mathbf{I}_p - \mathbf{B})^{-1'}\mathbf{D}^{-1} = \left\{(\mathbf{I}_p - \mathbf{B})\mathbf{D}\right\}^{-1}\Omega\left\{(\mathbf{I}_p - \mathbf{B})\mathbf{D}\right\}^{-1'} \\
&= \left\{\mathbf{D}(\mathbf{I}_p - \mathbf{D}^{-1}\mathbf{BD})\right\}^{-1}\Omega\left\{\mathbf{D}(\mathbf{I}_p - \mathbf{D}^{-1}\mathbf{BD})\right\}^{-1'} = (\mathbf{I}_p - \mathbf{B}_{\mathrm{S}})^{-1}\Omega_{\mathrm{S}}(\mathbf{I}_p - \mathbf{B}_{\mathrm{S}})^{-1'}.
\end{aligned}
\tag{9.23}
$$

Here, we have used (1.21), (4.15), and (4.16) with

$$\mathbf{B}_{\mathrm{S}} = \mathbf{D}^{-1}\mathbf{BD} \quad \text{and} \quad \Omega_{\mathrm{S}} = \mathbf{D}^{-1}\Omega\mathbf{D}^{-1}. \tag{9.24}$$

Substituting (9.23) in (9.21), it is rewritten as:

$$
\begin{aligned}
\log l(\mathbf{B}_{\mathrm{S}}, \Omega_{\mathrm{S}}) &= \frac{n}{2}\log\left|(\mathbf{I}_p - \mathbf{B}_{\mathrm{S}})'\Omega_{\mathrm{S}}^{-1}(\mathbf{I}_p - \mathbf{B}_{\mathrm{S}})\mathbf{R}\right| \\
&\quad - \frac{n}{2}\text{tr}(\mathbf{I}_p - \mathbf{B}_{\mathrm{S}})'\Omega_{\mathrm{S}}^{-1}(\mathbf{I}_p - \mathbf{B}_{\mathrm{S}})\mathbf{R},
\end{aligned}
\tag{9.25}
$$

which has the same form as (9.16).

The above results imply that the maximum of (9.25) equals that of (9.16) with the solution of $\mathbf{B}_{\mathrm{S}}$ and Ω_{S} maximizing (9.25) given by (9.24), in which $\hat{\mathbf{B}}$ and $\hat{\Omega}$ are substituted into $\mathbf{B}$ and Ω, respectively. The equivalence for the GFI between unstandardized and standardized solutions can be found by noting $\text{tr}\Sigma^{-1}\mathbf{V} = \text{tr}\Sigma^{-1}\mathbf{DRD} = \text{tr}\mathbf{D}\Sigma^{-1}\mathbf{DR} = \text{tr}(\mathbf{D}^{-1}\Sigma\mathbf{D}^{-1})^{-1}\mathbf{R} = \text{tr}\Sigma_{\mathrm{S}}^{-1}\mathbf{R}$, because of (9.20) and (9.22). Such properties of invariance between the two solutions are called *scale invariance*, and the solutions with this invariance are said to be *scale invariant*, as those properties imply that the solution remains essentially invariant, even if the scales of the variables change with standardization. The solutions of regression analysis and the procedures in Chaps. 10, 11, and 12 are also scale invariant.

9.8 Other and Extreme Models

Let us refer to the model in Fig. 9.1b as Model 1. Although this model was regarded as satisfactory, with a GFI exceeding 0.9, a model may exist that is better fitted to the data set in Table 9.1(B). This suggests that *other models* should be considered and *compared*; that is, the *model selection* shown in Fig. 8.5 (Sect. 8.7) is to be performed. Figure 9.3 shows two examples of other models, which we call Models 2 and 3. For *Model 2*, (9.6) is expressed as:

$$
\begin{array}{c}
\mathbf{x} \\
\begin{array}{|c|}
\hline
\text{IN} \\
\text{KN} \\
\text{AB} \\
\text{SH} \\
\text{RE} \\
\hline
\end{array}
\end{array}
=
\begin{array}{c}
\mathbf{(I-B)}^{-1} \\
\begin{array}{|ccccc|}
\hline
1 & & & & \\
 & 1 & & & \\
-b_1 & & 1 & & \\
-b_2 & -b_3 & & 1 & \\
-b_7 & -b_4 & -b_5 & -b_6 & 1 \\
\hline
\end{array}
\end{array}^{-1}
\begin{array}{c}
\mathbf{u} \\
\begin{array}{|c|}
\hline
\text{IN} \\
\text{KN} \\
e_3 \\
e_4 \\
e_5 \\
\hline
\end{array}
\end{array}
\qquad (9.26)
$$

Here, a parameter, b_7, is added to (9.7). The covariance matrix among explanatory variables is the same as that in (9.9). Except for the difference between (9.7) and

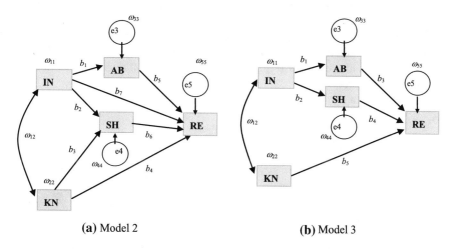

(a) Model 2   **(b)** Model 3

Fig. 9.3 Examples of models that differ from the one in Fig. 8.1b

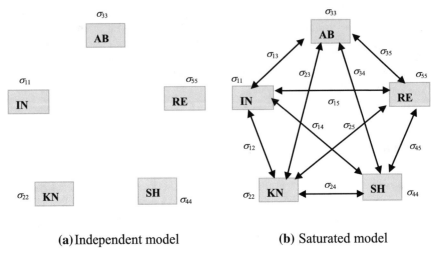

(a) Independent model   **(b)** Saturated model

Fig. 9.4 Two extreme models with the least and the most parameters

(9.26), the same procedure is performed for Model 2: The *maximum likelihood method* gives the solutions for Model 2 and other possible models.

Now, let us consider two types of extreme models. One is the *independent model* shown in Fig. 9.4a, where we find that no variable is linked to the others. It implies that all variables are assumed to be mutually independent. This model is the most restrictive, with its *number of parameters* the *least* among possible models. That number is p, i.e., the *number of variables*; only their variances are to be estimated.

The other extreme type is called the *saturated model*, whose number of parameters equals $p(p + 1)/2$, the number of the *distinctive covariances* in $\mathbf{V}$ or $\mathbf{\Sigma}$, this is, 15 for the data set in Table 9.1(B). This number is the *maximum* among those for all possible models. The saturated models contain several ones, and a typical saturated model is shown in Fig. 9.4b, where all variables are connected by double-headed arrows, implying that *all variables* are assumed to be *merely correlated*. That is, the model in Fig. 9.4b *states nothing* for the causal relationships among the variables.

The covariance structures of the independent and saturated models are expressed as:

$$
\mathbf{\Sigma} = \begin{array}{c} \\ \text{IN} \\ \text{KN} \\ \text{AB} \\ \text{SH} \\ \text{RE} \end{array}
\begin{array}{ccccc} \text{IN} & \text{KN} & \text{AB} & \text{SH} & \text{RE} \\ \sigma_{11} & & & & \\ & \sigma_{22} & & & \\ & & \sigma_{33} & & \\ & & & \sigma_{44} & \\ & & & & \sigma_{55} \end{array}
\quad , \quad
\mathbf{\Sigma} = \begin{array}{c} \\ \text{IN} \\ \text{KN} \\ \text{AB} \\ \text{SH} \\ \text{RE} \end{array}
\begin{array}{ccccc} \text{IN} & \text{KN} & \text{AB} & \text{SH} & \text{RE} \\ \sigma_{11} & & & & \\ \sigma_{21} & \sigma_{22} & & \textit{Sym} & \\ \sigma_{31} & \sigma_{32} & \sigma_{33} & & \\ \sigma_{41} & \sigma_{42} & \sigma_{43} & \sigma_{44} & \\ \sigma_{51} & \sigma_{52} & \sigma_{53} & \sigma_{54} & \sigma_{55} \end{array}
$$

respectively. The former is a *diagonal* matrix, while the latter is a simple unconstrained covariance matrix *without a special structure*.

Table 9.2 Number of parameters (NP) and the resulting index values for each model

Model	NP	GFI	AIC	BIC
Saturated	15	1.000	30.000	61.415
Model 2	16	0.987	28.035	55.262
Model 1	12	0.984	26.364	51.496
Model 3	11	0.908	39.792	62.830
Independent	5	0.389	231.480	241.952

9.9   Model Selection

So far, we have Models 1, 2, and 3 and two extreme models. For comparing those five models with respect to the *goodness of fit* to the data set, we *cannot use the GFI*, since the GFI values *increase* with the *number of parameters* in the models, and the GFI for the *saturated models* always attains the upper limit. This can be found in Table 9.2, where the models are arranged according to their numbers of parameters. This property of GFI is due to the fact that the number of parameters is not considered for defining the GFI, as found in (9.18). *GFI is thus only* useful for assessing whether *a considered model* is satisfactory or not.

The *information criteria* introduced in Sect. 8.7 are useful for *comparing models*, since the number of parameters is considered in the criteria. The values of typical information criteria, the *AIC* and *BIC*, for each model are shown in Table 9. 2. Since smaller values of the information criteria indicate better models, Model 1, for which both the AIC and BIC show the lowest values, is found to be the best of the five models. Different from this example, cases often arise when the AIC and BIC indicate other models are best.

9.10   Bibliographical Notes

It is difficult to find books in which path analysis is exclusively treated. It is, however, detailed in chapters of books for structural equation modeling, which include Bollen (1989) and Kaplan (2000).

Exercises

9.1. Present an example of a set of the variables V1, V2, and V3 whose relationships are represented as the following path diagram:

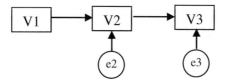

9.2. Present an example of a set of the variables V1, V2, V3, and V4 whose relationships are represented as the following path diagram:

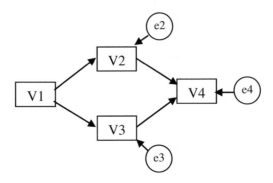

9.3. Present an example of a set of the variables V1, ..., V5 whose relationships are represented as the following path diagram:

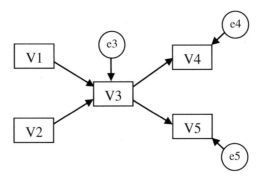

9.4. Let the elements of the $p \times 1$ vector $\mathbf{x}$ in (9.5) denoted as $\mathbf{x} = [z_1, ..., z_q, y_1, ..., y_r]' = [\mathbf{z}', \mathbf{y}']'$, with $\mathbf{z} = [z_1, ..., z_q]'$ the $q \times 1$ vector containing explanatory ones, $\mathbf{y} = [y_1, ..., y_r]'$ the $r \times 1$ vector consisting of explanatory variables, and $p = q + r$. Show that the path analysis model (9.5) can be rewritten as:

$$\mathbf{y} = \mathbf{A}\mathbf{y} + \mathbf{C}\mathbf{z} + \mathbf{e}, \qquad (9.27)$$

with $\mathbf{A}$ $(r \times r)$ and $\mathbf{C}$ $(r \times q)$ containing path coefficients.

9.5. Show the elements of $\mathbf{A}$ and $\mathbf{C}$ in model (9.27) represented as the diagram in Fig. 9.1b.

9.6. If the path analysis model is expressed as (9.27) with $\mathbf{z} \sim N_q(\mathbf{0}_q, \mathbf{\Phi})$, $\mathbf{e} \sim N_r(\mathbf{0}_r, \mathbf{\Psi})$, and no correlation found between $\mathbf{z}$ and $\mathbf{e}$, show that the covariance structure (9.12) can be rewritten as:

$$\Sigma = \begin{bmatrix} \boldsymbol{\Phi} & \boldsymbol{\Phi}\mathbf{C}'(\mathbf{I}_r - \mathbf{A})'^{-1} \\ (\mathbf{I}_r - \mathbf{A})^{-1}\mathbf{C}\boldsymbol{\Phi} & (\mathbf{I}_r - \mathbf{A})^{-1}(\mathbf{C}\boldsymbol{\Phi}\mathbf{C}' + \boldsymbol{\Psi})(\mathbf{I}_r - \mathbf{A})'^{-1} \end{bmatrix}, \quad (9.28)$$

where $\boldsymbol{\Psi}$ is an $r \times r$ diagonal element, and (9.28) is one of the block matrices which are detailed in Sect. 14.1.

9.7. For an $n \times p$ centered data matrix $\mathbf{X}$, the independent model can formally be expressed as $\mathbf{x} \sim N_p(\mathbf{0}_p, \Sigma)$ with $\Sigma = (\sigma_{jk})$ being constrained as a diagonal matrix. Show that the PDF of $\mathbf{x} = [x_1, \ldots, x_p]'$ in this model is expressed as:

$$P(\mathbf{x}|\Sigma) = \prod_{j=1}^{p} \frac{1}{\sqrt{2\pi\sigma_{jj}}} \exp\left\{ -\frac{1}{2}\frac{x_j^2}{\sigma_{jj}} \right\}, \quad (9.29)$$

using the fact that $|\Sigma| = \prod_{j=1}^{p} \sigma_{jj}$ if $\Sigma = (\sigma_{jk})$ is diagonal.

9.8. Show that the MLE of Σ in the independent model treated in (9.29) is given by the diagonal matrix whose diagonal elements are those of $\mathbf{V} = n^{-1}\mathbf{X}'\mathbf{X}$, with $\mathbf{X}$ the $n \times p$ centered data matrix whose rows are filled with $\mathbf{x}'$ for n individuals.

9.9. Let us consider model (9.27) with $\mathbf{z} \sim N_q(\mathbf{0}_q, \boldsymbol{\Phi})$ and $\mathbf{e} \sim N_r(\mathbf{0}_r, \boldsymbol{\Psi})$. If the jth variable in $\mathbf{y}$ cannot be a cause for the $(j-1)$th variable, the log likelihood of the parameters in (9.27) for the $n \times p$ centered data matrix $\mathbf{X} = [\mathbf{Z}, \mathbf{Y}]$, whose rows are filled with the observations of $\mathbf{x}' = [\mathbf{z}', \mathbf{y}']'$, is known to be given by

$$\log l(\mathbf{H}, \boldsymbol{\Psi}, \boldsymbol{\Phi}) = -\frac{n}{2}\left\{ \log|\boldsymbol{\Psi}| + \mathrm{tr}\mathbf{M}\boldsymbol{\Psi}^{-1} + \log|\boldsymbol{\Phi}| + \mathrm{tr}\mathbf{V}_{ZZ}\boldsymbol{\Phi}^{-1} \right\}$$
$$= -\frac{n}{2}\left\{ \log|\boldsymbol{\Psi}| + \mathrm{tr}(\mathbf{H}\mathbf{V}_{XX}\mathbf{H}' - 2\mathbf{V}_{YX}\mathbf{H}')\boldsymbol{\Psi}^{-1} + \mathrm{tr}\mathbf{V}_{YY}\boldsymbol{\Psi}^{-1} + \log|\boldsymbol{\Phi}| + \mathrm{tr}\mathbf{V}_{ZZ}\boldsymbol{\Phi}^{-1} \right\},$$
$$(9.28)$$

Here, $\mathbf{H} = [\mathbf{C}, \mathbf{A}]$ $(r \times p)$, $\mathbf{V}_{YY} = n^{-1}\mathbf{Y}'\mathbf{Y}$, $\mathbf{V}_{YX} = n^{-1}\mathbf{Y}'\mathbf{X}$, $\mathbf{V}_{XX} = n^{-1}\mathbf{X}'\mathbf{X}$, $\mathbf{V}_{ZZ} = n^{-1}\mathbf{Z}'\mathbf{Z}$, and $\mathbf{M} = \mathbf{H}\mathbf{V}_{XX}\mathbf{H}' - 2\mathbf{V}_{YX}\mathbf{H}' + \mathbf{V}_{YY}$, with $\mathbf{Z}$ $(n \times q)$ and $\mathbf{Y}$ $(n \times r)$ the blocks of $\mathbf{X}$. Show that the two terms $\mathrm{tr}(\mathbf{H}\mathbf{V}_{XX}\mathbf{H}' - 2\mathbf{V}_{YX}\mathbf{H}')\boldsymbol{\Psi}^{-1}$ and $\log|\boldsymbol{\Psi}| + \mathrm{tr}\mathbf{M}\boldsymbol{\Psi}^{-1}$ in (9.28) can be rewritten as:

$$\mathrm{tr}(\mathbf{H}\mathbf{V}_{XX}\mathbf{H}' - 2\mathbf{V}_{YX}\mathbf{H}')\boldsymbol{\Psi}^{-1} = \sum_{i=1}^{r}\frac{1}{\psi_i}\left(\sum_{j=1}^{p} v_{jj}h_{ij}^2 + 2\sum_{j=1}^{p}\sum_{k\neq j}^{p} v_{jk}h_{ij}h_{ik} - 2\sum_{j=1}^{p} w_{ij}h_{ij} \right) \quad (9.29)$$

$$\log|\boldsymbol{\Psi}| + \mathrm{tr}\mathbf{M}\boldsymbol{\Psi}^{-1} = \sum_{i=1}^{r}\left(\log\psi_i + \frac{m_{ii}}{\psi_i} \right), \quad (9.30)$$

with $\mathbf{V}_{XX} = (v_{jk})$, $\mathbf{V}_{YX} = (w_{ij})$, $\mathbf{H} = (h_{ij})$, m_{ii} the ith diagonal element of $\mathbf{M}$, and ψ_i that of $\boldsymbol{\Psi}$. For (9.30), use the fact that $|\mathbf{D}| = d_1 \times \cdots \times d_r$ if $\mathbf{D}$ is the $r \times r$ diagonal matrix whose diagonal elements are $d_1, \ldots, d_r$.

9.10. Let us consider maximizing (9.28). The MLE of $\mathbf{\Phi}$ is explicitly given by $\mathbf{\Phi} = \mathbf{V}_{ZZ}$, but the MLE of the nonzero elements in $\mathbf{H}$ and $\mathbf{\Psi}$ must be obtained by an iterative algorithm. Use (9.29) and (9.30) to show that the algorithm can be formed by the following steps:

Step 1. Initialize the nonzero elements of $\mathbf{H}$.
Step 2. Set $\psi_i = m_{ii}$ for $i = 1, \ldots, r$.
Step 3. Repeat updating h_{ij} as $h_{ij} = \frac{1}{v_{ij}} \left(w_{ij} - \sum_{k \neq j}^{p} v_{jk} h_{ik} \right)$ over all indexes i and j for the nonzero elements in $\mathbf{H}$.
Step 4. Finish if convergence is reached; otherwise, go back to Step 2.

The hint for Step 2 can be found in Exercise 8.1.

9.11. Show that the minimization of $\|\mathbf{XD} - \mathbf{FA}'\|^2$ over $\mathbf{F}$ and $\mathbf{A}$ gives an essentially different solutions from that of (5.4), which implies that the solutions of principal component analysis do not have scale invariance. Here, $\mathbf{X}$ is a data matrix, and $\mathbf{D}$ is a diagonal matrix whose diagonal elements are all positive and take mutually different values.

9.12. Show that the k-means clustering (KMC) for a data matrix $\mathbf{X}$ gives an essentially different solution from that for $\mathbf{XD}$, which implies that KMC solutions do not have scale invariance, with $\mathbf{D}$ a diagonal matrix whose diagonal elements are all positive and take mutually different values.

Chapter 10
Confirmatory Factor Analysis

Let the positive correlations be observed among the test scores for physics, chemistry, and biology. In order to investigate the causal relationships among the three variables, we can use the *path analysis* from the previous chapter. For example, we can evaluate the model in which a person's ability in physics influences his/her scores in chemistry and biology; ability in physics is a cause, while the scores in chemistry and biology are the results. However, it may be rather reasonable to assume that *all* of the scores for physics, chemistry, and biology are *the results of a single factor*, namely "an aptitude for the natural sciences." This is the idea underlying *factor analysis* (*FA*). British psychologist Spearman (1904) had such a conception in his studies of human intelligence, which is the origin of FA. Its key point is that all *observed variables* are regarded as the *results* caused by a few *unobserved latent variables* called *factors*, in contrast to path analysis, in which causal relationships among observed variables are modeled.

FA is classified into *exploratory FA* (*EFA*) and *confirmatory FA* (*CFA*). EFA refers to the FA procedures for exploring factors underlying observed variables for cases without prior knowledge of underlying factors (Thurstone 1935, 1947). In contrast, CFA refers to the procedures for confirming a model describing the relationships of factors to variables (Jöreskog 1969). Historically, the development of EFA preceded that of CFA, and EFA is often simply called "factor analysis." However, *CFA* is dealt with in this chapter, as introducing CFA before EFA suits the context of this book and CFA is easier to understand than EFA.

10.1 Example of Confirmatory Factor Analysis Model

We use the 100 (participants) × 8 (behavioral features) data matrix in Table 9.1(A) containing the self-ratings evaluating to what extent participants' behaviors are characterized by eight variables (features): **A** (Aggressive), **C** (Cheerful), **I** (Initiative), **B** (Blunt), **T** (Talkative), **V** (Vigor), **H** (tendency to Hesitate), and

© Springer Nature Singapore Pte Ltd. 2016
K. Adachi, *Matrix-Based Introduction to Multivariate Data Analysis*,
DOI 10.1007/978-981-10-2341-5_10

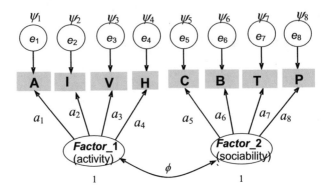

Fig. 10.1 Example of CFA model for the personality data

P (being Popular). For these eight variables, we consider the model with the assumption that A, I, V, and H are caused by *a factor* interpreted as an *activity*, while C, B, T, and P are caused by *another factor* that stands for *sociability*. The model is expressed as a set of eight equations:

$$
\begin{aligned}
\text{A} &= a_1 \times Factor\_1 & +c_1 + e_1 \\
\text{I} &= a_2 \times Factor\_1 & +c_2 + e_2 \\
\text{V} &= a_3 \times Factor\_1 & +c_3 + e_3 \\
\text{H} &= a_4 \times Factor\_1 & +c_4 + e_4 \\
\text{C} &= & a_5 \times Factor\_2 + c_5 + e_5 \\
\text{B} &= & a_6 \times Factor\_2 + c_6 + e_6 \\
\text{T} &= & a_7 \times Factor\_2 + c_7 + e_7 \\
\text{P} &= & a_8 \times Factor\_2 + c_8 + e_8
\end{aligned}
\tag{10.1}
$$

Here, c_j and e_j ($j = 1, \ldots, 8$) express an intercept and an error, respectively. Each equation in (10.1) is a model for *regression analysis*, though the *factor* is not an observed but rather an unobserved *latent* random variable. The model in (10.1) can be represented as the path diagram in Fig. 10.1.

10.2   Matrix Expression

Table 10.1(B) shows the centered scores for the raw data in (A). As in path analysis (Chap. 9), CFA for (A) and (B), with the assumption that the averages of factors are zeros, produces the same solution, except for the resulting *intercepts* ($c_1, \ldots, c_8$) being *zero* in the *latter* analysis. We thus *omit* the *intercepts* in CFA models, for the sake of simplicity, on the supposition that a data matrix to be analyzed contains

Table 10.1 Personality data: self-ratings for behavioral features (artificial example)

	A	C	I	B	T	V	H	P
(A) Raw data
1	9	7	9	2	9	8	3	8
2	2	3	5	8	1	3	7	3
3	5	6	7	6	8	4	6	6
4	4	6	6	3	8	5	7	7
5	6	5	7	6	6	5	6	6
6	4	5	5	5	6	3	5	5
7	6	7	6	5	8	3	6	8
8	6	6	7	4	8	7	6	7
9	7	6	8	5	6	5	3	4
10	4	4	6	8	4	3	6	3
11	5	6	6	4	6	4	5	7
12	6	4	6	5	5	6	4	6
13	7	5	8	5	5	6	7	6
14	4	5	6	7	5	4	7	5
15	3	6	5	6	4	2	6	5
16	5	6	7	3	9	5	7	7
17	4	5	8	4	8	5	5	6
18	7	6	8	4	6	6	4	6
19	5	7	7	4	9	5	4	7
20	5	5	7	5	4	4	6	8
21	6	7	7	4	8	6	6	6
22	4	6	6	4	5	3	7	5
23	3	6	5	7	4	3	8	2
24	7	7	8	5	7	8	5	6
25	4	6	7	6	5	4	5	6
26	5	6	5	4	8	5	4	7
27	3	5	4	6	5	4	6	5
28	4	6	6	5	5	4	7	5
29	6	5	9	5	8	6	5	6
30	5	6	8	5	6	6	6	6
31	4	3	7	7	4	5	5	5
32	3	3	6	9	3	4	7	3
33	5	5	6	7	5	6	6	5
34	5	6	7	5	4	4	8	8
35	2	5	5	5	5	3	6	5
36	4	6	7	7	5	5	6	5
37	7	7	9	5	6	8	3	5
38	5	7	5	3	8	5	5	5
39	6	5	7	4	7	6	2	7
40	5	5	4	7	5	3	8	4

(continued)

Table 10.1 (continued)

	A	C	I	B	T	V	H	P
41	3	4	6	9	4	3	6	4
42	5	7	7	6	6	4	6	7
43	6	4	7	7	5	5	5	4
44	5	7	5	4	9	6	5	6
45	5	7	9	3	8	7	3	7
46	2	5	4	7	3	4	6	5
47	5	9	6	4	9	4	4	8
48	5	5	7	8	4	5	3	5
49	7	6	8	5	7	7	2	6
50	6	5	6	7	4	6	5	4
51	6	6	7	3	6	5	6	7
52	6	6	8	4	8	8	3	9
53	7	6	8	4	8	6	4	7
54	7	6	7	5	7	6	5	7
55	4	4	6	8	2	3	7	3
56	7	6	8	3	7	8	2	7
57	7	5	8	6	7	6	2	6
58	6	6	6	3	6	4	3	4
59	6	5	7	6	5	5	3	7
60	4	4	7	6	4	5	5	4
61	4	6	6	5	6	5	9	6
62	6	4	5	7	4	3	5	4
63	4	5	6	6	6	4	7	5
64	2	6	5	5	5	2	9	5
65	5	6	6	6	7	6	5	6
66	4	5	4	7	5	3	7	5
67	4	6	6	4	7	4	6	6
68	5	4	7	6	6	8	4	4
69	6	6	7	4	7	4	5	8
70	4	6	5	7	5	4	6	6
71	6	5	7	5	5	6	4	6
72	6	7	8	5	7	7	4	7
73	6	5	8	5	6	6	2	5
74	5	6	6	6	6	2	8	5
75	6	6	8	5	6	6	3	6
76	4	4	5	6	5	4	5	3
77	7	5	7	6	5	7	3	6
78	8	6	9	6	6	8	1	7
79	8	5	8	4	6	9	2	7
80	4	5	3	6	5	3	6	5
81	4	6	5	6	5	4	6	5

(continued)

Table 10.1 (continued)

	A	C	I	B	T	V	H	P
82	5	6	7	6	5	4	6	5
83	6	6	7	6	4	4	3	5
84	8	6	6	7	6	8	2	7
85	4	6	6	6	6	5	6	6
86	5	7	7	4	8	6	5	9
87	4	4	5	9	2	2	8	4
88	3	5	6	6	4	3	7	4
89	4	6	6	5	6	4	9	6
90	5	6	7	6	5	4	6	6
91	3	5	6	7	5	4	9	5
92	3	3	6	7	3	3	7	4
93	6	6	8	5	5	7	3	7
94	4	5	5	6	6	4	8	5
95	5	7	6	4	6	6	3	5
96	4	4	6	7	3	5	6	5
97	3	7	6	4	7	3	7	5
98	4	7	5	6	7	3	7	4
99	5	4	5	9	4	5	6	7
100	4	5	7	7	3	5	4	3
(B) Centered data								
1	4.01	1.46	2.53	−3.51	3.28	3.12	−2.29	2.38
2	−2.99	−2.54	−1.47	2.49	−4.72	−1.88	1.71	−2.62
3	0.01	0.46	0.53	0.49	2.28	−0.88	0.71	0.38
4	−0.99	0.46	−0.47	−2.51	2.28	0.12	1.71	1.38
5	1.01	−0.54	0.53	0.49	0.28	0.12	0.71	0.38
6	−0.99	−0.54	−1.47	−0.51	0.28	−1.88	−0.29	−0.62
7	1.01	1.46	−0.47	−0.51	2.28	−1.88	0.71	2.38
8	1.01	0.46	0.53	−1.51	2.28	2.12	0.71	1.38
9	2.01	0.46	1.53	−0.51	0.28	0.12	−2.29	−1.62
10	−0.99	−1.54	−0.47	2.49	−1.72	−1.88	0.71	−2.62
11	0.01	0.46	−0.47	−1.51	0.28	−0.88	−0.29	1.38
12	1.01	−1.54	−0.47	−0.51	−0.72	1.12	−1.29	0.38
13	2.01	−0.54	1.53	−0.51	−0.72	1.12	1.71	0.38
14	−0.99	−0.54	−0.47	1.49	−0.72	−0.88	1.71	−0.62
15	−1.99	0.46	−1.47	0.49	−1.72	−2.88	0.71	−0.62
16	0.01	0.46	0.53	−2.51	3.28	0.12	1.71	1.38
17	−0.99	−0.54	1.53	−1.51	2.28	0.12	−0.29	0.38
18	2.01	0.46	1.53	−1.51	0.28	1.12	−1.29	0.38
19	0.01	1.46	0.53	−1.51	3.28	0.12	−1.29	1.38
20	0.01	−0.54	0.53	−0.51	−1.72	−0.88	0.71	2.38
21	1.01	1.46	0.53	−1.51	2.28	1.12	0.71	0.38

(continued)

Table 10.1 (continued)

	A	C	I	B	T	V	H	P
22	−0.99	0.46	−0.47	−1.51	−0.72	−1.88	1.71	−0.62
23	−1.99	0.46	−1.47	1.49	−1.72	−1.88	2.71	−3.62
24	2.01	1.46	1.53	−0.51	1.28	3.12	−0.29	0.38
25	−0.99	0.46	0.53	0.49	−0.72	−0.88	−0.29	0.38
26	0.01	0.46	−1.47	−1.51	2.28	0.12	−1.29	1.38
27	−1.99	−0.54	−2.47	0.49	−0.72	−0.88	0.71	−0.62
28	−0.99	0.46	−0.47	−0.51	−0.72	−0.88	1.71	−0.62
29	1.01	−0.54	2.53	−0.51	2.28	1.12	−0.29	0.38
30	0.01	0.46	1.53	−0.51	0.28	1.12	0.71	0.38
31	−0.99	−2.54	0.53	1.49	−1.72	0.12	−0.29	−0.62
32	−1.99	−2.54	−0.47	3.49	−2.72	−0.88	1.71	−2.62
33	0.01	−0.54	−0.47	1.49	−0.72	1.12	0.71	−0.62
34	0.01	0.46	0.53	−0.51	−1.72	−0.88	2.71	2.38
35	−2.99	−0.54	−1.47	−0.51	−0.72	−1.88	0.71	−0.62
36	−0.99	0.46	0.53	1.49	−0.72	0.12	0.71	−0.62
37	2.01	1.46	2.53	−0.51	0.28	3.12	−2.29	−0.62
38	0.01	1.46	−1.47	−2.51	2.28	0.12	−0.29	−0.62
39	1.01	−0.54	0.53	−1.51	1.28	1.12	−3.29	1.38
40	0.01	−0.54	−2.47	1.49	−0.72	−1.88	2.71	−1.62
41	−1.99	−1.54	−0.47	3.49	−1.72	−1.88	0.71	−1.62
42	0.01	1.46	0.53	0.49	0.28	−0.88	0.71	1.38
43	1.01	−1.54	0.53	1.49	−0.72	0.12	−0.29	−1.62
44	0.01	1.46	−1.47	−1.51	3.28	1.12	−0.29	0.38
45	0.01	1.46	2.53	−2.51	2.28	2.12	−2.29	1.38
46	−2.99	−0.54	−2.47	1.49	−2.72	−0.88	0.71	−0.62
47	0.01	3.46	−0.47	−1.51	3.28	−0.88	−1.29	2.38
48	0.01	−0.54	0.53	2.49	−1.72	0.12	−2.29	−0.62
49	2.01	0.46	1.53	−0.51	1.28	2.12	−3.29	0.38
50	1.01	−0.54	−0.47	1.49	−1.72	1.12	−0.29	−1.62
51	1.01	0.46	0.53	−2.51	0.28	0.12	0.71	1.38
52	1.01	0.46	1.53	−1.51	2.28	3.12	−2.29	3.38
53	2.01	0.46	1.53	−1.51	2.28	1.12	−1.29	1.38
54	2.01	0.46	0.53	−0.51	1.28	1.12	−0.29	1.38
55	−0.99	−1.54	−0.47	2.49	−3.72	−1.88	1.71	−2.62
56	2.01	0.46	1.53	−2.51	1.28	3.12	−3.29	1.38
57	2.01	−0.54	1.53	0.49	1.28	1.12	−3.29	0.38
58	1.01	0.46	−0.47	−2.51	0.28	−0.88	−2.29	−1.62
59	1.01	−0.54	0.53	0.49	−0.72	0.12	−2.29	1.38
60	−0.99	−1.54	0.53	0.49	−1.72	0.12	−0.29	−1.62
61	−0.99	0.46	−0.47	−0.51	0.28	0.12	3.71	0.38
62	1.01	−1.54	−1.47	1.49	−1.72	−1.88	−0.29	−1.62

(continued)

Table 10.1 (continued)

	A	C	I	B	T	V	H	P
63	−0.99	−0.54	−0.47	0.49	0.28	−0.88	1.71	−0.62
64	−2.99	0.46	−1.47	−0.51	−0.72	−2.88	3.71	−0.62
65	0.01	0.46	−0.47	0.49	1.28	1.12	−0.29	0.38
66	−0.99	−0.54	−2.47	1.49	−0.72	−1.88	1.71	−0.62
67	−0.99	0.46	−0.47	−1.51	1.28	−0.88	0.71	0.38
68	0.01	−1.54	0.53	0.49	0.28	3.12	−1.29	−1.62
69	1.01	0.46	0.53	−1.51	1.28	−0.88	−0.29	2.38
70	−0.99	0.46	−1.47	1.49	−0.72	−0.88	0.71	0.38
71	1.01	−0.54	0.53	−0.51	−0.72	1.12	−1.29	0.38
72	1.01	1.46	1.53	−0.51	1.28	2.12	−1.29	1.38
73	1.01	−0.54	1.53	−0.51	0.28	1.12	−3.29	−0.62
74	0.01	0.46	−0.47	0.49	0.28	−2.88	2.71	−0.62
75	1.01	0.46	1.53	−0.51	0.28	1.12	−2.29	0.38
76	−0.99	−1.54	−1.47	0.49	−0.72	−0.88	−0.29	−2.62
77	2.01	−0.54	0.53	0.49	−0.72	2.12	−2.29	0.38
78	3.01	0.46	2.53	0.49	0.28	3.12	−4.29	1.38
79	3.01	−0.54	1.53	−1.51	0.28	4.12	−3.29	1.38
80	−0.99	−0.54	−3.47	0.49	−0.72	−1.88	0.71	−0.62
81	−0.99	0.46	−1.47	0.49	−0.72	−0.88	0.71	−0.62
82	0.01	0.46	0.53	0.49	−0.72	−0.88	0.71	−0.62
83	1.01	0.46	0.53	0.49	−1.72	−0.88	−2.29	−0.62
84	3.01	0.46	−0.47	1.49	0.28	3.12	−3.29	1.38
85	−0.99	0.46	−0.47	0.49	0.28	0.12	0.71	0.38
86	0.01	1.46	0.53	−1.51	2.28	1.12	−0.29	3.38
87	−0.99	−1.54	−1.47	3.49	−3.72	−2.88	2.71	−1.62
88	−1.99	−0.54	−0.47	0.49	−1.72	−1.88	1.71	−1.62
89	−0.99	0.46	−0.47	−0.51	0.28	−0.88	3.71	0.38
90	0.01	0.46	0.53	0.49	−0.72	−0.88	0.71	0.38
91	−1.99	−0.54	−0.47	1.49	−0.72	−0.88	3.71	−0.62
92	−1.99	−2.54	−0.47	1.49	−2.72	−1.88	1.71	−1.62
93	1.01	0.46	1.53	−0.51	−0.72	2.12	−2.29	1.38
94	−0.99	−0.54	−1.47	0.49	0.28	−0.88	2.71	−0.62
95	0.01	1.46	−0.47	−1.51	0.28	1.12	−2.29	−0.62
96	−0.99	−1.54	−0.47	1.49	−2.72	0.12	0.71	−0.62
97	−1.99	1.46	−0.47	−1.51	1.28	−1.88	1.71	−0.62
98	−0.99	1.46	−1.47	0.49	1.28	−1.88	1.71	−1.62
99	0.01	−1.54	−1.47	3.49	−1.72	0.12	0.71	1.38
100	−0.99	−0.54	0.53	1.49	−2.72	0.12	−1.29	−2.62

centered scores, and the averages of the factors are zero. The model in (10.1) without intercepts can be expressed in the matrix form:

$$
\underset{8\times1}{\overset{\mathbf{x}}{\begin{bmatrix} A \\ I \\ V \\ H \\ C \\ B \\ T \\ P \end{bmatrix}}}
=
\underset{8\times2}{\overset{\mathbf{A}}{\begin{bmatrix} a_1 & \\ a_2 & \\ a_3 & \\ a_4 & \\ & a_5 \\ & a_6 \\ & a_7 \\ & a_8 \end{bmatrix}}}
\underset{2\times1}{\overset{\mathbf{f}}{\begin{bmatrix} Factor\_1 \\ Factor\_2 \end{bmatrix}}}
+
\underset{8\times1}{\overset{\mathbf{e}}{\begin{bmatrix} e_1 \\ e_2 \\ e_3 \\ e_4 \\ e_5 \\ e_6 \\ e_7 \\ e_8 \end{bmatrix}}}
\tag{10.2}
$$

with the blank cells in **A** occupied by zeros.

In any CFA model, a $p \times 1$ random variable vector **x** is expressed as

$$\mathbf{x} = \mathbf{Af} + \mathbf{e}, \tag{10.3}$$

where **A** is the p-variables $\times$ m-factors matrix whose elements are called *factor loadings* (or *path coefficients*), **f** is an $m \times 1$ vector whose elements are called *factor scores*, and **e** contains errors.

10.3 Distributional Assumptions for Factors

The factor vector **f** is assumed to have the multivariate normal (MVN) distribution whose average vector is $\mathbf{0}_m$ and whose covariance matrix is $\mathbf{\Phi}$, respectively:

$$\mathbf{f} \sim N_m(\mathbf{0}_m, \mathbf{\Phi}). \tag{10.4}$$

Here, the covariance matrix $\mathbf{\Phi}$ ($m \times m$) is constrained to be a *correlation matrix* with

$$
\mathbf{\Phi} = \begin{bmatrix}
1 & \phi_{12} & \cdots & \phi_{1p} \\
\phi_{12} & 1 & \ddots & \vdots \\
\vdots & \ddots & \ddots & \phi_{p-1,p} \\
\phi_{1p} & \cdots & \phi_{p-1,p} & 1
\end{bmatrix}.
\tag{10.5}
$$

It is equivalent to the assumption that factor scores are standard ones with their variances ones.

Let us consider the rationale of the above assumptions for averages and covariances. The average vector $\mathbf{0}_m$ is matched by supposing that a data set to be analyzed contains centered scores. The reason for assuming the factor scores to be standard ones is that factors are unobserved latent variables; thus, their variances can be freely determined; we may consider the values of a factor to be distributed over the range $[-100, 90]$, $[-50, 60]$, or $[-0.01, 0.01]$. For this reason, the variance is usually set to one, as it is a comprehensible value. This implies that the factor scores are standardized and the covariance matrix between factors is their correlation matrix. Thus, $\boldsymbol{\Phi}$ in (10.5) is called a *factor correlation* matrix.

10.4 Distributional Assumptions for Errors

The error vector $\mathbf{e}$ is assumed to follow the MVN distribution whose average vector is $\mathbf{0}_p$ and whose covariance matrix is $\boldsymbol{\Psi}$, respectively:

$$\mathbf{e} \sim N_p(\mathbf{0}_p, \boldsymbol{\Psi}), \tag{10.6}$$

with $\boldsymbol{\Psi}$ the $p \times p$ diagonal matrix, i.e.,

$$\boldsymbol{\Psi} = \begin{bmatrix} \psi_1 & & \\ & \ddots & \\ & & \psi_p \end{bmatrix}. \tag{10.7}$$

Assumption (10.7) implies that the errors for different variables are mutually uncorrelated, as found in Fig. 10.1, where each of the errors is only linked to the corresponding variable. This is an important feature of factor analysis. In contrast to the factors in vector $\mathbf{f}$ being the common causes for multiple variables, each error in $\mathbf{e}$ can be viewed as the factor that exclusively or uniquely contributes to the corresponding variable. For addressing this contrast, the factors in $\mathbf{f}$ are called *common factors*, while the errors in $\mathbf{e}$ are called *unique factors*. Further, the diagonal elements of $\boldsymbol{\Psi}$ are called *unique variances*, as they are the variances of unique factors.

10.5 Maximum Likelihood Method

We start with a property of the MVN distribution without its proof:

Note 10.1. A Property of MVN Distribution
If $\mathbf{u}_1 \sim N_r(\boldsymbol{\mu}_1, \boldsymbol{\Omega}_1)$, $\mathbf{u}_2 \sim N_r(\boldsymbol{\mu}_2, \boldsymbol{\Omega}_2)$, and $\mathbf{u}_1$ is distributed independently of $\mathbf{u}_2$, then

$$\mathbf{B}_1\mathbf{u}_1 + \mathbf{B}_2\mathbf{u}_2 \sim N_r\left(\mathbf{B}_1\boldsymbol{\mu}_1 + \mathbf{B}_2\boldsymbol{\mu}_2, \mathbf{B}_1\boldsymbol{\Omega}_1\mathbf{B}_1' + \mathbf{B}_2\boldsymbol{\Omega}_2\mathbf{B}_2'\right) \tag{10.8}$$

for fixed matrices $\mathbf{B}_1$ and $\mathbf{B}_2$.

In CFA, **f** and **e** are assumed to be distributed mutually independently. Using this assumption and (10.8) in (10.3), (10.4), and (10.6), the observation vector **x** in (10.3) is found to follow an MVN distribution, as follows:

$$\mathbf{x} \sim N_p(\mathbf{0}_p, \Sigma), \tag{10.9}$$

with its covariance matrix

$$\Sigma = \mathbf{A}\Phi\mathbf{A}' + \Psi, \tag{10.10}$$

which is called a *covariance structure*, as described in Note 9.2.

Let **X** denote the centered data matrix and $\mathbf{V} = n^{-1}\mathbf{X}'\mathbf{X}$ be the sample covariance matrix. As explained in Sect. 9.4, the log likelihood for CFA can be written in the form of (9.15), i.e., $l(\Sigma) = (n/2)\log|\Sigma^{-1}\mathbf{V}| - (n/2)\mathrm{tr}\Sigma^{-1}\mathbf{V}$. Substituting (10.10) into $l(\Sigma)$, we have

$$l(\mathbf{A}, \Psi, \Phi) = \frac{n}{2}\log\left|(\mathbf{A}\Phi\mathbf{A}' + \Psi)^{-1}\mathbf{V}\right| - \frac{n}{2}\mathrm{tr}(\mathbf{A}\Phi\mathbf{A}' + \Psi)^{-1}\mathbf{V}. \tag{10.11}$$

It is maximized over **A**, **Ψ**, and **Φ**, i.e., the 17 parameters $a_1, \ldots, a_8, v_{11}, \ldots, v_{88}$, c for the model in Fig. 10.1. Since the solution is not explicitly given, the minimization is attained by iterative algorithms. A popular algorithm is the one using a gradient method, which is illustrated in Appendix A.6.3. Setting the vector **θ** in A.6.3 to $[a_1, \ldots, a_8, v_{11}, \ldots, v_{88}, c]'$, the solution can be obtained. We express the resulting **A**, **Ψ**, and **Φ** as **Â**, **Ψ̂**, and **Φ̂**, respectively.

10.6   Solutions

The solution given by the maximum likelihood method is shown in Fig. 10.2a, where the estimated parameter values are presented at the corresponding parts. As in path analysis, the *GFI statistic* defined as (9.18) can be used for assessing whether a solution is satisfactory or not. A value of 0.9 is used as a benchmark, with GFI ≥ 0.9 indicating that a model is satisfactory. The GFI value for the solution in Fig. 10.2a was 0.953, which shows that the solution is to be accepted.

The solution in Fig. 10.2a is the *unstandardized* one obtained from variables with different variances. Thus, it is senseless to compare the largeness of the resulting parameter values. For the comparison to make sense, we must note the *standardized* solution obtained for the standard scores transformed from the original data set. That solution is shown in Fig. 10.2b. The unstandardized and standardized solutions in CFA can be considered two *different expressions of the same solution*, since CFA is scale invariant, as in path analysis.

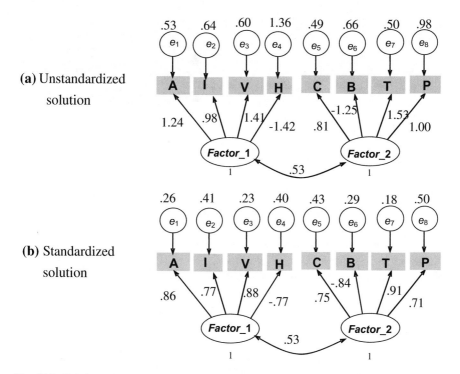

Fig. 10.2 Solution of the model in Fig. 9.1 for the data in Table 9.1

Note 10.2. Scale Invariance

Let us recall Note 9.3. Using (3.17), the sample covariance matrix $\mathbf{V} = n^{-1}\mathbf{X}'\mathbf{X}$ is rewritten as (9.20). Using it in log likelihood (9.15) for $\mathbf{V}$, it can be rewritten into the log likelihood for correlation matrix $\mathbf{R}$ as (9.21) with $\mathbf{\Sigma}_S = \mathbf{D}^{-1}\mathbf{\Sigma}\mathbf{D}^{-1}$. In this equation, we can use (10.10) to have

$$\mathbf{\Sigma}_S = \mathbf{D}^{-1}(\mathbf{A}\mathbf{\Phi}\mathbf{A}' + \mathbf{\Psi})\mathbf{D}^{-1} = \mathbf{A}_S\mathbf{\Phi}_S\mathbf{A}_S' + \mathbf{\Psi}_S, \qquad (10.12)$$

with

$$\mathbf{A}_S = \mathbf{D}^{-1}\mathbf{A}, \quad \mathbf{\Phi}_S = \mathbf{\Phi}, \quad \text{and} \quad \mathbf{\Psi}_S = \mathbf{D}^{-1}\mathbf{\Psi}\mathbf{D}^{-1}. \qquad (10.13)$$

Substituting (10.12) in (9.21), it is rewritten as follows:

$$\log l(\mathbf{A}_S, \mathbf{\Psi}_S, \mathbf{\Phi}_S) = \frac{n}{2}\log\left|\left(\mathbf{A}_S\mathbf{\Phi}_S\mathbf{A}_S' + \mathbf{\Psi}_S\right)^{-1}\mathbf{R}\right| = \frac{n}{2}\mathrm{tr}\left(\mathbf{A}_S\mathbf{\Phi}_S\mathbf{A}_S' + \mathbf{\Psi}_S\right)^{-1}\mathbf{R},$$

$$(10.14)$$

which shows the same form as (10.11). These results imply that the maximum of (10.14) equals that of (10.11), with the solution of $\mathbf{A}_S$, $\boldsymbol{\Psi}_S$, and $\boldsymbol{\Phi}_S$ maximizing (10.14) given by (10.13), in which $\hat{\mathbf{A}}$, $\hat{\boldsymbol{\Psi}}$, and $\hat{\boldsymbol{\Phi}}$ are substituted into $\mathbf{A}$, $\boldsymbol{\Psi}$, and $\boldsymbol{\Phi}$. This shows that CFA has *scale invariance*.

10.7   Other and Extreme Models

Let us refer to the model in Fig. 10.1 as "Two-factor Model 1." Though this model is regarded as satisfactory, with a GFI exceeding the benchmark value of 0.9, a model may exist that is better fitted to the data set in Table 10.1b. This suggests that *other models* should be considered and *compared*; that is, the *model selection* illustrated in Fig. 8.5 should be performed. Figure 10.3 shows two examples of other models. Figure 10.3a presents the *one-factor model* in which only one factor underlies the eight observed variables. For this model, the $\mathbf{A}$ and $\mathbf{f}$ in (10.3) are a vector and a scalar, respectively. Figure 10.3b shows the "*Two-factor Model 2*" in

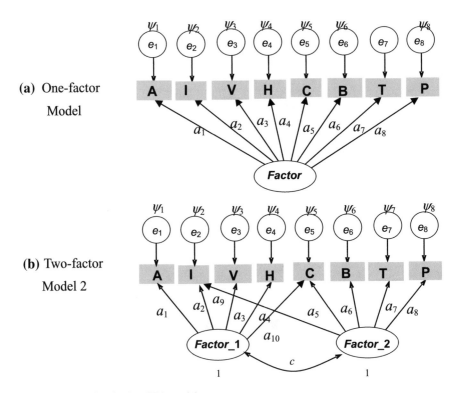

Fig. 10.3 Example of other CFA models

which the variables "Initiative" and "Cheerful" load both factors. This model is written as

$$
\begin{array}{cccc}
\mathbf{x} & \boldsymbol{\Lambda} & \mathbf{f} & \mathbf{e} \\
8\times1 & 8\times2 & 2\times1 & 8\times1
\end{array}
$$

$$
\begin{array}{c}
\begin{bmatrix} A \\ I \\ V \\ H \\ C \\ B \\ T \\ P \end{bmatrix} =
\begin{bmatrix} a_1 & \\ a_2 & a_9 \\ a_3 & \\ a_4 & \\ a_{10} & a_5 \\ & a_6 \\ & a_7 \\ & a_8 \end{bmatrix}
\begin{bmatrix} Factor\_1 \\ Factor\_2 \end{bmatrix}
+
\begin{bmatrix} e_1 \\ e_2 \\ e_3 \\ e_4 \\ e_5 \\ e_6 \\ e_7 \\ e_8 \end{bmatrix}
\end{array}
\tag{10.15}
$$

As in path analysis, the two types of extreme models are the *independent* and *saturated models*. In the former, all variables are mutually *independent*, without any factor. This is represented as the path diagram in which only eight variables are depicted, without any link among them. On the other hand, one of the *saturated models* is represented in the path diagram in which each of the eight variables are linked to the other seven by double-headed arrows, without any factors. This implies that *all variables are merely correlated*.

10.8 Model Selection

So far, we have the two-factor models (1 and 2), the one-factor model, and two extreme ones. For comparing those models with respect to the *goodness of fit* to the data set, we *cannot use the GFI* (9.18) for the reason explained in Sect. 9.9. On the other hand, the *information criteria* introduced in Sect. 8.7 are useful for *comparing models*, since the number of parameters is considered in the criteria. The values of typical information criteria, the *AIC* and *BIC*, for each model are shown in Table 10.2. There, the BIC shows that Two-factor Model 1 is the best, while it is found to be slightly worse than Two-factor Model 2 in the AIC values. This demonstrates that model selection statistics indicate different models as the best. For such a case, the model must be chosen by users' subjective consideration. This shows that no absolute index exists for model selection, which should be kept in mind.

Table 10.2 Number of parameters (NP) and the resulting index values for each model

Model	NP	GFI	AIC	BIC
Saturated	36	1.000	72.000	165.786
Two-factor Model 2	19	0.964	54.168	103.666
Two-factor Model 1	17	0.953	55.464	99.751
One-factor Model	16	0.642	193.494	235.176
Independent	8	0.354	504.322	525.163

10.9   Bibliographical Notes

It is difficult to find books in which CFA is exclusively treated. CFA is, however, described in chapters of books on structural equation modeling or factor analysis, which include Kaplan (2000), Mulaik (2010), and Wang and Wang (2012).

One drawback of CFA is that the model, i.e., the elements that are set to be zero in $\mathbf{A}$, must be selected by users. Such a drawback can be dealt with by the *sparse factor analysis* approaches in which the selection is computationally performed (e.g., Adachi and Trendafilov 2015a; Hirose and Yamamoto 2014).

Exercises

10.1. Let us consider the model $x = t + e$, with x an observed variable, e an error, and t a true score which is an unobserved latent variable. For example, t stands for the ability of mathematics possessed by an examinee, while x is the test score on mathematics shown by the examinee, and an error e must be considered, since t (ability) cannot be perfectly exactly measured by x (score). Present another example for a set of x, t, and e in the model.

10.2. Spearman (1904) hit upon the idea of factor analysis, by considering the scores of achievement tests as variables, and personality test scores have been used as an example in this chapter. Present an example of a data set that is not related to such tests and for which factor analysis is useful.

10.3. Consider another two-factor model for the data in Table 10.1.

10.4. Depict the path diagram of a saturated model for the data in Table 10.1 without a factor and a single-headed path.

10.5. Present an example of the CFA model for 15 observed variables with three factors.

10.6. Equation (10.3) can be rewritten as $\mathbf{x} = \mathbf{A}\mathbf{f} + \mathbf{e} = \mathbf{A}^*\mathbf{f}^* + \mathbf{e}$ with $\mathbf{f}^* = \mathbf{S}^{-1}\mathbf{f}$ and $\mathbf{A}^* = \mathbf{A}\mathbf{S}$. It suggests that $\mathbf{f}^*$ and $\mathbf{A}^*$ could also be regarded as a factor score vector and loading matrix, respectively, with $\mathbf{S}$ an $m \times m$ arbitrary nonsingular matrix. However, in CFA, except for special cases, it is not possible to regard $\mathbf{f}^*$ and $\mathbf{A}^*$ as above. Show the reason for this, noting that $\mathbf{A}$ is constrained in CFA.

10.7. Model (10.3) can be rewritten as $\mathbf{x} = \mathbf{Hg}$, with $\mathbf{H} = [\mathbf{A}, \ \mathbf{I}_p]$ being $p \times (m + p)$ and $\mathbf{g} = \begin{bmatrix} \mathbf{f} \\ \mathbf{e} \end{bmatrix} (m+p) \times 1$. If $\mathbf{A}$ and $\mathbf{x}$ are given, $\mathbf{x} = \mathbf{Hg}$ is regarded as a system of equations with $\mathbf{g}$ unknown. The necessary and sufficient condition of the system having the solutions of $\mathbf{g}$ is known to be $\mathbf{HH}^+\mathbf{x} = \mathbf{x}$. If this equation holds true, show that the solution of $\mathbf{g}$ satisfying $\mathbf{x} = \mathbf{Hg}$ is expressed as

$$\mathbf{g} = \mathbf{H}^+\mathbf{x} + \left(\mathbf{I}_{m+p} - \mathbf{H}^+\mathbf{H}\right)\mathbf{q}, \qquad (10.16)$$

with $\mathbf{H}^+$ the Moore–Penrose inverse of $\mathbf{H}$ defined in Exercise 5.10 and $\mathbf{q}$ an arbitrary $(m + p) \times 1$ vector.

10.8. Show the following: (10.16) implies that factor score vector $\mathbf{f}$ cannot be uniquely determined; i.e., we cannot select a single vector as $\mathbf{f}$ for given $\mathbf{A}$ and $\mathbf{x}$.

10.9. Let us consider the CFA model with intercept vector $\mathbf{c}$: $\mathbf{x} = \mathbf{Af} + \mathbf{c} + \mathbf{e}$, $\mathbf{f} \sim N_m(\mathbf{0}_m, \mathbf{\Phi})$, and $\mathbf{e} \sim N_p(\mathbf{0}_p, \mathbf{\Psi})$. Show that the MLE of transposed intercept vector $\mathbf{c}'$ is given by $n^{-1}\mathbf{1}_n'\mathbf{X}$ for the $n \times p$ data matrix $\mathbf{X}$ whose rows are the observations of $\mathbf{x}'$ for individuals $i = 1, \ldots, n$.

10.10. Let us consider a *confirmatory principal component analysis* (PCA) procedure formulated as minimizing $\|\mathbf{X} - \mathbf{FA}'\|^2$ over $\mathbf{F}$ and $\mathbf{A}$ subject to $n^{-1}\mathbf{F}'\mathbf{F} = \mathbf{I}_m$ and some elements of $\mathbf{A}$ constrained to be zero. Show that the function can be decomposed as

$$\|\mathbf{X} - \mathbf{FA}'\|^2 = \|\mathbf{X} - \mathbf{FB}'\|^2 + n\|\mathbf{B} - \mathbf{A}\|^2, \tag{10.17}$$

with $\mathbf{B} = n^{-1}\mathbf{X}'\mathbf{F}$ (Adachi and Trendafilov 2015b).

10.11. Show that an algorithm for the confirmatory PCA in Exercise 10.10 can be formed by the following steps:

Step 1. Initialize $\mathbf{F}$.
Step 2. Set the unconstrained elements of $\mathbf{A}$ to the corresponding ones of $n^{-1}\mathbf{X}'\mathbf{F}$.
Step 3. Obtain the SVD $\mathbf{XA} = \mathbf{K\Lambda L}'$ and set $\mathbf{F} = n^{1/2}\mathbf{KL}'$.
Step 4. Finish if convergence is reached; otherwise, go back to Step 2.

The hints for Steps 2 and 3 can be found in (10.17) and Theorem A.4.2 (Appendix A.4.2), respectively.

Chapter 11
Structural Equation Modeling

In confirmatory factor analysis (CFA), introduced in the previous chapter, all factors (latent variables) were the causes (explanatory variables). An extended *variant of CFA* is *structural equation modeling* (*SEM*), in which the *causal relationships among factors* are considered; i.e., *factors* appear that are *dependent variables*.

To the best of the author's knowledge, SEM was first presented by the Swedish statistician Jöreskog (1970), who *combined path analysis* and *factor analysis* to formulate SEM. This has been elaborated on and popularized, particularly with the developments of computer software, by the efforts of psychometricians including Bentler (1985).

11.1 Causality Among Factors

We will introduce structural equation modeling (SEM) by starting with the formation of a model, which is followed by the description of the data to be observed.

Let us consider a *model of the causality among four factors*, depicted in Fig. 11.1, with the factors as follows:

[**F1**] Prior achievements before postgraduate school (PGS);
[**F2**] Adaptation to PGS;
[**F3**] Achievements in PGS;
[**F4**] Satisfaction with having gone to PGS.

© Springer Nature Singapore Pte Ltd. 2016
K. Adachi, *Matrix-Based Introduction to Multivariate Data Analysis*,
DOI 10.1007/978-981-10-2341-5_11

Fig. 11.1 Path diagram for a
structural equation model

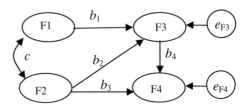

The path diagram in the figure is expressed as a set of formulas:

$$F3 = b_1 F1 + b_2 F2 + e_{F3},$$
$$F4 = b_3 F2 + b_4 F3 + e_{F4}, \tag{11.1}$$

with e_{F3} and e_{F4} being errors. Here, *intercepts* are *omitted*, since it is known that they may be zero, assuming that the averages of the factors are zeros. The set of formulas is a kind of *path analysis* model, though the variables are not observed but rather latent *factors*, which differs from the ordinary path analysis in Chap. 9. A model such as (11.1) is called a *structural equation model* for latent variables.

11.2   Observed Variables as Indicator of Factors

It is reasonable to consider that the above four *factors are difficult to measure directly*, but each of them (F1, F2, F3, F4) *can be measured with several indices* (*observed variables*). Then, let us suppose that each factor can be measured by the four variables shown in Table 11.1. For example, we suppose that X9 (scores for lecture courses), X10 (scores for practice courses), X11 (evaluation of a Master's thesis), and X12 (self-rating of achievement) can be used as the *indicators* for F3 (achievements in PGS).

The four path diagrams in Fig. 11.2 represent the fact that F1, F2, F3, and F4 are indicated by the variables in Table 11.1. Each diagram can be expressed by a set of formulas; for example, the third diagram is expressed as the set of four equations

$$X9 = a_9 F3 + e_9,$$
$$X10 = a_{10} F3 + e_{10},$$
$$X11 = a_{11} F3 + e_{11}, \tag{11.2}$$
$$X12 = a_{12} F3 + e_{12}.$$

This is just a *factor analysis* model, which is also called a *measurement equation model*, as (11.2) stands for how an *unobserved factor* (F3), which *cannot be*

Table 11.1 Variables indicating factors

F	Variable	
F1	X1	Scores for languages when one was a student in a faculty
	X2	Scores for sciences when one was a student in a faculty
	X3	Scores for the entrance examination for a postgraduate school
	X4	Evaluation of a graduation thesis
F2	X5	Goodness of fit to the education in the postgraduate school
	X6	Goodness of fit to the atmosphere in the postgraduate school
	X7	Goodness of fit to the facilities in the postgraduate school
	X8	Inconvenience found in the systems of the postgraduate school
F3	X9	Scores for lecture courses in the postgraduate school
	X10	Scores for practice courses in the postgraduate school
	X11	Evaluation of a Master's thesis
	X12	Self-rating of achievement
F4	X13	Fulfillment felt from life in the postgraduate school
	X14	How well one enjoyed life in the postgraduate school
	X15	Regret that one went to the postgraduate school
	X16	Hope for the future

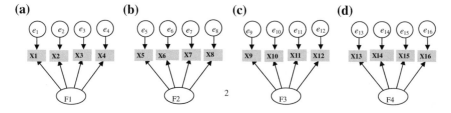

Fig. 11.2 Path diagrams for four measurement equation models

measured directly, is *measured* using *several observed variables* as the *indicators of the factor*.

Let 300×16 data matrix **X** contains the centered scores of 300 postgraduate students for the 16 items in Table 11.1 with covariance matrix $\mathbf{V} = n^{-1}\mathbf{X}'\mathbf{X}$ for the 16 variables shown in Table 11.2. The data matrix **X** is too big to be presented; in place of it, the sample covariance matrix **V** is presented here. As described in Note 9.2, the procedures in Chaps. 9–12 can be feasible only with **V**, even if **X** is not available.

Table 11.2 Data set observed for 300 postgraduate students, which is an artificial example found in Adachi (2006)

	X1	X2	X3	X4	X5	X6	X7	X8	X9	X10	X11	X12	X13	X14	X15	X16
X1	40.323	27.475	21.883	2.669	1.685	1.405	1.478	-0.820	15.131	19.270	2.506	2.757	2.161	1.686	-1.685	1.306
X2	27.475	55.696	28.439	3.104	1.320	1.469	1.410	-0.822	14.911	17.290	3.065	2.602	2.816	2.091	-2.117	1.726
X3	21.883	28.439	52.305	2.662	0.792	1.119	0.549	-0.502	12.839	12.088	2.551	1.977	2.040	1.847	-1.889	1.415
X4	2.669	3.104	2.662	0.620	0.155	0.132	0.164	-0.109	1.557	2.272	0.294	0.268	0.284	0.162	-0.204	0.163
X5	1.685	1.320	0.792	0.155	0.810	0.484	0.454	-0.439	2.026	2.437	0.359	0.247	0.371	0.293	-0.311	0.305
X6	1.405	1.469	1.119	0.132	0.484	0.961	0.468	-0.477	2.108	2.698	0.374	0.305	0.436	0.344	-0.414	0.374
X7	1.478	1.410	0.549	0.164	0.454	0.468	1.128	-0.458	2.177	2.545	0.380	0.273	0.407	0.277	-0.368	0.286
X8	-0.820	-0.822	-0.502	-0.109	-0.439	-0.477	-0.458	1.011	-1.597	-2.363	-0.356	-0.262	-0.385	-0.262	0.331	-0.270
X9	15.131	14.911	12.839	1.557	2.026	2.108	2.177	-1.597	44.213	28.670	4.163	3.809	3.566	2.500	-3.144	3.072
X10	19.270	17.290	12.088	2.272	2.437	2.698	2.545	-2.363	28.670	58.632	4.513	4.123	4.567	2.349	-3.595	3.083
X11	2.506	3.065	2.551	0.294	0.359	0.374	0.380	-0.356	4.163	4.513	1.176	0.703	0.747	0.417	-0.600	0.553
X12	2.757	2.602	1.977	0.268	0.247	0.305	0.273	-0.262	3.809	4.123	0.703	1.077	0.575	0.389	-0.506	0.542
X13	2.161	2.816	2.040	0.284	0.371	0.436	0.407	-0.385	3.566	4.567	0.747	0.575	1.377	0.573	-0.727	0.672
X14	1.686	2.091	1.847	0.162	0.293	0.344	0.277	-0.262	2.500	2.349	0.417	0.389	0.573	0.924	-0.473	0.551
X15	-1.685	-2.117	-1.889	-0.204	-0.311	-0.414	-0.368	0.331	-3.144	-3.595	-0.600	-0.506	-0.727	-0.473	1.141	-0.592
X16	1.306	1.726	1.415	0.163	0.305	0.374	0.286	-0.270	3.072	3.083	0.553	0.542	0.672	0.551	-0.592	1.349

11.3 SEM Model

The *structural equation model* in Fig. 11.1 and the four *measurement equation models* in Fig. 11.2 are *integrated into a single model* in Fig. 11.3. This is a *SEM model* for the covariance matrix in Table 11.2. The outer parts of the diagram in Fig. 11.3 are a–d in Fig. 11.2, while the inner part in Fig. 11.3 is the diagram in Fig. 11.1. In other words, the *outer* parts stand for *measurement* equation models (i.e., *factor analysis* models), while the *inner* part represents a *structural* equation model (i.e., a *path analysis* model with latent factors). That is, SEM is an analysis procedure with a model into which *structural* and *measurement* equation models are *integrated*. However, the procedure is called structural equation modeling, without the use of the term "measurement."

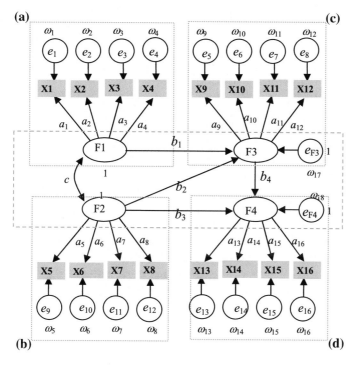

Fig. 11.3 A SEM model

11.4 Matrix Expression

The path diagram in Fig. 11.3 is formally expressed using the two equations in (11.1) and the four sets of measurement equations, with an example of a set presented in (11.2). Those equations can be written as a single equation in matrix form:

$$
\begin{array}{c}
\mathbf{t} \\
\begin{bmatrix}
F1 \\ F2 \\ F3 \\ F4 \\ X1 \\ X2 \\ X3 \\ X4 \\ X5 \\ X6 \\ X7 \\ X8 \\ X9 \\ X10 \\ X11 \\ X12 \\ X13 \\ X14 \\ X15 \\ X16
\end{bmatrix}
\end{array}
=
\begin{array}{c}
\mathbf{B} \\
\begin{bmatrix}
0 & 0 & 0 & 0 & 0 & 0 & 0 & 0 & 0 & 0 & 0 & 0 & 0 & 0 & 0 & 0 & 0 & 0 & 0 & 0 \\
0 & 0 & 0 & 0 & 0 & 0 & 0 & 0 & 0 & 0 & 0 & 0 & 0 & 0 & 0 & 0 & 0 & 0 & 0 & 0 \\
b_1 & b_2 & 0 & 0 & 0 & 0 & 0 & 0 & 0 & 0 & 0 & 0 & 0 & 0 & 0 & 0 & 0 & 0 & 0 & 0 \\
0 & b_3 & b_4 & 0 & 0 & 0 & 0 & 0 & 0 & 0 & 0 & 0 & 0 & 0 & 0 & 0 & 0 & 0 & 0 & 0 \\
a_1 & 0 & 0 & 0 & 0 & 0 & 0 & 0 & 0 & 0 & 0 & 0 & 0 & 0 & 0 & 0 & 0 & 0 & 0 & 0 \\
a_2 & 0 & 0 & 0 & 0 & 0 & 0 & 0 & 0 & 0 & 0 & 0 & 0 & 0 & 0 & 0 & 0 & 0 & 0 & 0 \\
a_3 & 0 & 0 & 0 & 0 & 0 & 0 & 0 & 0 & 0 & 0 & 0 & 0 & 0 & 0 & 0 & 0 & 0 & 0 & 0 \\
a_4 & 0 & 0 & 0 & 0 & 0 & 0 & 0 & 0 & 0 & 0 & 0 & 0 & 0 & 0 & 0 & 0 & 0 & 0 & 0 \\
0 & a_5 & 0 & 0 & 0 & 0 & 0 & 0 & 0 & 0 & 0 & 0 & 0 & 0 & 0 & 0 & 0 & 0 & 0 & 0 \\
0 & a_6 & 0 & 0 & 0 & 0 & 0 & 0 & 0 & 0 & 0 & 0 & 0 & 0 & 0 & 0 & 0 & 0 & 0 & 0 \\
0 & a_7 & 0 & 0 & 0 & 0 & 0 & 0 & 0 & 0 & 0 & 0 & 0 & 0 & 0 & 0 & 0 & 0 & 0 & 0 \\
0 & a_8 & 0 & 0 & 0 & 0 & 0 & 0 & 0 & 0 & 0 & 0 & 0 & 0 & 0 & 0 & 0 & 0 & 0 & 0 \\
0 & 0 & a_9 & 0 & 0 & 0 & 0 & 0 & 0 & 0 & 0 & 0 & 0 & 0 & 0 & 0 & 0 & 0 & 0 & 0 \\
0 & 0 & a_{10} & 0 & 0 & 0 & 0 & 0 & 0 & 0 & 0 & 0 & 0 & 0 & 0 & 0 & 0 & 0 & 0 & 0 \\
0 & 0 & a_{11} & 0 & 0 & 0 & 0 & 0 & 0 & 0 & 0 & 0 & 0 & 0 & 0 & 0 & 0 & 0 & 0 & 0 \\
0 & 0 & a_{12} & 0 & 0 & 0 & 0 & 0 & 0 & 0 & 0 & 0 & 0 & 0 & 0 & 0 & 0 & 0 & 0 & 0 \\
0 & 0 & 0 & a_{13} & 0 & 0 & 0 & 0 & 0 & 0 & 0 & 0 & 0 & 0 & 0 & 0 & 0 & 0 & 0 & 0 \\
0 & 0 & 0 & a_{14} & 0 & 0 & 0 & 0 & 0 & 0 & 0 & 0 & 0 & 0 & 0 & 0 & 0 & 0 & 0 & 0 \\
0 & 0 & 0 & a_{15} & 0 & 0 & 0 & 0 & 0 & 0 & 0 & 0 & 0 & 0 & 0 & 0 & 0 & 0 & 0 & 0 \\
0 & 0 & 0 & a_{16} & 0 & 0 & 0 & 0 & 0 & 0 & 0 & 0 & 0 & 0 & 0 & 0 & 0 & 0 & 0 & 0
\end{bmatrix}
\end{array}
\begin{array}{c}
\mathbf{t} \\
\begin{bmatrix}
F1 \\ F2 \\ F3 \\ F4 \\ X1 \\ X2 \\ X3 \\ X4 \\ X5 \\ X6 \\ X7 \\ X8 \\ X9 \\ X10 \\ X11 \\ X12 \\ X13 \\ X14 \\ X15 \\ X16
\end{bmatrix}
\end{array}
+
\begin{array}{c}
\mathbf{u} \\
\begin{bmatrix}
F1 \\ F2 \\ e_{F3} \\ e_{F4} \\ e_1 \\ e_2 \\ e_3 \\ e_4 \\ e_5 \\ e_6 \\ e_7 \\ e_8 \\ e_9 \\ e_{10} \\ e_{11} \\ e_{12} \\ e_{13} \\ e_{14} \\ e_{15} \\ e_{16}
\end{bmatrix}
\end{array}
$$

(11.3)

Here, $\mathbf{t}$ is the random vector whose first elements are *factors* and the remaining ones are *observed variables*, while $\mathbf{u}$ is the vector whose first elements are the factors being explanatory variables and the remaining ones are the errors for the dependent factors and observed variables. Matrix $\mathbf{B}$ is filled with zeros except for the *path coefficients* corresponding to the links between factors and the links of factors to variables. The first and second rows in the left- and right-hand sides of (11.3) stand for "F1 = F1" and "F2 = F2," which obviously hold true; the third and fourth rows express (11.1), and the remaining ones stand for the measurement equation models (Fig. 11.2), with the rows for X9 to X12 corresponding to (11.2).

Any *SEM model* is expressed as

$$\mathbf{t} = \mathbf{Bt} + \mathbf{u}.$$

(11.4)

Here, $\mathbf{B}$ is an $(m + p) \times (m + p)$ path coefficient matrix, with m and p being the numbers of factors and observed variables, respectively. Vector $\mathbf{t}$ is $(m + p) \times 1$ with

$$\mathbf{t} = \begin{bmatrix} \mathbf{f} \\ \mathbf{x} \end{bmatrix}.$$

(11.5)

Its first m elements are those of an $m \times 1$ *factor* vector $\mathbf{f}$ and the $(m + 1)$th, ..., $(m + p)$th elements of $\mathbf{t}$ are the 1st, ..., pth *observed variables* in $\mathbf{x}$. Vector $\mathbf{u}$ is $(m + p) \times 1$ with

$$\mathbf{u} = \begin{bmatrix} \mathbf{f}_E \\ \mathbf{e}_D \\ \mathbf{e}_X \end{bmatrix}. \tag{11.6}$$

Its first m_E elements are those of the $m_E \times 1$ vector $\mathbf{f}_E$ containing factors being explanatory variables; the next m_D elements are those of the $m_D \times 1$ vector $\mathbf{e}_D$ consisting of the errors for dependent factors, and the remaining p ones are the elements of the $p \times 1$ vector $\mathbf{e}_X$ containing the errors for $\mathbf{x}$.

Equation (11.4) can be rewritten as $(\mathbf{I}_{m+p} - \mathbf{B})\mathbf{t} = \mathbf{u}$ with $\mathbf{I}_{m+p}$ the $(m + p) \times (m + p)$ identity matrix. It can be further rewritten as:

$$\mathbf{t} = (\mathbf{I}_{m+p} - \mathbf{B})^{-1}\mathbf{u}, \quad \text{i.e.,} \quad \begin{bmatrix} \mathbf{f} \\ \mathbf{x} \end{bmatrix} = (\mathbf{I}_{m+p} - \mathbf{B})^{-1}\mathbf{u}, \tag{11.7}$$

where we have supposed the existence of $(\mathbf{I}_{m+p} - \mathbf{B})^{-1}$. Now, let us define a $p \times (m + p)$ matrix as:

$$\mathbf{H} = \begin{bmatrix} 0 & \cdots & 0 & 1 & 0 & \cdots & 0 \\ 0 & \cdots & 0 & 0 & 1 & \ddots & \vdots \\ \vdots & \vdots & \vdots & \vdots & \ddots & \ddots & 0 \\ 0 & \cdots & 0 & 0 & \cdots & 0 & 1 \end{bmatrix}, \tag{11.8}$$

whose first m columns are filled with zeros and whose remaining p columns are the *those* of $\mathbf{I}_p$. We find that

$$\mathbf{x} = \mathbf{H}\begin{bmatrix} \mathbf{f} \\ \mathbf{x} \end{bmatrix}, \quad \text{i.e.,} \quad \mathbf{x} = \mathbf{Ht}. \tag{11.9}$$

Using (11.7) in (11.9), it is expressed as:

$$\mathbf{x} = \mathbf{H}(\mathbf{I}_{m+p} - \mathbf{B})^{-1}\mathbf{u}. \tag{11.10}$$

This is the SEM model for the observation vector $\mathbf{x}$.

11.5 Distributional Assumptions

Let us assume that vector $\mathbf{u}$ is distributed according to the multivariate normal (MVN) distribution, with its mean vector $\mathbf{0}_{m+p}$ and covariance matrix $\mathbf{\Omega}$:

$$\mathbf{u} \sim N_p(\mathbf{0}_{m+p}, \boldsymbol{\Omega}). \tag{11.11}$$

The elements of the covariance matrix are described as:

$$
\boldsymbol{\Omega} \;=\;
\begin{array}{c}
\\ F1 \\ F2 \\ e_{F3} \\ e_{F4} \\ e_1 \\ \cdots \\ e_{16}
\end{array}
\!\!
\begin{array}{|cccccccc|}
\hline
F1 & F2 & e_{F3} & e_{F4} & e_1 & \cdots & e_{16} \\
1 & c & & & & & \\
c & 1 & & & & & \\
& & 1 & & & & \\
& & & 1 & & & \\
& & & & \omega_1 & & \\
& & & & & \cdots & \\
& & & & & & \omega_{16} \\
\hline
\end{array}
\tag{11.12}
$$

for the model in Fig. 11.3, where the blanks (=zeros) indicate *no correlation between errors* and *no correlation of errors to explanatory variables*. They are not linked by paths, as found in the figure.

In (11.12), we should note the following constraints:

$$V(F1) = V(F2) = V(e_{F3}) = V(e_{F4}) = 1, \tag{11.13}$$

with $V(F1)$ denoting the variance of F1. The reason for constraining the variances of factors to be one with $V(F1) = V(F2) = 1$ is the same in factor analysis (Sect. 10.3); the variances can be set to one, as the factors are unobserved latent variables and their variances can be freely determined. The errors e_{F3} and e_{F4} for factors F3 and F4, respectively, are also unobserved and their variances can be freely determined. Thus, $V(e_{F3})$ and $V(e_{F4})$ can be set to one. The constraint $V(F1) = V(F2) = 1$ implies that factors F1 and F2 are standardized; thus, their covariance c is a correlation coefficient.

Because of (11.10), (11.11), and (9.10), observed variable vector $\mathbf{x}$ is found to follow an MVN distribution as

$$\mathbf{x} \sim N_p(\mathbf{0}_p, \Sigma), \tag{11.14}$$

with the covariance matrix

$$\Sigma = \mathbf{H}(\mathbf{I}_{m+p} - \mathbf{B})^{-1}\boldsymbol{\Omega}(\mathbf{I}_{m+p} - \mathbf{B})^{-1'}\mathbf{H}'. \tag{11.15}$$

11.6 Maximum Likelihood Method

Let $\mathbf{X}$ denote the centered data matrix and $\mathbf{V} = n^{-1}\mathbf{X}'\mathbf{X}$ be the sample covariance matrix. As explained in Sect. 9.4, the log likelihood for SEM can be written in the form of (9.15), i.e., $l(\boldsymbol{\Sigma}) = (n/2)\log|\boldsymbol{\Sigma}^{-1}\mathbf{V}| - (n/2)\mathrm{tr}\boldsymbol{\Sigma}^{-1}\mathbf{V}$. Substituting (11.15) into $l(\boldsymbol{\Sigma})$, we have the *log likelihood* of *parameter* matrices $\mathbf{B}$ and $\boldsymbol{\Omega}$:

$$
\begin{aligned}
\log l(\mathbf{B}, \boldsymbol{\Omega}) = {} & \frac{n}{2}\log\left|\{\mathbf{H}(\mathbf{I}_{m+p} - \mathbf{B})^{-1}\boldsymbol{\Omega}(\mathbf{I}_{m+p} - \mathbf{B})'^{-1}\mathbf{H}'\}^{-1}\mathbf{V}\right| \\
& - \frac{n}{2}\mathrm{tr}\{\mathbf{H}(\mathbf{I}_{m+p} - \mathbf{B})^{-1}\boldsymbol{\Omega}(\mathbf{I}_{m+p} - \mathbf{B})'^{-1}\mathbf{H}'\}^{-1}\mathbf{V}.
\end{aligned}
\tag{11.16}
$$

This is maximized over $\mathbf{B}$ and $\boldsymbol{\Omega}$, that is, the 37 parameters $a_1, \ldots, a_{16}, b_1, b_2, b_3, b_4, \omega_1, \ldots, \omega_{16}, c$. Since the solution is not explicitly given, the maximization is attained by iterative algorithms. A popular algorithm is the one using a gradient method, illustrated in Appendix A.6.3; setting the vector $\boldsymbol{\theta}$ in A.6.3 to $[a_1, \ldots, a_{16}, b_1, b_2, b_3, b_4, \omega_1, \ldots, \omega_{16}, c]'$, the solution can be obtained. We express the resulting $\mathbf{A}$, $\boldsymbol{\Psi}$, and $\boldsymbol{\Phi}$ as $\hat{\mathbf{A}}$, $\hat{\boldsymbol{\Psi}}$, and $\hat{\boldsymbol{\Phi}}$, respectively.

11.7 Solutions

The solution given by the maximum likelihood method is shown in Fig. 11.4; the estimated parameter values are presented at the corresponding parts. As in path analysis and confirmatory factor analysis, the *GFI statistic* defined in (9.18) can be

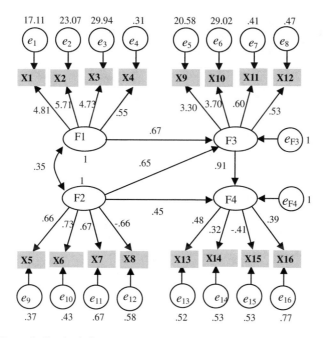

Fig. 11.4 Unstandardized solution

used for assessing whether a solution is satisfactory or not. A value of 0.9 is used as a benchmark with a GFI ≥ 0.9 showing a model that is satisfactory. The GFI value for the solution was 0.96, which shows that the solution is to be accepted.

The solution in Fig. 11.4 is the *unstandardized* one obtained from those variables with different variances. Thus, it is senseless to compare the largeness of the resulting parameter values. For the comparison to make sense, we must obtain the *standardized* solution obtained for the standard scores transformed from the original data set. The standardized solution is shown in Fig. 11.5. In this solution, the variances of the latent variables (*factors*) are also adjusted to be unity. Here, we should recall (11.13): for the factors F3 and F4 being dependent variables, the variances of the corresponding errors are unity with $\mathrm{Var}(e_{F3}) = \mathrm{Var}(e_{F4}) = 1$, but the variances of F3 and F4 are not constrained to be one. Their *variances* have been adjusted so as to become one in the standardized solution. How that is done will not be discussed here.

As in path analysis and confirmatory factor analysis, SEM also has scale invariance, though its proof is omitted here. Thus, the attained value of the maximum of log likelihood (11.15) is equivalent for unstandardized and standardized solutions, and so is the GFI. Further, the standardized solution is easily transformed from the unstandardized one. We may thus consider unstandardized and standardized solutions to be two *different expressions of the same solution*.

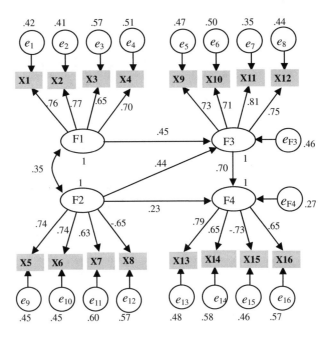

Fig. 11.5 Standardized solution

11.8   Model Selection

As in path analysis and confirmatory factor analysis, several models, including two extreme (independent and saturated) models, should be compared in SEM. For the comparison, information criteria such as the AIC and BIC are useful for selecting a good model, although the GFI cannot be used.

An example of SEM models differing from the model in Fig. 11.3 is shown in Fig. 11.5, where the path connecting F2 and F4 in Fig. 11.3 has been deleted. In Table 11.3, the AIC and BIC for the model in Fig. 11.3 are found to be the least, which shows that model to be the best among the four considered.

Table 11.3 Number of parameters (NP) and the resulting index values for each model

Model	NP	GFI	AIC	BIC
Saturated	136	1.000	272.000	775.714
Figure 11.3	37	0.960	175.052	312.092
Figure 11.6	36	0.957	183.703	317.039
Independent	16	0.332	2034.899	2036.828

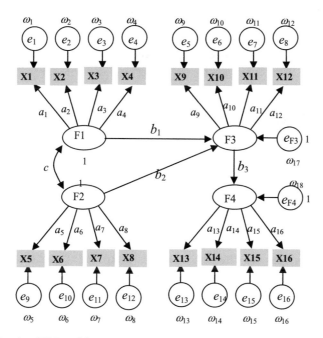

Fig. 11.6 Another SEM model

11.9   Bibliographical Notes

The books in which SEM is exhaustively detailed include Bollen (1989), Kaplan (2000), and Wang and Wang (2012). SEM is also illustrated in a chapter of Lattin et al. (2003). The formulation of SEM in this chapter is based on Toyoda (1988), which is a very excellent book, but written in Japanese.

Exercises

11.1. Present an example of a set of the variables (V1–V5) and factors (F1 and F2) whose relationships are represented as the following path diagram:

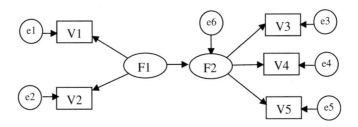

11.2. The above diagram can be changed into the one for CFA by changing a few parts. Show those changes.
11.3. Present another SEM model for the covariance matrix in Table 11.2.
11.4. Describe what is implied by removing the double-headed path between F1 and F2 from Fig. 11.3.
11.5. Show that the structural equation model describing the causal relationships among factors can be expressed as

$$\mathbf{f}_D = \mathbf{C}\mathbf{f}_D + \mathbf{Q}\mathbf{f}_E + \mathbf{e}_D. \tag{11.17}$$

Here, $\mathbf{f}_D$ ($m_D \times 1$) contains the factors as dependent variables, $\mathbf{f}_E$ ($m_E \times 1$) contains the factors as explanatory variables, and $\mathbf{e}_D$ consists of errors, as defined in Sect. 11.4, with $\mathbf{C}$ and $\mathbf{Q}$ path coefficient matrices

11.6. Show the elements of $\mathbf{C}$ and $\mathbf{Q}$ in (11.17) corresponding to the diagram in Fig. 11.1
11.7. Show that the measurement equations describing the relationships of the factor vectors $\mathbf{f}_D$ and $\mathbf{f}_E$ in (11.17) to observed variables can be expressed as

$$\mathbf{y} = \mathbf{A}_Y \mathbf{f}_D + \mathbf{e}_Y, \tag{11.18}$$

$$\mathbf{z} = \mathbf{A}_Z\mathbf{f}_E + \mathbf{e}_Z. \tag{11.19}$$

Here, $\mathbf{y}$ ($p_D \times 1$) and $\mathbf{z}$ ($p_E \times 1$) are the observed variable vectors corresponding to $\mathbf{f}_D$ and $\mathbf{f}_E$, respectively; $\mathbf{y}$, $\mathbf{z}$, $\mathbf{e}_Y$, and $\mathbf{e}_Z$ form the $p \times 1$ vectors $\mathbf{x} = [\mathbf{z}', \mathbf{y}']$ and $\mathbf{e}_X = [\mathbf{e}_Z, \mathbf{e}_Y]$ in (11.5) and (11.6), respectively, with $p = p_D + p_E$; and $\mathbf{A}_Y$ and $\mathbf{A}_Z$ are loading matrices.

11.8. Show that model (11.4) is equivalent to a set of (11.17), (11.18), and (11.19).

11.9. If $\mathbf{f}_E \sim N_{m_E}(\mathbf{0}_{m_E}, \boldsymbol{\Phi})$, $\mathbf{e}_D \sim N_{m_D}(\mathbf{0}_{m_D}, \boldsymbol{\Psi})$, $\mathbf{e}_Y \sim N_{p_D}(\mathbf{0}_{p_D}, \boldsymbol{\Theta}_Y)$, and $\mathbf{e}_Z \sim N_{p_E}(\mathbf{0}_{p_E}, \boldsymbol{\Theta}_Z)$, show that we can use (11.17), (11.18), and (11.19) to rewrite covariance structure (11.15) as a block matrix $\begin{bmatrix} \Sigma_{ZZ} & \Sigma_{ZY} \\ \Sigma'_{ZY} & \Sigma_{YY} \end{bmatrix}$, with

$$\Sigma_{ZZ} = \mathbf{A}_Z\boldsymbol{\Phi}\mathbf{A}_{Z'} + \boldsymbol{\Theta}_Z,$$

$$\Sigma_{ZY} = \mathbf{A}_Z\boldsymbol{\Phi}\mathbf{Q}'(\mathbf{I}_{m_D} - \mathbf{C})^{'-1}\mathbf{A}_{Y'},$$

$$\Sigma_{YY} = \mathbf{A}_Y(\mathbf{I}_{m_D} - \mathbf{C})^{-1}(\mathbf{Q}\boldsymbol{\Phi}\mathbf{Q}' + \boldsymbol{\Psi})(\mathbf{I}_{m_D} - \mathbf{C})^{'-1}\mathbf{A}'_Y + \boldsymbol{\Theta}_Y.$$

Block matrices are detailed in Sect. 14.1.

Chapter 12
Exploratory Factor Analysis

As described in Chap. 10, factor analysis (FA) is classified into *exploratory FA* (*EFA*) and confirmatory FA (CFA). EFA refers to the procedures for exploring factors underlying observed variables for cases without prior knowledge of what factors explain the variables. EFA is introduced in this chapter. Two features of *EFA* are that [1] *all factors* are assumed to be linked to *all variables*, and [2] *multiple solutions* exist for a data set.

The FA model conceived by Spearman (1904), the originator of FA, was restricted to one factor. In the single-factor case, CFA is not distinguished from EFA, as only that model can be considered in which the factor is linked to all variables. Spearman's single-factor FA was extended to FA with *multiple factors* by Thurstone (1935, 1947). Then, he chose the EFA approach with all factors linked to all variables. That was the origin of EFA.

12.1 Example of Exploratory Factor Analysis Model

We use the same data set as that in Chap. 10, the 100 participants × 8 behavioral features data matrix in Table 10.1(A). It contains the self-ratings regarding what extent participants' behaviors are characterized by eight variables (features): **A** (Aggressive), **C** (Cheerful), **I** (Initiative), **B** (Blunt), **T** (Talkative), **V** (Vigor), **H** (tendency to Hesitate), and **P** (being Popular).

Let us suppose that two factors underlie the eight variables, though it is unknown what variables are related to each factor. Thus, those links are considered which connect all variables to all factors, as illustrated in Fig. 12.1. This is a key point in EFA. The model in Fig. 12.1 can be written as the set of eight equations:

© Springer Nature Singapore Pte Ltd. 2016
K. Adachi, *Matrix-Based Introduction to Multivariate Data Analysis*,
DOI 10.1007/978-981-10-2341-5_12

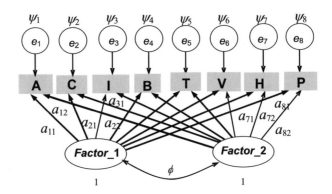

Fig. 12.1 EFA model with two factors for personality data

$$A = a_{11} \times Factor\_1 + a_{12} \times Factor\_2 + c_1 + e_1$$
$$C = a_{21} \times Factor\_1 + a_{22} \times Factor\_2 + c_2 + e_2$$
$$I = a_{31} \times Factor\_1 + a_{32} \times Factor\_2 + c_3 + e_3$$
$$B = a_{41} \times Factor\_1 + a_{42} \times Factor\_2 + c_4 + e_4$$
$$T = a_{51} \times Factor\_1 + a_{52} \times Factor\_2 + c_5 + e_5 \qquad (12.1)$$
$$V = a_{61} \times Factor\_1 + a_{62} \times Factor\_2 + c_6 + e_6$$
$$H = a_{71} \times Factor\_1 + a_{72} \times Factor\_2 + c_7 + e_7$$
$$P = a_{81} \times Factor\_1 + a_{82} \times Factor\_2 + c_8 + e_8.$$

Here, c_j and e_j ($j = 1, \ldots, 8$) express an intercept and an error, respectively; the first subscript j and the second k in a_{jk} indicate a variable and a factor, respectively. The path coefficients a_{jk} are also called factor loadings.

In Fig. 12.1, we can find that each *factor* is a *common cause* for variables, while each *error* is a *cause* for *a single* variable. For this reason, a factor is called a *common factor*, while an error is also called a *unique factor* (a factor *uniquely* influencing a single variable) with its variance called a *unique variance*, as already mentioned in Chap. 10.

12.2 Matrix Expression

Table 10.1(B) shows the centered scores of the raw data in (A). EFA for (A) and that for (B), on the assumption of the averages of factors being zeros, produce the same solution except for the resulting *intercepts* being *zero* in the *latter* analysis. We thus *omit the intercepts* in EFA models, for the sake of simplicity, by supposing that a data matrix to be analyzed contains *centered scores*. Model (12.1) without intercepts can be expressed in matrix form:

$$
\begin{array}{c}
\mathbf{x} \\
{\scriptstyle 8\times 1}
\end{array}
\begin{bmatrix}
A \\ C \\ I \\ B \\ T \\ V \\ H \\ P
\end{bmatrix}
=
\begin{array}{c}
\mathbf{A} \\
{\scriptstyle 8\times 2}
\end{array}
\begin{bmatrix}
a_{11} & a_{12} \\
a_{21} & a_{22} \\
a_{31} & a_{32} \\
a_{41} & a_{42} \\
a_{51} & a_{52} \\
a_{61} & a_{62} \\
a_{71} & a_{72} \\
a_{81} & a_{82}
\end{bmatrix}
\begin{array}{c}
\mathbf{f} \\
{\scriptstyle 2\times 1}
\end{array}
\begin{bmatrix}
Factor\_1 \\
Factor\_2
\end{bmatrix}
+
\begin{array}{c}
\mathbf{e} \\
{\scriptstyle 8\times 1}
\end{array}
\begin{bmatrix}
e_1 \\ e_2 \\ e_3 \\ e_4 \\ e_5 \\ e_6 \\ e_7 \\ e_8
\end{bmatrix}
\tag{12.2}
$$

In any EFA model, a $p \times 1$ random variable vector $\mathbf{x}$ is expressed as:

$$
\mathbf{x} = \mathbf{A}\mathbf{f} + \mathbf{e}, \tag{12.3}
$$

where $\mathbf{A} = (a_{jk})$ is the p-variables $\times$ m-factors matrix containing *factor loadings*, $\mathbf{f}$ is an $m \times 1$ vector whose elements are called *factor scores*, and $\mathbf{e}$ contains errors. This is the same as the CFA model in Chap. 10 except that $\mathbf{A}$ is unconstrained in EFA.

12.3   Distributional Assumptions

The error vector $\mathbf{e}$ is assumed to be distributed according to the multivariate normal (MVN) distribution whose average vector and covariance matrix are $\mathbf{0}_p$ and $\mathbf{\Psi}$, respectively:

$$
\mathbf{e} \sim N_p(\mathbf{0}_p, \mathbf{\Psi}), \tag{12.4}
$$

with $\mathbf{\Psi}$ the diagonal matrix including *unique variances*, i.e.,

$$
\mathbf{\Psi} =
\begin{bmatrix}
\psi_1 & & & \\
 & \psi_2 & & \\
 & & \ddots & \\
 & & & \psi_8
\end{bmatrix}. \tag{12.5}
$$

The factor vector $\mathbf{f}$ is supposed to be distributed according to the MVN distribution whose average vector and covariance matrix are $\mathbf{0}_m$ and $\mathbf{I}_m$, respectively:

$$
\mathbf{f} \sim N_m(\mathbf{0}_m, \mathbf{I}_m). \tag{12.6}
$$

This differs from assumption (10.4) for CFA in Chap. 10 in that the covariance matrix is the identity matrix. However, this can be transformed into a *factor correlation* matrix $\mathbf{\Phi}$, as in (10.4), for the reason described in Sect. 12.5.

Because of (10.8), assumptions (12.4) and (12.6) imply that the observed variable vector $\mathbf{x}$ is distributed according to the following MVN distribution:

$$\mathbf{x} \sim N_p(\mathbf{0}_p, \Sigma), \tag{12.7}$$

with its covariance matrix

$$\Sigma = \mathbf{A}\mathbf{A}' + \mathbf{\Psi}. \tag{12.8}$$

12.4   Maximum Likelihood Method

Let $\mathbf{X}$ denote the centered data matrix in Table 10.1(B) and $\mathbf{V} = n^{-1}\mathbf{X}'\mathbf{X}$ be the sample covariance matrix. As explained in Sect. 9.4, the *log likelihood* is written in the form of (9.15). By substituting (12.8) into (9.15), we have the following:

$$l(\mathbf{A}, \mathbf{\Psi}) = \frac{n}{2}\log\left|(\mathbf{A}\mathbf{A}' + \mathbf{\Psi})^{-1}\mathbf{V}\right| - \frac{n}{2}\mathrm{tr}(\mathbf{A}\mathbf{A}' + \mathbf{\Psi})^{-1}\mathbf{V}. \tag{12.9}$$

This is maximized over $\mathbf{A}$ and $\mathbf{\Psi}$. Since the solution is not explicitly given, the maximization is attained by iterative algorithms. One of the algorithms is a gradient algorithm, which is illustrated in Appendix A.6.3. Another one is called an *EM algorithm* (Dempster et al. 1977). The principle for this algorithm is omitted here, but its formulas are shown next:

Note 12.1. EM algorithm for EFA
Let $\mathbf{A}_{[t]}$ and $\mathbf{\Psi}_{[t]}$ be $\mathbf{A}$ and $\mathbf{\Psi}$ at the t-th iteration, respectively. The monotonic increase of the log likelihood (12.9) with iteration, $l(\mathbf{A}_{[t+1]}, \mathbf{\Psi}_{[t+1]}) \geq l(\mathbf{A}_{[t]}, \mathbf{\Psi}_{[t]})$, is guaranteed if $\mathbf{A}_{[t]}$ and $\mathbf{\Psi}_{[t]}$ are updated to $\mathbf{A}_{[t+1]}$ and $\mathbf{\Psi}_{[t+1]}$ with

$$\mathbf{A}_{[t+1]} = \mathbf{C}_{[t]}\mathbf{S}_{[t]}^{-1} \quad \text{and} \quad \mathbf{\Psi}_{[t+1]} = \mathrm{diag}\left\{\mathbf{V} - \mathbf{C}_{[t]}\mathbf{S}_{[t]}^{-1}\mathbf{C}_{[t]}'\right\}.$$

Here, $\mathrm{diag}\{\mathbf{M}\}$ denotes the diagonal matrix whose diagonal elements are those of square $\mathbf{M}$,

$$\mathbf{C}_{[t]} = \mathbf{V}\mathbf{Q}_{[t]}' \quad \text{and} \quad \mathbf{S}_{[t]} = \mathbf{Q}_{[t]}\mathbf{V}\mathbf{Q}_{[t]}' + \mathbf{I}_p - \mathbf{Q}_{[t]}\mathbf{A}_{[t]},$$

with

$$\mathbf{Q}_{[t]} = \mathbf{A}_{[t]}'(\mathbf{A}_{[t]}\mathbf{A}_{[t]}' + \mathbf{\Psi}_{[t]})^{-1}.$$

The algorithm is detailed in Rubin and Thayer (1982) and its properties are discussed in Adachi (2013).

We express the resulting $\mathbf{A}$ and $\mathbf{\Psi}$ as $\hat{\mathbf{A}}$ and $\hat{\mathbf{\Psi}}$, respectively.

12.5 Indeterminacy of EFA Solutions

An important property of EFA is that its solution of $\mathbf{A}$ is *not unique* (i.e., is not single). This is true because the FA model (12.3) can be rewritten as:

$$\mathbf{x} = \mathbf{Af} + \mathbf{e} = \mathbf{ATT'f} + \mathbf{e} = \mathbf{A_T f_T} + \mathbf{e}. \tag{12.10}$$

Here,

$$\mathbf{A_T} = \mathbf{AT} \quad \text{and} \quad \mathbf{f_T} = \mathbf{T'f}, \tag{12.11}$$

with $\mathbf{T}$ an $m \times m$ matrix satisfying

$$\mathbf{T'T} = \mathbf{TT'} = \mathbf{I}_m. \tag{12.12}$$

This $\mathbf{T}$ is called an *orthonormal* matrix, which is detailed in Appendix A.1.2. Because of (9.10) and (12.12), (12.6) leads to

$$\mathbf{f_T} = \mathbf{T'f} \sim N_m(\mathbf{0}_m, \mathbf{I}_m) \tag{12.13}$$

where $\mathbf{f_T}$ follows the same distribution as that for $\mathbf{f}$. That is, (12.11) satisfies the assumptions of EFA, which implies that $\hat{\mathbf{A}}_T = \hat{\mathbf{A}}\mathbf{T}$ is also the solution of $\mathbf{A}$ if $\hat{\mathbf{A}}$ is the solution.

We can also *relax* the condition (12.12) for $\mathbf{T}$ as:

$$\mathbf{T'T} = \mathbf{\Phi} = \begin{bmatrix} 1 & & \# \\ & \ddots & \\ \# & & 1 \end{bmatrix}, \tag{12.14}$$

where the right-hand side stands for the diagonal elements of $\mathbf{\Phi}$ being restricted to one, but its off-diagonal elements are unconstrained. Then, the EFA model (12.3) can be rewritten as:

$$\mathbf{x} = \mathbf{Af} + \mathbf{e} = \mathbf{AT'^{-1}T'f} + \mathbf{e} = \mathbf{A_T f_T} + \mathbf{e}. \tag{12.15}$$

with

$$\mathbf{A_T} = \mathbf{AT'^{-1}} \quad \text{and} \quad \mathbf{f_T} = \mathbf{T'f}. \tag{12.16}$$

Because of (9.10), (12.6) and (12.14) imply

$$\mathbf{f_T} = \mathbf{T'f} \sim N_m(\mathbf{0}_m, \mathbf{\Phi}). \tag{12.17}$$

Though it differs from (12.6), (12.17) is a reasonable assumption, if factors are assumed to be correlated, since (12.14) implies that $\mathbf{\Phi}$ is a *factor correlation* matrix with its diagonal elements ones. This shows that $\hat{A}T'^{-1}$ is also the solution of A with (12.14) providing the corresponding factor correlation matrix.

As discussed above, the EFA solution of A is not unique. But, the solution of the diagonal matrix Ψ is *uniquely determined*; the solution of Ψ is single.

12.6 Two-Stage Procedure

As described in the last section, if $\hat{A}$ is the solution of A, $\hat{A}T$ is that with (12.12), and, further, $\hat{A}T'^{-1}$ is also a solution with (12.14). Thus, EFA involves the following *two-stage* procedure:

Stage 1. A set of solutions for A and Ψ, i.e., $\hat{A}$ and $\hat{\Psi}$, is obtained.
Stage 2. A suitable T is found to have a solution $\hat{A}T$ with (12.12) or $\hat{A}T'^{-1}$ (and $\Phi = T'T$) with (12.14).

Indeed, the procedure in Sect. 12.4 corresponds to Stage 1. On the other hand, the procedure in Stage 2, which is called "*rotation*," is not treated in this chapter, but is detailed in the next chapter. In the next two sections, we illustrate the interpretation of the solution after Stage 2.

12.7 Interpretation of Loadings

The EFA solutions obtained with the above procedures also have *scale invariance*, though its proof is omitted here. Thus, the *unstandardized* and *standardized* solutions of EFA can be viewed as *two expressions of the same solution*. In this chapter, only the standardized one is shown. For the data in Table 10.1(B), Stage 1 in the last section, the EFA procedure with $m = 2$ in Sect. 12.4, provides the solution in Table 12.1(A), where the inter-factor correlation is found to be zero, as shown in (12.6). For this solution, Stage 2 in the last section provides the result in Table 12.1(B), where a rotation technique called "*oblique geomin rotation*," which will be detailed in Sect. 13.5, has been used for finding T.

Let us consider interpreting what each factor in Table 12.1(B) stands for. For facilitating the interpretation, bold font is used for the loadings of large absolute values in Table 12.1(B). Further, the absolute values can be visually captured in Fig. 12.2, where the widths of paths are proportional to the absolute values of the corresponding loadings and their signs are distinguished by solid and dotted lines. In the figure, we can find that the variables A, I, and V load *Factor_1* heavily and positively, while H loads that factor negatively; i.e., the higher *Factor_1* leads to less H (tendency to hesitate). This result allows us to call *Factor_1* "*the factor of activity*." On the other hand, the variables C, T, and P load *Factor_2* heavily and positively, while V loads that factor negatively. This allows us to interpret *Factor_2*

Table 12.1 Loadings, unique variances, and factor correlation obtained for the data set in Table 11.1

j	(A) Before rotation $\hat{\mathbf{A}}$		ψ_j	(B) After rotation $\mathbf{A}_T$		ψ_j
A	0.77	−0.38	0.26	**0.82**	0.08	0.26
C	0.61	0.50	0.38	−0.13	**0.84**	0.38
I	0.67	−0.36	0.41	**0.74**	0.04	0.41
B	−0.74	−0.40	0.30	−0.04	**−0.82**	0.30
T	0.79	0.43	0.18	0.04	**0.88**	0.18
V	0.76	−0.44	0.22	**0.87**	0.01	0.22
H	−0.63	0.46	0.39	**−0.82**	0.08	0.39
P	0.70	0.18	0.47	0.23	**0.58**	0.47
φ_{12}	0.00			0.48		

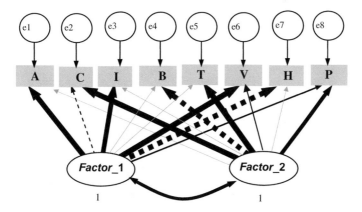

Fig. 12.2 Path diagram in which the widths of paths are proportional to the absolute values of loadings, with *solid* and *dotted lines* indicating positive and negative loadings, respectively

as "*the factor of sociability*." Table 12.1(B) shows that the correlation between those two factors is 0.48, which implies the factor of activity is *positively correlated* with the sociability factor.

12.8 Interpretation of Unique Variances

Unique variances are *uniquely determined*, as found in Table 12.1, as their values are equivalent between (A) and (B). The table shows that the unique variance for A (aggressive) is 0.26. This implies that 26 % of the variance of variable A remains unexplained by the two common factors; in other words, 74 % (= [1 − 0.26] × 100 %) of the activities of individuals are accounted for by the two common factors.

The proportion (one minus a unique variance in the standardized solution) is called communality. It makes sense to compare the largeness of unique variances among variables in Table 12.1, since the solution is standardized. The largest is that of P (popular) (0.47). It is least explained by the factors; in other words, P is characterized by a feature unique to that variable beside the two common factors.

12.9   Selecting the Number of Factors

When EFA is used, the suitable number of factors (m) is often unknown for a data set. In order to select m, *information criteria* such as AIC and BIC can be used. For a data set, we can carry out EFA with m set at some candidate numbers, so as to choose the solution with the least AIC or BIC as the solution with the suitable m. In EFA, the number of parameters η used for obtaining AIC and BIC is given by

$$\eta = p + pm - \frac{m(m-1)}{2}. \tag{12.18}$$

Here, $p + pm$ is the number of unique variances and loadings, from which m $(m-1)/2$ must be *subtracted* for the reason described next:

Note 12.2. Loadings Set at Zero

EFA loadings have indeterminacy shown by (12.10) to (12.13). It is known that an orthonormal matrix $\mathbf{T}_0$ exists that leads to

$$\mathbf{A}_0 = \hat{\mathbf{A}}\mathbf{T}_0, \tag{12.19}$$

where the elements of $\mathbf{A}_0 = \left(a_{ij}^{[0]}\right)$ satisfy

$$a_{ij}^{[0]} = 0 \quad \text{for} \quad i \leq m-1 \quad \text{and} \quad i \leq j, \tag{12.20}$$

implying $m(m-1)/2$ elements are zero. This is illustrated in Table 12.2. The post-multiplication of the left $\hat{\mathbf{A}}$ by $\mathbf{T}_0 = \begin{bmatrix} 0.47 & 0.20 & 0.86 \\ -0.40 & 0.92 & 0.00 \\ -0.79 & -0.35 & 0.51 \end{bmatrix}$ leads to the right $\mathbf{A}_0 = \hat{\mathbf{A}}\mathbf{T}_0$ whose upper left $3(3-1)/2$ elements are zero.

The above fact implies the following: "If $\hat{\mathbf{A}}$ is a solution, then $\mathbf{A}_0$ is also so." This can be rewritten as "If $\mathbf{A}_0$ is a solution, $\hat{\mathbf{A}} = \mathbf{A}_0\mathbf{T}_0'$ is so," which follows from (12.19), leading to $\mathbf{A}_0\mathbf{T}_0' = \hat{\mathbf{A}}\mathbf{T}_0\mathbf{T}_0'$, with $\mathbf{T}_0\mathbf{T}_0' = \mathbf{I}$. Further, $\hat{\mathbf{A}} = \mathbf{A}_0\mathbf{T}_0'$ and (12.11) imply that $\mathbf{A}_T = \hat{\mathbf{A}}\mathbf{T} = \mathbf{A}_0\mathbf{T}_0'\mathbf{T}$ is a solution, where $\mathbf{T}_0'\mathbf{T}$ is orthonormal: $(\mathbf{T}_0'\mathbf{T})'\mathbf{T}_0'\mathbf{T} = \mathbf{T}_0'\mathbf{T}(\mathbf{T}_0'\mathbf{T})' = \mathbf{I}_m$. That is, if the

Table 12.2 Two solutions of loading matrices for the correlation matrix in Table 12.4

$\hat{\mathbf{A}}$: maximum likelihood estimate			$\mathbf{A}_0 = \hat{\mathbf{A}}\,\mathbf{T}_0'$ with 3 zero elements		
0.60961	0.00029	0.36018	0.00000	0.00000	0.70806
0.38404	−0.03143	0.24305	0.00000	−0.03454	0.45426
0.42288	−0.13882	0.37412	−0.04198	−0.17035	0.55433
0.49556	−0.09901	0.22700	0.09179	−0.06804	0.54208
0.68417	−0.28993	−0.31198	0.68124	−0.01740	0.43022
0.67711	−0.40867	−0.24367	0.67191	−0.15122	0.45884
0.66746	−0.38456	−0.32760	0.72385	−0.10202	0.40785
0.67352	−0.17645	−0.12146	0.48046	0.01826	0.51801
0.68557	−0.45757	−0.25595	0.70521	−0.19000	0.45986
0.48125	0.58511	−0.46732	0.35695	0.79602	0.17686
0.55191	0.30992	−0.12420	0.23023	0.43959	0.41212
0.48094	0.51620	−0.10070	0.09563	0.60577	0.36305
0.60801	0.24713	0.03432	0.15671	0.33861	0.54103
0.40242	0.00397	0.00947	0.17839	0.08261	0.35128
0.37857	0.03413	0.08833	0.09299	0.07800	0.37088
0.50229	0.03928	0.32167	−0.03533	0.02711	0.59609
0.44799	0.13903	0.00273	0.15057	0.21790	0.38714
0.51477	0.26950	0.16531	0.00104	0.29462	0.52739
0.44212	0.06012	0.11279	0.09286	0.10630	0.43804
0.61658	−0.13514	0.11094	0.25417	−0.03614	0.58722
0.60193	0.23149	0.06913	0.13274	0.31099	0.55349
0.61152	−0.12704	0.11541	0.24503	−0.03131	0.58515
0.69611	−0.05113	0.12739	0.24444	0.05132	0.66410
0.65299	0.18337	−0.20214	0.38966	0.37142	0.45944

Table 12.3 AIC and BIC values as functions of m for the data in Table 12.4 with the number of parameters

m	1	2	3	4	5	6	7
η	48	71	93	114	134	153	171
AIC	4247.5	4077.3	3987.9	3956.1	3968.3	3962.9	3972.4
BIC	4390.4	4288.6	4264.7	4295.4	4367.2	4418.3	4481.5

$\{pm - m(m-1)/2\}$ nonzero elements in $\mathbf{A}_0$ are estimated, we can obtain any solution of the loading matrix for a data set. This leads to (12.18).

The AIC and BIC values for some m are shown in Table 12.3, which were obtained by the EFA solutions for the correlation matrix in Table 12.4. This is a famous data set known as the 24 psychological test data (Holzinger and Swineford 1939).

The least AIC and BIC are found for $m = 4$ and $m = 3$, which suggest that the best number of factors is 3 or 4. The loading matrices with $m = 3$ in Table 12.2 have been obtained by EFA for the correlations in Table 12.4.

12.10   Difference to Principal Component Analysis

For an n-individuals $\times$ p-variables data matrix $\mathbf{X}$, the EFA model (12.3) can be rewritten in matrix form as:

$$\mathbf{X} = \mathbf{FA}' + \mathbf{E}. \tag{12.21}$$

Here, $\mathbf{F}$ is an n-individuals $\times$ m-factors matrix and $\mathbf{E}$ is an $n \times p$ matrix of errors whose ith rows are the vectors $\mathbf{f}'$ and $\mathbf{e}'$ for individual i, respectively. We find that (12.21) is same as model (5.1) for *principal component analysis* (*PCA*). This begs the question "In what points does EFA differ from PCA?" One might answer that EFA is a maximum likelihood (ML) procedure, while PCA is a least squares (LS) one. But, this is incorrect, since EFA can be formulated as an LS procedure (Harman 1976; Mulaik 2011), and PCA can be formulated as an ML procedure (Bishop 2006; Tipping and Bishop 1999).

A crucial difference between EFA and PCA is found in the *errors*. *No assumption* is made for $\mathbf{E}$ in *PCA*. Thus, it can be formulated simply as minimizing (5.4), i.e., $\|\mathbf{E}\|^2 = \|\mathbf{X} - \mathbf{FA}'\|^2$. In contrast, the covariance matrix for errors in EFA is constrained to be a *diagonal matrix* $\mathbf{\Psi}$, as in (12.4). That is, the error for a variable is assumed to be *uncorrelated* with those for the other variables. Thus, errors are called *unique factors*, and its variances (i.e., the diagonal elements of $\mathbf{\Psi}$) are called *unique variances* in *EFA*. On the other hand, the errors in *PCA* are not unique to variables; the *correlations* are found among the columns of the resulting $\mathbf{E} = \mathbf{X} - \mathbf{FA}'$.

Table 12.5 shows the EFA and PCA solutions for the data set in Table 12.4 with $m = 3$. Here, the PCA solution has been given by (5.28), which can be obtained if only a covariance or correlation matrix is available, as found in Note 6.1. The varimax rotation has been performed for the EFA and PCA loading matrices. The PCA loading matrix can be rotated if constraint (5.26) is removed, as explained in Note 5.5. In Table 12.4, var($\mathbf{e}_j$) for EFA expresses the diagonal elements of $\mathbf{\Psi}$, while var($\mathbf{e}_j$) for PCA are those of the resulting $n^{-1}\mathbf{E}'\mathbf{E}$. There, we can find the similarity between EFA and PCA solutions. The difference is the interpretation for errors. For example, var($\mathbf{e}_1$) = 0.5 for EFA can be interpreted as 50 % of the variance in "visual perception" being explained by the corresponding unique factor, but var($\mathbf{e}_1$) = 0.44 for PCA *cannot be interpreted so*.

Table 12.4 Correlation matrix for 24 psychological test data

	1	2	3	4	5	6	7	8	9	10	11	12	13	14	15	16	17	18	19	20	21	22	23	24
1	1.000	0.318	0.403	0.468	0.321	0.335	0.304	0.332	0.326	0.116	0.308	0.314	0.489	0.125	0.238	0.414	0.176	0.368	0.270	0.365	0.369	0.413	0.474	0.282
2	0.318	1.000	0.317	0.230	0.285	0.234	0.157	0.157	0.195	0.057	0.150	0.145	0.239	0.103	0.131	0.272	0.005	0.255	0.112	0.292	0.306	0.232	0.348	0.211
3	0.403	0.317	1.000	0.305	0.247	0.268	0.223	0.382	0.184	-0.075	0.091	0.140	0.321	0.177	0.065	0.263	0.177	0.211	0.312	0.297	0.165	0.250	0.383	0.203
4	0.468	0.230	0.305	1.000	0.227	0.327	0.335	0.391	0.325	0.099	0.110	0.160	0.327	0.066	0.127	0.322	0.187	0.251	0.137	0.339	0.349	0.380	0.335	0.248
5	0.321	0.285	0.247	0.227	1.000	0.622	0.656	0.578	0.723	0.311	0.344	0.215	0.344	0.280	0.229	0.187	0.208	0.263	0.190	0.398	0.318	0.441	0.435	0.420
6	0.335	0.234	0.268	0.327	0.622	1.000	0.722	0.527	0.714	0.203	0.353	0.095	0.309	0.292	0.251	0.291	0.273	0.167	0.251	0.435	0.263	0.386	0.431	0.433
7	0.304	0.157	0.223	0.335	0.656	0.722	1.000	0.619	0.685	0.246	0.232	0.181	0.345	0.236	0.172	0.180	0.228	0.159	0.226	0.451	0.314	0.396	0.405	0.437
8	0.332	0.157	0.382	0.391	0.578	0.527	0.619	1.000	0.532	0.170	0.280	0.271	0.395	0.252	0.175	0.296	0.255	0.250	0.274	0.427	0.362	0.357	0.501	0.388
9	0.326	0.195	0.184	0.325	0.723	0.714	0.685	0.532	1.000	0.170	0.280	0.113	0.280	0.260	0.248	0.242	0.274	0.208	0.274	0.446	0.266	0.483	0.504	0.424
10	0.116	0.057	-0.075	0.099	0.311	0.203	0.246	0.170	0.170	1.000	0.484	0.484	0.408	0.172	0.154	0.124	0.289	0.317	0.190	0.173	0.405	0.160	0.262	0.531
11	0.308	0.150	0.091	0.110	0.344	0.353	0.232	0.280	0.280	0.484	1.000	0.428	0.535	0.350	0.240	0.314	0.362	0.350	0.290	0.202	0.399	0.304	0.251	0.412
12	0.314	0.145	0.140	0.160	0.215	0.095	0.181	0.271	0.113	0.484	0.428	1.000	0.512	0.131	0.173	0.119	0.278	0.349	0.110	0.246	0.355	0.193	0.350	0.414
13	0.489	0.239	0.321	0.327	0.344	0.309	0.345	0.395	0.280	0.408	0.535	0.512	1.000	0.195	0.139	0.281	0.194	0.323	0.263	0.241	0.425	0.279	0.382	0.358
14	0.125	0.103	0.177	0.066	0.280	0.292	0.236	0.252	0.260	0.172	0.350	0.131	0.195	1.000	0.370	0.412	0.341	0.201	0.206	0.302	0.183	0.243	0.242	0.304
15	0.238	0.131	0.065	0.127	0.229	0.251	0.172	0.175	0.248	0.154	0.240	0.173	0.139	0.370	1.000	0.325	0.345	0.334	0.192	0.272	0.232	0.246	0.256	0.165
16	0.414	0.272	0.263	0.322	0.187	0.291	0.180	0.296	0.242	0.124	0.314	0.119	0.281	0.412	0.325	1.000	0.324	0.344	0.258	0.388	0.348	0.283	0.360	0.262
17	0.176	0.005	0.177	0.187	0.208	0.273	0.228	0.255	0.274	0.289	0.362	0.278	0.194	0.341	0.345	0.324	1.000	0.448	0.324	0.262	0.173	0.273	0.287	0.326
18	0.368	0.255	0.211	0.251	0.263	0.167	0.159	0.250	0.208	0.317	0.350	0.349	0.323	0.201	0.334	0.344	0.448	1.000	0.358	0.301	0.357	0.317	0.272	0.405
19	0.270	0.112	0.312	0.137	0.190	0.251	0.226	0.274	0.274	0.190	0.290	0.110	0.263	0.206	0.192	0.258	0.324	0.358	1.000	0.167	0.331	0.342	0.303	0.374
20	0.365	0.292	0.297	0.339	0.398	0.435	0.451	0.427	0.446	0.173	0.202	0.246	0.241	0.302	0.272	0.388	0.262	0.301	0.167	1.000	0.413	0.463	0.509	0.366
21	0.369	0.306	0.165	0.349	0.318	0.263	0.314	0.362	0.266	0.405	0.399	0.355	0.425	0.183	0.232	0.348	0.173	0.357	0.331	0.413	1.000	0.374	0.451	0.448
22	0.413	0.232	0.250	0.380	0.441	0.386	0.396	0.357	0.483	0.160	0.304	0.193	0.279	0.243	0.246	0.283	0.273	0.317	0.342	0.463	0.374	1.000	0.503	0.375
23	0.474	0.348	0.383	0.335	0.435	0.431	0.405	0.501	0.504	0.262	0.251	0.350	0.382	0.242	0.256	0.360	0.287	0.272	0.303	0.509	0.451	0.503	1.000	0.434
24	0.282	0.211	0.203	0.248	0.420	0.433	0.437	0.388	0.424	0.531	0.412	0.414	0.358	0.304	0.165	0.262	0.326	0.405	0.374	0.366	0.448	0.375	0.434	1.000

12.11   Bibliographical Notes

Various subjects on EFA are exhaustively detailed in Bartholomew et al. (2011), Harman (1976) and Mulaik (2011). Papers reviewing EFA well include Unkel and Trendafilov (2010) and Yanai and Ichikawa (2007).

Exercises

12.1. Present another example of a set of the variables for which EFA is useful.

12.2. Show that EFA model (12.3) with (12.4) can be rewritten as $\mathbf{x} = \mathbf{Af} + \mathbf{\Psi}^{1/2}\mathbf{u}$ with $\mathbf{u} \sim N_p(\mathbf{0}_p, \mathbf{I}_p)$.

12.3. In model (12.3), factor vector $\mathbf{f}$ is regarded as a random variable. In contrast to this, the EFA model also exists in which $\mathbf{f}$ is regarded as a fixed parameter vector. This model is called a *fixed factor model*, which is expressed as:

$$\mathbf{x}_i = \mathbf{Af}_i + \mathbf{e}_i \tag{12.22}$$

by attaching subscript i to $\mathbf{x}$, $\mathbf{f}$, and $\mathbf{e}$ in (12.3) for explicitly showing that they are related to individual i. If $\mathbf{A}$ is given, show that the squared norm of the error for i, $\|\mathbf{e}_i\|^2 = \|\mathbf{x}_i - \mathbf{Af}_i\|^2$, is minimized for $\mathbf{f}_i = (\mathbf{A}'\mathbf{A})^{-1}\mathbf{A}'\mathbf{x}_i$.

12.4. Model (12.22) with $\mathbf{e}_i \sim N_p(\mathbf{0}_m, \mathbf{\Psi})$ implies $\mathbf{x}_i \sim N_p(\mathbf{Af}_i, \mathbf{\Psi})$. Show that it leads to the log likelihood for $\mathbf{X} = \begin{bmatrix} \mathbf{x}'_n \\ \vdots \\ \mathbf{x}'_n \end{bmatrix} = [\mathbf{y}_1, \ldots, \mathbf{y}_p]$ being expressed as:

$$
\begin{aligned}
\log l(\mathbf{F}, \mathbf{A}, \mathbf{\Psi}) &= -\frac{n}{2}\log|\mathbf{\Psi}| - \sum_{i=1}^{n}(\mathbf{x}_i - \mathbf{Af}_i)'\mathbf{\Psi}^{-1}(\mathbf{x}_i - \mathbf{Af}_i) \\
&= -\frac{n}{2}\log|\mathbf{\Psi}| - \frac{1}{2}\mathrm{tr}(\mathbf{X} - \mathbf{FA}')\mathbf{\Psi}^{-1}(\mathbf{X} - \mathbf{FA}')' \quad (12.23) \\
&= -\frac{1}{2}\left(n\sum_{j=1}^{p}\log\psi_j + \sum_{j=1}^{p}\frac{1}{\psi_j}\|\mathbf{y}_j - \mathbf{Fa}_j\|^2 \right),
\end{aligned}
$$

with $\mathbf{F} = [\mathbf{f}_1, \ldots, \mathbf{f}_n]'$, $\mathbf{A} = [\mathbf{a}_1, \ldots, \mathbf{a}_p]'$, ψ_j the jth diagonal element of $\mathbf{\Psi}$, and the term irrelevant to $\mathbf{X}$, $\mathbf{F}$, $\mathbf{A}$, and $\mathbf{\Psi}$ omitted.

12.5. Show that $\mathbf{f}_i = (\mathbf{A}'\mathbf{\Psi}^{-1}\mathbf{A})^{-1}\mathbf{A}'\mathbf{\Psi}^{-1}\mathbf{x}_i$ $(i = 1, \ldots, n)$ maximizes (12.23) for given $\mathbf{A}$ and $\mathbf{\Psi}$, using $\|\mathbf{\Psi}^{-1/2}\mathbf{x}_i - \mathbf{\Psi}^{-1/2}\mathbf{Af}_i\|^2 = (\mathbf{x}_i - \mathbf{Af}_i)'\mathbf{\Psi}^{-1}(\mathbf{x}_i - \mathbf{Af}_i)'$.

12.6. Show that $\psi_j = n^{-1}\|\mathbf{y}_j - \mathbf{Fa}_j\|^2$ maximizes (12.23) for given $\mathbf{F}$ and $\mathbf{A}$, by noting the fact in Exercise 8.1.

12.7. Show that the MLE of $\mathbf{F}$, $\mathbf{A}$, and $\mathbf{\Psi}$ does not exist for (12.23), since it diverges to infinity when $\mathbf{F}$, $\mathbf{A}$, and $\mathbf{\Psi}$ are jointly estimated.

12.8. Let us consider a restrictive version of the model (12.22) with $\mathbf{e}_i \sim N_p(\mathbf{Af}_i, \psi\mathbf{I}_m)$, i.e., the error variance for every variable equaling ψ. Show that its log likelihood for the data matrix $\mathbf{X}$ is expressed as:

$$\log l(\mathbf{F}, \mathbf{A}, \psi) = -\frac{np}{2}\log\psi - \frac{1}{2\psi}\|\mathbf{X} - \mathbf{FA}'\|^2. \tag{12.24}$$

12.9. The maximization of log likelihood (12.24) has been introduced as a maximum likelihood estimation for principal component analysis (PCA) in Bishop (2006, p. 571). Show that maximizing (12.24) over $\mathbf{F}$, $\mathbf{A}$, and ψ is equivalent to minimizing (5.4) over $\mathbf{F}$ and $\mathbf{A}$, i.e., PCA.

12.10. A least squares method for EFA is formulated as minimizing $\|\mathbf{R} - (\mathbf{AA}'+\mathbf{\Psi})\|^2$ over $p \times m$ loading matrix $\mathbf{A} = [\mathbf{a}_1, \ldots, \mathbf{a}_m]'$ and $p \times p$ diagonal matrix

$$\mathbf{\Psi} = \begin{bmatrix} \psi_1 & & \\ & \ddots & \\ & & \psi_p \end{bmatrix}$$ for correlation coefficient matrix $\mathbf{R} = (r_{jk})$. Let $\mathbf{r}_j$ be the

$(p - 1) \times 1$ vector obtained by deleting r_{jj} from the jth column of $\mathbf{R}$. Show that the minimization can be attained by the following algorithm (Harman and Jones 1966):

Step 1. Initialize $\mathbf{A}$.
Step 2. Repeat the update of the ith row of $\mathbf{A}$ by the transpose of $\mathbf{a}_j = (\mathbf{A}_j'\mathbf{A}_j)^{-1}\mathbf{A}_j'\mathbf{r}_j$ for $j = 1, \ldots, m$, where $\mathbf{A}_j$ is the $(p - 1) \times m$ matrix obtained by deleting $\mathbf{a}_j'$ from the current $\mathbf{A}$
Step 3. Set $\psi_j = 1 - \mathbf{a}_j'\mathbf{a}_j$ to finish if convergence is reached; otherwise, go back to Step 2.

12.11. In another least squares method for EFA, which differs from the one in Exercise 12.7, a loss function is defined as:

$$\left\|\mathbf{X} - (\mathbf{FA}' + \mathbf{U\Psi}^{1/2})\right\|^2 = \|\mathbf{X} - \mathbf{ZB}'\|^2, \tag{12.25}$$

for an $n \times p$ centered data matrix $\mathbf{X}$ (Unkel and Trendafilov 2010). Here, $\mathbf{F}$, $\mathbf{A}$, $\mathbf{U}$, and $\mathbf{\Psi}$ are $n \times m$, $p \times m$, $n \times p$, and $p \times p$ matrices, respectively, $\mathbf{\Psi}$ is diagonal, $\mathbf{Z} = [\mathbf{F}, \mathbf{U}]$ is an $n \times (m + p)$ block matrix, and $\mathbf{B} = [\mathbf{A}, \mathbf{\Psi}^{1/2}]$ is a $p \times (m + p)$ block matrix, with $\mathbf{Z}$ constrained as:

$$\frac{1}{n}\mathbf{Z}'\mathbf{Z} = \mathbf{I}_{m+p}. \tag{12.26}$$

Use Theorem A.4.1 in Appendix A.4.1 to show that (12.25) is minimized over $\mathbf{Z}$ for

Table 12.5 Solutions obtained by the varimax rotation for the data in Table 12.4

	EFA				PCA			
	1	2	3	Var (e_j)	1	2	3	Var (e_j)
Visual perception	0.64	0.14	0.26	0.50	0.69	0.17	0.23	0.44
Cubes	0.41	0.06	0.18	0.79	0.55	0.01	0.14	0.67
Paper form board	0.54	−0.06	0.22	0.66	0.64	−0.08	0.20	0.54
Flags	0.46	0.06	0.30	0.69	0.56	0.01	0.30	0.60
General information	0.11	0.22	0.77	0.35	0.10	0.20	0.80	0.30
Paragraph comprehension	0.15	0.09	0.81	0.32	0.16	0.10	0.83	0.28
Sentence completion	0.08	0.14	0.82	0.30	0.07	0.12	0.86	0.24
Word classification	0.27	0.23	0.61	0.50	0.26	0.20	0.67	0.44
Word meaning	0.14	0.06	0.85	0.26	0.14	0.09	0.86	0.23
Addition	−0.06	0.87	0.16	0.21	−0.12	0.82	0.19	0.28
Code	0.24	0.55	0.23	0.58	0.13	0.71	0.22	0.44
Counting dots	0.23	0.67	0.05	0.49	0.15	0.72	0.05	0.45
Straight-curved capitals	0.39	0.47	0.25	0.57	0.36	0.53	0.23	0.54
Word recognition	0.24	0.19	0.27	0.84	0.20	0.32	0.27	0.79
Number recognition	0.29	0.17	0.20	0.85	0.29	0.30	0.16	0.80
Figure recognition	0.55	0.14	0.18	0.64	0.62	0.22	0.13	0.55
Object number	0.27	0.32	0.22	0.78	0.22	0.49	0.18	0.68
Number figure	0.45	0.39	0.12	0.64	0.46	0.52	0.04	0.52
Figure word	0.35	0.21	0.22	0.79	0.36	0.31	0.18	0.74
Deduction	0.43	0.14	0.45	0.59	0.46	0.14	0.48	0.53
Numerical puzzles	0.42	0.44	0.24	0.58	0.42	0.47	0.23	0.55
Problem reasoning	0.43	0.14	0.44	0.60	0.44	0.17	0.47	0.56
Series completion	0.50	0.23	0.45	0.50	0.51	0.23	0.47	0.46
Arithmetic problems	0.22	0.53	0.41	0.50	0.17	0.57	0.43	0.45

$$\mathbf{Z} = \sqrt{n}\mathbf{K}\mathbf{L}' = \sqrt{n}\mathbf{K}_1\mathbf{L}'_1 + \sqrt{n}\mathbf{K}_2\mathbf{L}'_2, \tag{12.27}$$

subject to (12.26) when $\mathbf{B}$ is fixed. Here, $\mathbf{K}_1$ ($n \times p$) and $\mathbf{L}_1$ ($(m + p) \times p$) are obtained from the SVD of $\mathbf{XB}$ defined as $\mathbf{XB} = \mathbf{K}_1\boldsymbol{\Lambda}\mathbf{L}'_1$, while $\mathbf{K}_2$ ($n \times m$) and $\mathbf{L}_2$ ($(m + p) \times m$) form $\mathbf{K} = [\mathbf{K}_1, \mathbf{K}_2]$ and $\mathbf{L} = [\mathbf{L}_1, \mathbf{L}_2]$ satisfying $\mathbf{K}'\mathbf{K} = \mathbf{L}'\mathbf{L} = \mathbf{I}_{m+p}$. A hint is the fact that the minimization of (12.25) is equivalent to the maximization of tr$(\mathbf{XB})'\mathbf{Z}$ under (12.26).

12.12. The least squares EFA in Exercise 12.10 with $\mathbf{R}$ replaced by the corresponding covariance matrix gives the essentially different solution from the one obtained for $\mathbf{R}$ being a correlation matrix. The EFA in Exercise 12.11 with $\mathbf{X}$ replaced by $\mathbf{XD}$ also gives an essentially different solution from the one obtained without the replacement, where $\mathbf{D}$ is a diagonal matrix whose diagonal elements take mutually different positive values. Show that those facts imply that least squares EFA solutions are not scale invariant, though the maximum likelihood EFA described in main text is scale invariant.

12.13. *Independent component analysis (ICA)* refers to a class of procedures, the most general form of whose models can be expressed as $\mathbf{x} = \mathbf{f}(\mathbf{s}) + \mathbf{e}$ (Izenman 2008, p. 558). Here, $\mathbf{x}$ is a $p \times 1$ observed variable vector, $\mathbf{e}$ is an error vector, $\mathbf{s}$ is an $m \times 1$ vector containing unobserved signals originating from m mutually independent sources, and $\mathbf{f}(\mathbf{s})$ is a function of $\mathbf{s}$ providing a $p \times 1$ vector. Discuss the relationships of ICA to EFA.

Part IV
Miscellaneous Procedures

The types of matrices to be analyzed by the procedures in this part differ from those in Parts II and III. The techniques in Chap. 13 are not procedures for analyzing data, but rather for transforming solutions. The data sets to be analyzed by the procedures in Chap. 14 are given as block and categorical data matrices. In Chap. 15, data sets are treated in which individuals are classified into some groups, while data are considered which describe the quasi-distances among objects in Chap. 16.

Chapter 13
Rotation Techniques

In some analysis procedures, the solution for a data set is *not uniquely determined*; multiple solutions exist. An example of such procedures is exploratory factor analysis (EFA). In that procedure, one of the solutions is first found, and then it is transformed into a useful solution that is included in multiple solutions. A family of such transformations is the *rotation* treated in this section. The rotation for EFA solutions in particular is called *factor rotation*, although the rotation can be used for solutions of procedures other than EFA. This chapter starts with illustrating why the term "rotation" is used, before explaining which solutions are useful in Sect. 13.3. This is followed by the introduction of some rotation techniques.

13.1 Geometric Illustration of Factor Rotation

As discussed with (12.16) in Sect. 12.5, when loading matrix $\hat{\mathbf{A}}$ is an EFA solution of a loading matrix, its transformed version,

$$\mathbf{A}_{\mathrm{T}} = \hat{\mathbf{A}} \mathbf{T}'^{-1}, \tag{13.1}$$

is *also a solution*. Here, $\mathbf{T}$ is an $m \times m$ matrix that satisfies (12.14), which is written again here:

$$\mathbf{T}'\mathbf{T} = \begin{bmatrix} 1 & & \# \\ & \ddots & \\ \# & & 1 \end{bmatrix}, \tag{13.2}$$

where # stands for the fact that the off-diagonal elements are unconstrained. In this section, we geometrically illustrate the transformation of $\hat{\mathbf{A}}$ into $\mathbf{A}_{\mathrm{T}} = \hat{\mathbf{A}} \mathbf{T}'^{-1}$, supposing that $\mathbf{T}$ is given.

© Springer Nature Singapore Pte Ltd. 2016
K. Adachi, *Matrix-Based Introduction to Multivariate Data Analysis*,
DOI 10.1007/978-981-10-2341-5_13

Let us use $\mathbf{a}'_j$ for the jth row of the original matrix $\hat{\mathbf{A}}$ and use $\mathbf{a}_j^{(\mathrm{T})\prime}$ for that of the transformed $\mathbf{A}_{\mathrm{T}}$. Then, $\mathbf{A}_{\mathrm{T}} = \hat{\mathbf{A}}\mathbf{T}'^{-1}$ is rewritten as:

$$\mathbf{a}_j^{(\mathrm{T})\prime} = \mathbf{a}'_j\mathbf{T}'^{-1} \quad (j = 1, \ldots, p). \tag{13.3}$$

Post-multiplying both sides of (13.3) by $\mathbf{T}'$ leads to $\mathbf{a}_j^{(\mathrm{T})\prime}\mathbf{T}' = \mathbf{a}'_j$, i.e.,

$$\mathbf{a}'_j = \mathbf{a}_j^{(\mathrm{T})\prime}\mathbf{T}' \quad (j = 1, \ldots, p), \tag{13.4}$$

which shows that the original loading vector $\mathbf{a}'_j$ for variable j is expressed by the post-multiplication of the transformed $\mathbf{a}_j^{(\mathrm{T})\prime}$ by $\mathbf{T}'$. We suppose $m = 2$ and define the columns of $\mathbf{T}$ as:

$$\mathbf{T} = [\mathbf{t}_1, \mathbf{t}_2], \quad \text{with} \quad ||\mathbf{t}_1|| = ||\mathbf{t}_2|| = 1, \tag{13.5}$$

which satisfies (13.2). Using (13.5) and $\mathbf{a}_j^{(\mathrm{T})\prime} = [a_{j1}^{(\mathrm{T})}, a_{j2}^{(\mathrm{T})}]$, (13.4) is rewritten as:

$$\mathbf{a}'_j = a_{j1}^{(\mathrm{T})}\mathbf{t}'_1 + a_{j2}^{(\mathrm{T})}\mathbf{t}'_2. \tag{13.6}$$

It shows that the *original loading vector* for variable j is equal to the *sum* of $\mathbf{t}_k$ ($k = 1, 2$) *multiplied* by the *transformed loadings*. Its geometric implications are illustrated in the next two paragraphs.

In Table 13.1(A), we again show the original loading matrix $\hat{\mathbf{A}}$ in Table 12.1(A) obtained by EFA. Its row vectors $\mathbf{a}'_j$ ($j = 1, \ldots, 8$) corresponding to variables are shown in Fig. 13.1a; the vector $\mathbf{a}'_7$ for H is depicted by the line extending to [−0.63, 0.46], and the other vectors are done in parallel manners. Now, let us consider transforming $\hat{\mathbf{A}}$ into $\mathbf{A}_{\mathrm{T}} = \hat{\mathbf{A}}\mathbf{T}'^{-1}$ by

$$\mathbf{T}'^{-1} = \begin{bmatrix} 1.18 & -0.42 \\ -0.32 & 1.14 \end{bmatrix}, \text{ following from } \mathbf{T} = [\mathbf{t}_1, \mathbf{t}_2] = \begin{bmatrix} 0.94 & 0.26 \\ 0.34 & 0.97 \end{bmatrix}. \tag{13.7}$$

This $\mathbf{T}'^{-1}$ leads to $\mathbf{A}_{\mathrm{T}} = \hat{\mathbf{A}}\mathbf{T}'^{-1}$ in Table 13.1(B). There, we find that the vector for H is $\mathbf{a}_7^{(\mathrm{T})\prime} = \mathbf{a}'_7\mathbf{T}'^{-1} = [-0.89, 0.79]$, transformed from $\mathbf{a}'_7 = [-0.63, 0.46]$ in (A). Those two vectors satisfy the relationship in (13.6):

$$[-0.63, 0.46] = -0.89\mathbf{t}'_1 + 0.79\mathbf{t}'_2, \tag{13.8}$$

with $\mathbf{t}'_1 = [0.94, 0.34]$ and $\mathbf{t}'_2 = [0.26, 0.97]$.

The geometric implication of (13.8), which is an example of (13.6), is illustrated in Fig. 13.1b. There, the axes extending in the directions of $\mathbf{t}'_1 = [0.94, 0.34]$ and $\mathbf{t}'_2 = [0.26, 0.98]$ are depicted, together with the original loading vectors $\mathbf{a}'_1, \ldots, \mathbf{a}'_8$ whose locations are the same as in (A). Let us note that vector $\mathbf{a}'_7$ for H satisfies

Table 13.1 A solution
obtained with EFA
(Table 11.2A) and
an example of its
rotated version

		(A) Before rotation			(B) After rotation		
		$\hat{\mathbf{A}}$		ψ_j	$\mathbf{A}_T$		ψ_j
A		0.77	−0.38	0.26	1.03	−0.76	0.26
C		0.61	0.50	0.38	0.56	0.32	0.38
I		0.67	−0.36	0.41	0.90	−0.69	0.41
B		−0.74	−0.40	0.30	−0.75	−0.15	0.30
T		0.79	0.43	0.18	0.80	0.16	0.18
V		0.76	−0.44	0.22	1.04	−0.82	0.22
H		−0.63	0.46	0.39	−0.89	0.79	0.39
P		0.70	0.18	0.47	0.77	−0.09	0.47
ϕ_{12}		0.00			0.57		

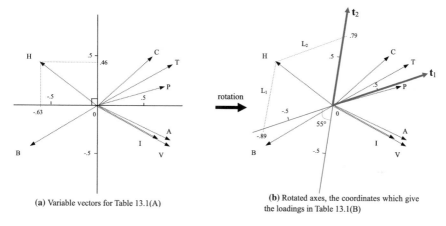

(a) Variable vectors for Table 13.1(A)   (b) Rotated axes, the coordinates which give
                                             the loadings in Table 13.1(B)

Fig. 13.1 Illustration of rotation as that of axes

(13.8), i.e., the −0.89 times of $\mathbf{t}_1'$ plus the 0.79 times of $\mathbf{t}_2$ is equivalent to $\mathbf{a}_7' = [-0.63, 0.64]$. Here, the transformed loadings −0.89 and 0.79 can be viewed as the coordinates of point H on $\mathbf{t}_1'$ and $\mathbf{t}_2'$ axes, as shown by the dotted lines L_1 and L_2 in Fig. 13.1b, where L_1 and L_2 extend in *parallel* to $\mathbf{t}_2$ and $\mathbf{t}_1$, respectively. This relationship holds for the other loading vectors.

In summary, transformation (13.1) implies the rotation of the original horizontal and vertical axes in Fig. 13.1a to the *new axes* extending in the direction of the column vectors of $\mathbf{T}$ as in Fig. 13.1b, where the transformed loadings are the *coordinates* on the new axes. The reason (13.1) is called rotation is found above.

13.2   Oblique and Orthogonal Rotation

Rotation is classified into oblique and orthogonal. The transformation illustrated in the last section is *oblique rotation*, since the new axes are intersected obliquely, as in Fig. 13.1b. On the other hand, *orthogonal rotation* refers to the rotation of *axes* by keeping their *orthogonal intersection*, whose example is described later in Fig. 13.2a. In orthogonal rotation, constraint (13.2) is strengthened so that it is the $m \times m$ identity matrix:

$$\mathbf{T'T} = \mathbf{I}_m. \tag{13.9}$$

The matrix $\mathbf{T}$ satisfying (13.9) is said to be *orthonormal,* and its properties are detailed in Appendix A.1.2. Customarily, the rotation made by orthonormal $\mathbf{T}$ is not called orthonormal rotation, but rather orthogonal rotation. Using (13.9), (13.1) is simplified as

$$\mathbf{A}_T = \hat{\mathbf{A}}\mathbf{T} \tag{13.10}$$

in orthogonal rotation.

In summary, rotation is classified into two types:

[1] *Oblique* rotation (13.1) with $\mathbf{T}$ constrained as (13.2)
[2] *Orthogonal* rotation (13.10) with $\mathbf{T}$ constrained as (13.9)

Orthogonal rotation can be viewed as a special case of oblique rotation in which (13.2) is strengthened as (13.9).

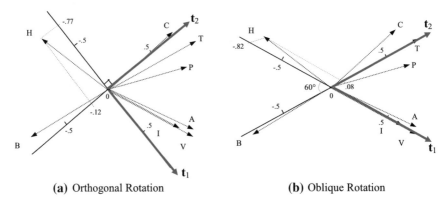

(a) Orthogonal Rotation   (b) Oblique Rotation

Fig. 13.2 Illustrations of rotation to a simple structure

13.3 Rotation to Simple Structure

The transformed loading matrix in Table 13.1(B) is not a useful one. That matrix is merely an example for illustrating rotation. A "good rotation procedure" is one that gives a useful matrix. Here, we have the question: "What matrix is *useful?*" A variety of answers exist; which answer is good varies from case to case.

When a matrix is a variables × factors loading matrix, usefulness can be defined as "*interpretability,*" i.e., being easily interpreted. What matrix is interpretable? An ideal example is shown in Table 13.2(A), where # indicates a nonzero (positive or negative) value. This matrix has two features:

[1] *Sparse*, i.e., a number of elements are *zero*
[2] *Well classified*, i.e., different variables load different factors

Feature [1] allows us to focus on the nonzero elements to capture the relationships of variables to factors. Feature [2] clarifies the differences between the two factors. The matrix in Table 13.2(A) is said to have a *simple structure* in the literature for psychometrics (statistics for psychology).

Table 13.2(A) shows an ideally simple structure, but it is almost impossible to have such a matrix; $\mathbf{T}$ cannot be chosen so that some elements of $\mathbf{A_T} = \hat{\mathbf{A}}\mathbf{T}'^{-1}$ are exactly zero as in (A). However, it is feasible to obtain $\mathbf{A_T} = \hat{\mathbf{A}}\mathbf{T}'^{-1}$ that approximates the ideal. It is illustrated in Table 13.2(B). There, "small" stands for a value *close to zero*, but not exactly being zero, while "Large" expresses a value with a large absolute value. A matrix, which is not ideal but *approximates ideal structure*, is also said to have a simple structure.

Let us remember that $\mathbf{A_T} = \hat{\mathbf{A}}\mathbf{T}'^{-1}$ can be viewed as the coordinates on rotated axes. How should the axes be rotated so as to make the loading matrix $\mathbf{A_T}$ be of a simple structure? One answer is found in Fig. 13.2, where the useful orthogonal and oblique rotation for the variable vectors in Fig. 13.1a is illustrated. First, let us note the axes of $\mathbf{t}_1$ and $\mathbf{t}_2$ in Fig. 13.2b. The former axis is approximately *parallel* to the vectors for a group of variables {C, T, B, P} (Group 1), while the latter is almost

Variable	(A) Ideally simple		(B) Simple	
	F1	F2	F1	F2
1	#	0	**Large**	Small
2	0	#	Small	**Large**
3	0	#	Small	**Large**
4	#	0	**Large**	Small
5	0	#	Small	**Large**
6	#	0	**Large**	Small
7	#	0	**Large**	Small
8	0	#	Small	**Large**

Table 13.2 Simple structure in a matrix of variables × factors

Table 13.3 The loadings, unique variances, and factor correlation obtained for the data set in Table 11.1

	(A) Before rotation			(B) After varimax rotation			(C) After geomin rotation		
	A		ψ_{jj}	$\mathbf{A_T}$		ψ_{jj}	$\mathbf{A_T}$		ψ_{jj}
A	**0.77**	−0.38	0.26	**0.81**	0.28	0.26	**0.82**	0.08	0.26
C	**0.61**	0.50	0.38	0.07	**0.78**	0.38	−0.13	**0.84**	0.38
I	**0.67**	−0.36	0.41	**0.73**	0.22	0.41	**0.74**	0.04	0.41
B	**−0.74**	−0.40	0.30	−0.24	**−0.80**	0.30	−0.04	**−0.82**	0.30
T	**0.79**	0.43	0.18	0.25	**0.87**	0.18	0.04	**0.88**	0.18
V	**0.76**	−0.44	0.22	**0.85**	0.23	0.22	**0.87**	0.01	0.22
H	**−0.63**	0.46	0.39	**−0.77**	−0.12	0.39	**−0.82**	0.08	0.39
P	**0.70**	0.18	0.47	0.37	**0.63**	0.47	0.23	**0.58**	0.47
c	0.00			0.00			0.48		

parallel to those for another group {A, V, I, H} (Group 2). Thus, Group 1 has the coordinates of large absolute values on the t_1 axis, but shows those of small absolutes on the t_2 axis. On the other hand, Group 2 shows the coordinates of large and small absolutes for t_2 and t_1 axes, respectively. The resulting loading matrix is presented in Table 13.3(C). There, the matrix successfully attains the simple structure as in Table 13.2(B). Orthogonal rotation is illustrated in Fig. 13.2a, where t_1 and t_2 are orthogonally intersected; (13.9) is satisfied. On the other hand, the axes are obliquely intersected in Fig. 13.2b. Also in (A), the t_1 and t_2 axes are almost parallel to Groups 1 and 2, respectively, which provide the matrix for a simple structure in Table 13.3(B).

   In the above paragraph, we visually illustrated how $\mathbf{T} = [t_1, t_2]$ is set to be parallel to groups of variable vectors, so that $\mathbf{A_T} = \mathbf{\hat{A}T}^{\prime-1}$ has a simple structure. But, this task can only be attained by human vision and is impossible even by that when m exceeds three dimensions. Indeed, the optimal $\mathbf{T}$ is obtained not visually but *computationally* with

$$\text{maximize Simp}(\mathbf{A_T}) = \text{Simp}(\mathbf{\hat{A}T}^{\prime-1}) \text{ over } \mathbf{T} \text{ subject to } (13.2) \text{ or } (13.9). \quad (13.11)$$

Here, $\text{Simp}(\mathbf{\hat{A}T}^{\prime-1})$ is the abbreviation for the simplicity of $\mathbf{\hat{A}T}^{\prime-1}$ and is a function of $\mathbf{T}$ that stands for how well $\mathbf{A_T} = \mathbf{\hat{A}T}^{\prime-1}$ approximates the ideal simple structure, that is, how simple the structure in $\mathbf{A_T}$ is. The procedures formulated as (13.11) are generally called (algebraic) *rotation techniques*. In exactness, we should call them *simple structure rotation procedures* in order to distinguish them from the rotation that does not involve a simple structure. A number of simple structure rotation techniques have been proposed so far, which differ in terms of how to define Simp $(\mathbf{\hat{A}T}^{\prime-1})$. Two popular techniques are introduced in the next two sections.

13.4 Varimax Rotation

The rotation techniques with (13.9) chosen as the constraint in (13.11) are called *orthogonal* rotation techniques. Among them, the *varimax* rotation method presented by Kaiser (1958) is well known. In this method, the simplicity of $\mathbf{A}_T = \hat{\mathbf{A}}\mathbf{T}$ is defined as:

$$\mathrm{Simp}(\mathbf{A}_T) = \mathrm{Simp}(\hat{\mathbf{A}}\mathbf{T}) = \sum_{k=1}^{m} \mathrm{var}\left(a_{1k}^{(\mathrm{T})2} \ldots a_{pk}^{(\mathrm{T})2}\right) \tag{13.12}$$

to be maximized. Here, we have used the fact that (13.1) is simplified as (13.10) and $\mathrm{var}(a_{1k}^{(\mathrm{T})2} \ldots a_{pk}^{(\mathrm{T})2})$ stands for the *variance of the squared elements* in the kth column of $\mathbf{A}_T = (a_{jk}^{(\mathrm{T})})$:

$$\mathrm{var}\left(a_{1k}^{(\mathrm{T})2} \ldots a_{pk}^{(\mathrm{T})2}\right) = \frac{1}{m} \sum_{j=1}^{p} \left(a_{jk}^{(\mathrm{T})2} - \bar{a}_{.k}^{(\mathrm{T})2}\right)^2, \tag{13.13}$$

with $\bar{a}_{.k}^{(\mathrm{T})2}$ the average of $a_{1k}^{(\mathrm{T})2}, \ldots, a_{pk}^{(\mathrm{T})2}$. That is, the varimax rotation is formulated as:

$$\text{maximize } \mathrm{Simp}(\hat{\mathbf{A}}\mathbf{T}) = \frac{1}{m} \sum_{k=1}^{m} \sum_{j=1}^{p} \left(a_{jk}^{(\mathrm{T})2} - \bar{a}_{.k}^{(\mathrm{T})2}\right)^2 \text{ over } \mathbf{T} \text{ subject to } \mathbf{T}'\mathbf{T} = \mathbf{I}_m.$$

$$\tag{13.14}$$

For this maximization, an iterative algorithm is needed. One of the algorithms can be included in the gradient methods introduced in Appendix A.6.3 (Jennrich 2001). However, that is out of the scope of this book.

We should note that variance (13.13) is not defined for loadings $a_{jk}^{(\mathrm{T})}$ but for its squares $a_{jk}^{(\mathrm{T})2}$; they are irrelevant to whether $a_{jk}^{(\mathrm{T})}$ are positive or negative, but are relevant to the absolute values of $a_{jk}^{(\mathrm{T})}$. If variance (13.13) is larger, the *absolute values* of the loadings in each column of $\mathbf{A}_T$ would take a *variety* of values so that

some absolute values are larger, but others are small, (13.15)

as illustrated in Table 13.2(B).

The sum of the above variances over m columns defines the simplicity as in (13.12). By maximizing the sum, all m columns can have loadings with (13.15). This allows us to consider the two different $\mathbf{A}_T$ results illustrated in Table 13.4(A) and (B). There, we find that (A) is equivalent to Table 13.2(B), i.e., it shows a simple structure, while Table 13.4(B) is not simple, in that the same variables

Table 13.4 Variables × factors matrices with and without a simple structure

Variable	(A) Simple		(B) Not simple	
	F1	F2	F1	F2
1	Large	Small	Large	Large
2	Small	Large	Small	Small
3	Small	Large	Small	Small
4	Large	Small	Large	Large
5	Small	Large	Small	Small
6	Large	Small	Large	Large
7	Large	Small	Large	Large
8	Small	Large	Small	Small

heavily load two factors. However, (13.14) hardly provides a loading matrix $\mathbf{A}_T$, as in Table 13.4(B), since it necessitates $\mathbf{t}_1$ and $\mathbf{t}_2$ extending almost in parallel, which contradicts constraint (13.9).

The varimax rotation for loading matrix $\hat{\mathbf{A}}$ in Table 13.3(A) provides the rotation matrix:

$$\mathbf{T} = \begin{bmatrix} 0.705 & 0.710 \\ -0.711 & 0.704 \end{bmatrix}, \tag{13.16}$$

which is the solution for (13.14). Post-multiplication of $\hat{\mathbf{A}}$ in Table 13.3(A) by (13.16) yields the matrix $\mathbf{A}_T = \hat{\mathbf{A}}\mathbf{T}$ in Table 13.3(B) that shows a simple structure. Indeed, Fig. 13.2a has been depicted according to Table 13.3(B).

Let us compare $\hat{\mathbf{A}}$ in Table 13.3(A) and $\mathbf{A}_T$ in (B). It is difficult to reasonably interpret the former loadings in (A), as all variables show the loadings of large absolute values for Factor 1 and those of rather small absolutes for Factor 2. It obliges one to consider that Factor 1 loads all variables, while Factor 2 is irrelevant to all variables, which implies that Factor 2 is trivial. On the other hand, $\mathbf{A}_T = \hat{\mathbf{A}}\mathbf{T}$ can be reasonably interpreted in the same manner as described in Sect. 12.7.

13.5 Geomin Rotation

The phrase "*maximize* Simp($\mathbf{A}_T$)" in (13.11) is equivalent to "*minimize* $-1 \times$ Simp ($\mathbf{A}_T$)". Here, $-1 \times$ Simp($\mathbf{A}_T$) can be rewritten as Comp($\mathbf{A}_T$) which abbreviates the *complexity* of $\mathbf{A}_T$ and represents to what extent $\mathbf{A}_T$ deviates from a simple structure. Some rotation techniques are formulated as substituting "minimize Comp($\mathbf{A}_T$)" for "maximize Simp($\mathbf{A}_T$)" in (13.11). One of them is Yates's (1987) *geomin* rotation method, in which complexity is defined as:

$$\text{Comp}(\mathbf{A_T}) = \text{Comp}(\hat{\mathbf{A}}\mathbf{T}'^{-1}) = \sum_{j=1}^{p}\left\{\prod_{k=1}^{m}(a_{jk}^{(T)\,2}+\varepsilon)\right\}^{1/m}, \tag{13.17}$$

with ε, a specified small positive value such as 0.01. The geomin rotation method has orthogonal and oblique versions. In this section, we treat the latter, i.e., the *oblique* geomin rotation, which is formulated as:

$$\text{minimize Comp}(\hat{\mathbf{A}}\mathbf{T}'^{-1}) = \sum_{j=1}^{p}\left\{\prod_{k=1}^{m}(a_{jk}^{(T)\,2}+\varepsilon)\right\}^{1/m}\ \text{over } \mathbf{T}\text{ subject to (13.2).}$$
$$\tag{13.18}$$

For this minimization, an iterative algorithm is needed. One of the algorithms can be included in the gradient methods introduced in Appendix A.6.3 (Jennrich 2002). However, that is beyond the scope of this book.

Let us note the parenthesized part in the right-hand side of (13.17):

$$\prod_{k=1}^{m}\left(a_{jk}^{(T)\,2}+\varepsilon\right)=\left(a_{j1}^{(T)\,2}+\varepsilon\right)\times\cdots\times\left(a_{jm}^{(T)\,2}+\varepsilon\right). \tag{13.19}$$

It is close to zero, if some of $a_{jk}^{(T)}$ are close to zero, which would give a matrix approximating that in Table 13.2(A). The sum of (13.19) over p variables is minimized as in (13.18). This minimization for $\hat{\mathbf{A}}$ in Table 13.3(A) provides the rotation matrix:

$$\mathbf{T}'^{-1} = \begin{bmatrix} 0.581 & 0.582 \\ -0.979 & 0.979 \end{bmatrix}. \tag{13.20}$$

Post-multiplication of $\hat{\mathbf{A}}$ in Table 13.3(A) by (13.20) yields $\mathbf{A_T}=\hat{\mathbf{A}}\mathbf{T}'^{-1}$ in Table 13.3(C). This has also been presented in Table 12.1(B), as described in Sect. 12.7.

The reason for adding a small positive constant ε to loadings, as in (13.19), is as follows: (13.19) would be $\prod_{k=1}^{K}a_{jk}^{(T)\,2}=a_{j1}^{(T)\,2}\times\cdots\times a_{jm}^{(T)\,2}$ without ε. Then, the solution which allows $\prod_{k=1}^{K}a_{jk}^{(T)\,2}$ to attain the lower bound 0 is not uniquely determined; multiple solutions could exist. For example, let m be 2. If $a_{j1}^{(T)}=0$, then $a_{j1}^{(T)\,2}\times a_{j2}^{(T)\,2}=0$ whatever value $a_{j2}^{(T)}$ takes. This existence of multiple solutions is avoided by adding ε as in (13.19).

13.6   Orthogonal Procrustes Rotation

In this section, we introduce *Procrustes* rotation, whose purpose is *different* from the procedures treated so far. Procrustes rotation generally refers to a class of rotation techniques to rotate $\hat{\mathbf{A}}$, so that the resulting $\mathbf{A}_T$ is *matched* with a *target* matrix $\mathbf{B}$. The rotation was originally conceived by Mosier (1939) and named by Hurley and Cattell (1962) after a figure appearing in Greek mythology.

Let us consider *orthogonal Procrustes rotation* with (13.9), i.e., $\mathbf{T}$ ($m \times m$) constrained to be orthonormal. This is formulated as:

$$\text{minimize} f(\mathbf{T}) = ||\mathbf{B} - \hat{\mathbf{A}}\mathbf{T}||^2 \text{ over } \mathbf{T} \text{ subject to } \mathbf{T}'\mathbf{T} = \mathbf{I}_m. \qquad (13.21)$$

This is useful for every case, in which one wishes to match $\hat{\mathbf{A}}\mathbf{T}$ to target $\mathbf{B}$ and examine how *similar* the resulting matrix $\mathbf{A}_T = \hat{\mathbf{A}}\mathbf{T}$ is to the target, under constraint (13.9).

The function $f(\mathbf{T})$ in (13.21) can be expanded as:

$$f(\mathbf{T}) = ||\mathbf{B}||^2 - 2\text{tr}\mathbf{B}'\hat{\mathbf{A}}\mathbf{T} + \text{tr}\mathbf{T}'\hat{\mathbf{A}}'\hat{\mathbf{A}}\mathbf{T} = ||\mathbf{B}||^2 - 2\text{tr}\mathbf{B}'\hat{\mathbf{A}}\mathbf{T} + ||\hat{\mathbf{A}}||^2, \qquad (13.22)$$

where we have used $\mathbf{T}\mathbf{T}' = \mathbf{I}_m$ following from (13.9). In the right-hand side of (13.22), only $-2\text{tr}\mathbf{T}'\hat{\mathbf{A}}'\mathbf{B}$ is relevant to $\mathbf{T}$. Thus, the minimization of (13.22) amounts to:

$$\text{maximize } g(\mathbf{T}) = \text{tr}\mathbf{B}'\hat{\mathbf{A}}\mathbf{T} \text{ over } \mathbf{T} \text{ subject to } \mathbf{T}'\mathbf{T} = \mathbf{I}_m. \qquad (13.23)$$

This problem is equivalent to the one in Theorem A.4.2 (Appendix A.4.2). As found there, the solution of $\mathbf{T}$ is given through the singular value decomposition of $\hat{\mathbf{A}}'\mathbf{B}$.

A numerical example is given in Table 13.5. The matrices $\mathbf{B}$ and $\hat{\mathbf{A}}$ presented there seem to be very different. The orthogonal Procrustes rotation for them provide $\mathbf{T} = \begin{bmatrix} 0.53 & 0.85 \\ -0.85 & 0.53 \end{bmatrix}$. The resulting $\hat{\mathbf{A}}\mathbf{T}$ is shown in the right-hand side of Table 13.5, where $\hat{\mathbf{A}}\mathbf{T}$ is found to be very similar to $\mathbf{B}$.

Table 13.5 Example of orthogonal Procrustes rotation

B		$\hat{\mathbf{A}}$		$\hat{\mathbf{A}}\mathbf{T}$	
0.0	0.8	0.6	0.4	−0.02	0.72
0.3	0.7	0.8	0.1	0.34	0.73
0.6	0.6	0.8	−0.2	0.59	0.57
0.8	0.1	0.5	−0.6	0.77	0.11
0.9	0.0	0.5	−0.8	0.94	0.00

13.7 Bibliographical Notes

Simple structure rotation techniques are exhaustively described in Browne (2001) and Mulaik (2011). Procrustes rotation techniques are detailed in Gower and Dijksterhuis (2004), with its special extended version presented by Adachi (2009). A simple structure can be related to the sparse analysis mentioned in Sect. 6.7, as discussed in Trendafilov (2014).

Exercises

13.1. Show that $\mathbf{T} = \mathbf{S} \, \text{diag}(\mathbf{S}'\mathbf{S})^{-1/2}$ satisfies (13.2), where $\text{diag}(\mathbf{S}'\mathbf{S})$ denotes the $m \times m$ diagonal matrix whose diagonal elements $d_1, \ldots, d_m$ are those of $\mathbf{S}'\mathbf{S}$, and $\text{diag}(\mathbf{S}'\mathbf{S})^{-1/2}$ is the $m \times m$ diagonal matrix whose diagonal elements are $1/d_1^{1/2}, \ldots, 1/d_m^{1/2}$.

13.2. Show that $\text{diag}(\mathbf{T}'\mathbf{T}) = \mathbf{I}_m$ is rewritten as: $\mathbf{T}'\mathbf{T} = \begin{bmatrix} 1 & & \# \\ & \ddots & \\ \# & & 1 \end{bmatrix}$.

13.3. Show that a 2×2 orthonormal matrix $\mathbf{T}$ is expressed as $\mathbf{T} = \begin{bmatrix} \cos\theta & -\sin\theta \\ \sin\theta & \cos\theta \end{bmatrix}$.

13.4. Thurstone (1947) defined simple structure with provisions, which have been rewritten more clearly by Browne (2001, p. 115) as follows:

[1] Each row should contain at least one zero.
[2] Each column should contain at least m zeros, with m the number of factors.
[3] Every pair of columns should have several rows with a zero in one column but not the other.
[4] If $m \geq 4$, every pair of columns should have several rows with zeros in both columns.
[5] Every pair of columns should have a few rows with nonzero loadings in both columns.

Present an example of a 20×4 matrix meeting provisions [1]–[5].

13.5. Minimizing $\frac{1}{m}\sum_{k<l}\sum_{j=1}^{p}\left(a_{jk}^{(T)2} - \bar{a}_{.k}^{(T)2}\right)\left(a_{jl}^{(T)2} - \bar{a}_{.l}^{(T)2}\right)$ over $\mathbf{T}$ subject to $\text{diag}(\mathbf{T}'\mathbf{T}) = \mathbf{I}_m$ is included in a family of oblique rotation called *oblimin rotation* (Jennrich and Sampson 1966), where $a_{jk}^{(T)}$ is the (j, k) element of the rotated loading matrix $\hat{\mathbf{A}}\mathbf{T}'^{-1}$. Discuss the purpose of the above minimization. The symbol $\sum_{k<l}$ is explained in Note 16.1.

13.6. Oblique rotation tends to give a matrix of a simpler structure than orthogonal rotation. Explain the reason for this.

13.7. Show that orthogonal rotation is feasible for the $p \times m$ matrix $\mathbf{A}$ that minimizes $\|\mathbf{V} - \mathbf{AA}\prime\|^2$ subject to $\mathbf{A}\prime\mathbf{A} = \mathbf{I}_m$ for given $\mathbf{V}$.

13.8. Show that oblique rotation is feasible for the solution of principal component analysis, if constraint (5.25) is relaxed as $n^{-1}\mathrm{diag}(\mathbf{F}\prime\mathbf{F}) = \mathbf{I}_m$ without (5.26).

13.9. Show the objective function (13.12) in the varimax rotation can be rewritten as:

$$f = \frac{1}{n}\mathrm{tr}\,\mathbf{T}\prime\hat{\mathbf{A}}\prime\{(\mathbf{T}\prime\hat{\mathbf{A}}\prime) * (\hat{\mathbf{A}}\mathbf{T}) * (\hat{\mathbf{A}}\mathbf{T})\} - \frac{1}{n^2}\mathrm{tr}\,\mathbf{T}\prime\hat{\mathbf{A}}\prime\hat{\mathbf{A}}\mathbf{T}\{\mathrm{diag}(\mathbf{T}\prime\hat{\mathbf{A}}\prime\hat{\mathbf{A}}\mathbf{T})\}.$$

(ten Berge et al. 1988). Here, $\mathrm{diag}(\mathbf{T}\prime\hat{\mathbf{A}}\prime\hat{\mathbf{A}}\mathbf{T})$ is defined in Exercise 13.1, and the * denotes the element-wise product called the *Hadamard product* (e.g., Schott 2005, pp. 295–305): $\mathbf{X} * \mathbf{Y} = \begin{bmatrix} x_{11}y_{11} & \cdots & x_{1p}y_{1p} \\ & \vdots & \\ x_{n1}y_{n1} & \cdots & x_{np}y_{np} \end{bmatrix} = (x_{ij}y_{ij})\ (n \times p)$ for

$n \times p$ matrices $\mathbf{X} = \begin{bmatrix} x_{11} & \cdots & x_{1p} \\ & \vdots & \\ x_{n1} & \cdots & x_{np} \end{bmatrix}$ and $\mathbf{Y} = \begin{bmatrix} y_{11} & \cdots & y_{1p} \\ & \vdots & \\ y_{n1} & \cdots & y_{np} \end{bmatrix}$.

13.10. *Generalized orthogonal rotation* is formulated as minimizing $\sum_{k=1}^{K}\|\mathbf{H} - \mathbf{A}_k\mathbf{T}_k\|^2$ over $\mathbf{H}$, $\mathbf{T}_1, \ldots, \mathbf{T}_K$ subject to $\mathbf{T}_k\prime\mathbf{T}_k = \mathbf{T}_k\mathbf{T}_k\prime = \mathbf{I}_m$, $k = 1, \ldots, K$, for given $p \times m$ matrices $\mathbf{A}_1, \ldots, \mathbf{A}_K$. Show that the minimization can be attained by the following algorithm:

Step 1. Initialize $\mathbf{T}_1, \ldots, \mathbf{T}_K$.
Step 2. Set $\mathbf{H} = \frac{1}{K}\sum_{k=1}^{K}\mathbf{A}_k\mathbf{T}_k$.
Step 3. Compute the SVD $\mathbf{A}_k\prime\mathbf{H} = \mathbf{K}_k\mathbf{\Lambda}_k\mathbf{L}_k\prime$ to set $\mathbf{T}_k = \mathbf{K}_k\mathbf{L}_k\prime$ for $k = 1, \ldots, K$.
Step 4. Finish if convergence is reached; otherwise, go back to Step 2.

13.11. Show that $K\sum_{k=1}^{K}\|\mathbf{H} - \mathbf{A}_k\mathbf{T}_k\|^2 = \sum_{k=1}^{K-1}\sum_{l=k+1}^{K}\|\mathbf{A}_k\mathbf{T}_k - \mathbf{A}_l\mathbf{T}_l\|^2$ for the $\mathbf{H}$ set in Step 2 of Exercise 13.10.

13.12. Let us consider the minimization of $\|[\mathbf{M}, \mathbf{c}] - \mathbf{AT}\|^2$ over $\mathbf{T}$ ($m \times m$) and $\mathbf{c}$ ($p \times 1$) subject to $\mathbf{T}\prime\mathbf{T} = \mathbf{TT}\prime = \mathbf{I}_m$ for given $\mathbf{M}$ ($p \times (m - 1)$) and $\mathbf{A}$ ($p \times m$). Show that the minimization can be attained by the following algorithm:

Step 1. Initialize $\mathbf{T}$.
Step 2. Set $\mathbf{c}$ to the final column of $\mathbf{AT}$.
Step 3. Compute the SVD $\mathbf{A}\prime[\mathbf{M}, \mathbf{c}] = \mathbf{K}\mathbf{\Lambda}\mathbf{L}\prime$ to set $\mathbf{T} = \mathbf{KL}\prime$.
Step 4. Finish if convergence is reached; otherwise, go back to Step 2.

13.13. Kier's (1994) *simplimax rotation* is a generalization of Procrustes procedures, which is used for having a matrix of simple structure. In simplimax rotation, target matrix **B** is unknown except that **B** is constrained to have a specified number of zero elements: $||\mathbf{B} - \hat{\mathbf{A}}\mathbf{T}'^{-1}||^2$ is minimized over **B** and **T** subject to (13.2) or (13.9) and s elements in **B** being zero, though the locations of the s zero elements are unknown. Show that, for fixed **T**, the optimal $\mathbf{B} = (b_{jk})$ is given by $b_{jk} = \begin{cases} 0 & \text{if } a_{jk}^{[T]2} \leq a_{<s>}^{[T]2} \\ a_{jk}^{[T]} & \text{otherwise} \end{cases}$, where $a_{jk}^{[T]}$ is the (j, k) element of $\hat{\mathbf{A}}\mathbf{T}'^{-1}$ and $a_{<s>}^{[T]2}$ is the sth smallest value among the squares of the elements in $\hat{\mathbf{A}}\mathbf{T}'^{-1}$.

Chapter 14
Canonical Correlation and Multiple Correspondence Analyses

In this chapter, we treat procedures for the data set in which *variables* are classified into some *groups*. Such a data set is expressed as a *block matrix*, introduced in Sect. 14.1. Then, we describe *canonical correlation analysis (CCA)* for data with two groups of variables, which is followed by the introduction of *generalized CCA (GCCA)* for more than two groups of variables in Sect. 14.3. GCCA provides a foundation for a procedure analyzing the *multivariate categorical data* described in Sect. 14.4. This procedure is called *multiple correspondence analysis (MCA)*, whose purpose is to *quantify unnumerical categories*, i.e., finding the optimal scores to be given to the categories, as shown in Sect. 14.5.

CCA was originally formulated by Hotelling (1936), and some types of GCCA have been presented (Gifi 1990; Kettenring 1984; van de Geer 1984), among which Gifi's (1990) approach is chosen for describing GCCA and MCA in this chapter.

14.1 Block Matrices

We start with introducing the blocks of a matrix by the following note:

Note 14.1. Blocks of a Matrix
We can rewrite a 5 × 4 matrix **Y** as follows:

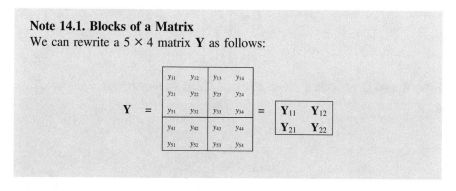

© Springer Nature Singapore Pte Ltd. 2016
K. Adachi, *Matrix-Based Introduction to Multivariate Data Analysis*,
DOI 10.1007/978-981-10-2341-5_14

where

$$\mathbf{Y}_{11} = \begin{bmatrix} y_{11} & y_{12} \\ y_{21} & y_{22} \\ y_{31} & y_{32} \end{bmatrix}, \quad \mathbf{Y}_{12} = \begin{bmatrix} y_{13} & y_{14} \\ y_{23} & y_{24} \\ y_{33} & y_{34} \end{bmatrix}, \quad \mathbf{Y}_{21} = \begin{bmatrix} y_{41} & y_{42} \\ y_{51} & y_{52} \end{bmatrix}, \quad \mathbf{Y}_{22} = \begin{bmatrix} y_{43} & y_{44} \\ y_{53} & y_{54} \end{bmatrix}$$

$\mathbf{Y}_{11}$, $\mathbf{Y}_{12}$, $\mathbf{Y}_{21}$, and $\mathbf{Y}_{22}$ are called the *blocks* of $\mathbf{Y}$, while $\mathbf{Y}$ is called a *block matrix* consisting of $\mathbf{Y}_{11}$, $\mathbf{Y}_{12}$, $\mathbf{Y}_{21}$, and $\mathbf{Y}_{22}$.

This example is generalized as follows: An $n \times p$ matrix $\mathbf{Y}$ can be rewritten as:

$$\mathbf{Y} = \begin{bmatrix} \mathbf{Y}_{11} & \cdots & \mathbf{Y}_{1j} & \cdots & \mathbf{Y}_{1J} \\ & & \vdots & & \\ \mathbf{Y}_{i1} & \cdots & \mathbf{Y}_{ij} & \cdots & \mathbf{Y}_{iJ} \\ & & \vdots & & \\ \mathbf{Y}_{I1} & \cdots & \mathbf{Y}_{Ij} & \cdots & \mathbf{Y}_{IJ} \end{bmatrix}. \tag{14.1}$$

Here, $\mathbf{Y}_{ij}$ is called the (i, j) *block* of $\mathbf{Y}$, while $\mathbf{Y}$ is called a *block matrix* containing $\mathbf{Y}_{ij}$ ($i = 1, \ldots, I$; $j = 1, \ldots, J$). If $\mathbf{Y}_{ij}$ is $n_i \times p_j$, then $n = \sum_{i=1}^{I} n_i$ and $p = \sum_{j=1}^{J} p_j$.

In this chapter, a block matrix of data is treated in which blocks $\mathbf{X}_1, \ldots, \mathbf{X}_J$ are arranged horizontally:

$$\mathbf{X} = \begin{bmatrix} \mathbf{X}_1, \ldots, \mathbf{X}_j, \ldots, \mathbf{X}_J \end{bmatrix}, \tag{14.2}$$

while a block matrix of parameters is considered in which $\mathbf{C}_1, \ldots, \mathbf{C}_J$ are stacked vertically:

$$\mathbf{C} = \begin{bmatrix} \mathbf{C}_1 \\ \vdots \\ \mathbf{C}_j \\ \vdots \\ \mathbf{C}_J \end{bmatrix}. \tag{14.3}$$

Here, $\mathbf{X}_j$ and $\mathbf{C}_j$ are called the jth block of $\mathbf{X}$ and $\mathbf{C}$, respectively.

A weighted sum of matrices can be expressed block-wise as follows:

Note 14.2. Weighted Sum of Block Matrices

Let the block matrices $\mathbf{A} = \begin{bmatrix} \mathbf{A}_{11} & \cdots & \mathbf{A}_{1J} \\ & \vdots & \\ \mathbf{A}_{I1} & \cdots & \mathbf{A}_{IJ} \end{bmatrix}$ and $\mathbf{B} = \begin{bmatrix} \mathbf{B}_{11} & \cdots & \mathbf{B}_{1J} \\ & \vdots & \\ \mathbf{B}_{I1} & \cdots & \mathbf{B}_{IJ} \end{bmatrix}$

be of the same order and their blocks $\mathbf{A}_{ij}$ and $\mathbf{B}_{ij}$ ($i = 1, \ldots, I; j = 1, \ldots, J$) also be so. Then, the sum of $\mathbf{A}$ and $\mathbf{B}$ multiplied by scalars s and t is defined as:

$$s\mathbf{A} + t\mathbf{B} = \begin{bmatrix} s\mathbf{A}_{11} + t\mathbf{B}_{11} & \cdots & s\mathbf{A}_{1J} + t\mathbf{B}_{1J} \\ & \vdots & \\ s\mathbf{A}_{I1} + t\mathbf{B}_{I1} & \cdots & s\mathbf{A}_{IJ} + t\mathbf{B}_{IJ} \end{bmatrix}, \qquad (14.4)$$

whose (i, j) block is $s\mathbf{A}_{ij} + t\mathbf{B}_{ij}$.

The product of the matrices can also be expressed block-wise:

Note 14.3. Product of Block Matrices

Let us define $n \times p$ and $p \times m$ block matrices as $\mathbf{A} = \begin{bmatrix} \mathbf{A}_{11} & \cdots & \mathbf{A}_{1J} \\ & \vdots & \\ \mathbf{A}_{I1} & \cdots & \mathbf{A}_{IJ} \end{bmatrix}$

and $\mathbf{Q} = \begin{bmatrix} \mathbf{Q}_{11} & \cdots & \mathbf{Q}_{1K} \\ & \vdots & \\ \mathbf{Q}_{J1} & \cdots & \mathbf{Q}_{JK} \end{bmatrix}$, respectively, with $\mathbf{A}_{ij}$ being the (i, j) block of

$\mathbf{A}$, $\mathbf{Q}_{jk}$ the (j, k) one of $\mathbf{Q}$ and the number of the columns of $\mathbf{A}_{ij}$ equaling the number of rows of $\mathbf{Q}_{jk}$. Post-multiplication of $\mathbf{A}$ by $\mathbf{Q}$ provides the $n \times m$ matrix

$$\mathbf{V} = \mathbf{AQ} = \begin{bmatrix} \mathbf{V}_{11} & \cdots & \mathbf{V}_{1K} \\ & \vdots & \\ \mathbf{V}_{I1} & \cdots & \mathbf{V}_{IK} \end{bmatrix}, \qquad (14.5)$$

whose (i, k) block is:

$$\mathbf{V}_{ik} = \sum_{j=1}^{J} \mathbf{A}_{ij}\mathbf{Q}_{jk} = \mathbf{A}_{i1}\mathbf{Q}_{1k} + \mathbf{A}_{i2}\mathbf{Q}_{2k} + \cdots + \mathbf{A}_{iJ}\mathbf{Q}_{Jk}. \qquad (14.6)$$

In this chapter, the special case of (14.5),

$$\mathbf{XC} = \sum_{j=1}^{J} \mathbf{X}_j \mathbf{C}_j = \mathbf{X}_1 \mathbf{C}_1 + \mathbf{X}_2 \mathbf{C}_2 + \cdots + \mathbf{X}_J \mathbf{C}_J, \qquad (14.7)$$

is often used with $\mathbf{X} = [\mathbf{X}_1, \ldots, \mathbf{X}_J]$ and $\mathbf{C} = \begin{bmatrix} \mathbf{C}_1 \\ \vdots \\ \mathbf{C}_J \end{bmatrix}$.

14.2 Canonical Correlation Analysis

Let us consider an n-individuals $\times$ p-variables data matrix $\mathbf{X} = [\mathbf{X}_1, \mathbf{X}_2]$ consisting of the *two blocks* $\mathbf{X}_1 = [\mathbf{x}_{11}, \ldots, \mathbf{x}_{1p_1}] (n \times p_1)$ and $\mathbf{X}_2 = [\mathbf{x}_{21}, \ldots, \mathbf{x}_{2p_2}] (n \times p_2)$. That is, the p-variables in $\mathbf{X}$ are classified into a group of p_1-variables and into a group of p_2-variables. It is supposed that $\mathbf{X}$ is centered with $\mathbf{1}'_n \mathbf{X} = \mathbf{0}'_p$. An example of such data is presented in Table 14.1.

For $\mathbf{X} = [\mathbf{X}_1, \mathbf{X}_2]$, *canonical correlation analysis* (*CCA*) is formulated as minimizing

$$f(\mathbf{C}_1, \mathbf{C}_2) = ||\mathbf{X}_1 \mathbf{C}_1 - \mathbf{X}_2 \mathbf{C}_2||^2 \qquad (14.8)$$

over $p_1 \times m$ coefficient matrix $\mathbf{C}_1$ and $p_2 \times m$ coefficient matrix $\mathbf{C}_2$ subject to the constraints

$$\frac{1}{n} \mathbf{C}'_1 \mathbf{X}'_1 \mathbf{X}_1 \mathbf{C}_1 = \frac{1}{n} \mathbf{C}'_2 \mathbf{X}'_2 \mathbf{X}_2 \mathbf{C}_2 = \mathbf{I}_m, \qquad (14.9)$$

with $m \leq \mathrm{rank}(\mathbf{X}'_1 \mathbf{X}_2)$. That is, the purpose of CCA is to obtain the coefficient matrices $\mathbf{C}_1$ and $\mathbf{C}_2$ that allow $\mathbf{X}_1 \mathbf{C}_1$ and $\mathbf{X}_2 \mathbf{C}_2$ to be mutually best *matched*. Loss function (14.8) can be rewritten using (14.9) as $\mu = \mathrm{tr}\mathbf{C}'_1 \mathbf{X}'_1 \mathbf{X}_1 \mathbf{C}_1 + \mathrm{tr}\mathbf{C}'_2 \mathbf{X}'_2 \mathbf{X}_2 \mathbf{C}_2 - 2\mathrm{tr}\ \mathbf{C}'_1 \mathbf{X}'_1 \mathbf{X}_2 \mathbf{C}_2 = 2m - 2\mathrm{tr}\mathbf{C}'_1 \mathbf{X}'_1 \mathbf{X}_2 \mathbf{C}_2$, whose minimization is equivalent to maximizing

$$\frac{1}{n} \mathrm{tr}\mathbf{C}'_1 \mathbf{X}'_1 \mathbf{X}_2 \mathbf{C}_2. \qquad (14.10)$$

Its maximization subject to (14.9) is attained as in Theorem A.4.8 (Appendix A.4.5, where we can set $\mathbf{V}_{11} = n^{-1}\mathbf{X}'_1 \mathbf{X}_1 / \ \mathbf{V}_{22} = n^{-1}\mathbf{X}'_2 \mathbf{X}_2 /$ and $\mathbf{V}_{12} = n^{-1}\mathbf{X}'_1 \mathbf{X}_2$ to find the solution for the above CCA problem).

We illustrate CCA by performing it to the data set in Table 14.1, setting $m = 1$. In this unidimensional case, $\mathbf{C}_1$ and $\mathbf{C}_2$ are rewritten as vectors $\mathbf{c}_1 = [c_{11}, \ldots, c_{1,p_1}]'$ and $\mathbf{c}_2 = [c_{21}, \ldots, c_{2p_2}]'$, respectively; $\mathbf{X}_1 \mathbf{C}_1$ and $\mathbf{X}_2 \mathbf{C}_2$ are expressed as

Table 14.1 Standard scores for strength and athletic test data (Tanaka and Tarumi 1995)

Ind.	X_1: Strength test[a]							X_2: Athletic test[a]				
	RJ	VJ	DM	GP	SM	DB	BW	SP	LJ	LT	CE	MA
1	−0.42	−0.68	0.74	1.19	0.56	1.62	1.51	−0.96	1.13	−0.30	0.11	−0.22
2	1.36	−0.68	−1.24	−0.49	0.99	0.45	−1.20	0.18	0.54	0.79	−0.68	−0.60
3	−0.42	1.35	−0.48	−1.24	2.15	0.45	1.75	−0.96	−0.26	1.52	0.38	0.62
.	.	.	.	.	.	.	.	.	.	.	.	.
.	.	.	.	.	.	.	.	.	.	.	.	.
.	.	.	.	.	.	.	.	.	.	.	.	.
37	1.36	0.91	0.99	0.44	−1.45	−0.23	−1.81	−0.96	1.51	0.07	0.90	−0.38
38	0.17	1.20	−0.92	0.07	−0.92	1.29	1.26	0.18	1.91	0.07	0.38	−0.47

[a]Variable names are abbreviated as follows: RJ = repetition of jump, VJ = vertical jump, DM = dorsal muscles, GP = grasping power, SM = step motion, DB = deep forward bow, BW = body warping, SP = sprint, LJ = long jump, LT = long throw, CE = chinning exercises, MA = marathon

$\mathbf{X}_1\mathbf{c}_1 = c_{11}\mathbf{x}_{11} + \cdots + c_{1,p_1}\mathbf{x}_{1p_1}$ and $\mathbf{X}_2\mathbf{c}_2 = c_{21}\mathbf{x}_{21} + \cdots + c_{2p_2}\mathbf{x}_{2p_2}$, respectively. The CCA for the data set gives the following solution:

$$
\begin{aligned}
\mathbf{X}_1\mathbf{c}_1 = {} & 0.442 \times \text{RJ} + 0.267 \times \text{VJ} + 0.588 \times \text{DM} \\
& + 0.061 \times \text{GP} + 0.222 \times \text{SM} + 0.091 \times \text{DB} + 0.014 \times \text{BW},
\end{aligned}
\tag{14.11}
$$

$$
\mathbf{X}_2\mathbf{c}_2 = -0.426 \times \text{SP} + 0.233 \times \text{LJ} + 0.370 \times \text{LT} + 0.004 \times \text{CE} - 0.356 \times \text{MA},
\tag{14.12}
$$

where the resulting coefficient for each variable is followed by the abbreviation of its name in Table 14.1. The solutions in (14.11) and (14.12) stand for the weighted sums of strength and athletic test scores that are best matched.

Since $\mathbf{1}'_n\mathbf{X} = \mathbf{0}'_p$, the correlation coefficient between $\mathbf{X}_1\mathbf{c}_1$ and $\mathbf{X}_2\mathbf{c}_2$ is expressed as:

$$
\frac{n^{-1}\mathbf{c}_{11}\mathbf{X}_1'\mathbf{X}_2\mathbf{c}_{21}}{\sqrt{n^{-1}\mathbf{c}_{11}\mathbf{X}_1'\mathbf{X}_1\mathbf{c}_{11}}\sqrt{n^{-1}\mathbf{c}_{21}\mathbf{X}_2'\mathbf{X}_2\mathbf{c}_{21}}},
\tag{14.13}
$$

whose denominator equals one because of (14.9): (14.10) with $m = 1$ is equivalent to (14.13). This particular coefficient is called a *canonical correlation coefficient* between the variables in $\mathbf{X}_1$ and those in $\mathbf{X}_2$. The CCA solution for the data set in Table 14.1 gives the (14.13) value equaling 0.85, which shows that the items in the strength test are strongly related to those in the athletic test.

14.3 Generalized Canonical Correlation Analysis

Let us compare the CCA loss function (14.8) and the function

$$
f(\mathbf{F}, \mathbf{C}_1, \mathbf{C}_2) = ||\mathbf{F} - \mathbf{X}_1\mathbf{C}_1||^2 + ||\mathbf{F} - \mathbf{X}_2\mathbf{C}_2||^2
\tag{14.14}
$$

with a new matrix $\mathbf{F}(n \times m)$ whose rows correspond to individuals. The minimization of (14.8) is equivalent to minimizing (14.14) over $\mathbf{F}$, $\mathbf{C}_1$, and $\mathbf{C}_2$. It follows from the fact that the solution of $\mathbf{F}$ must satisfy $\mathbf{F} = 2^{-1}(\mathbf{X}_1\mathbf{C}_1 + \mathbf{X}_2\mathbf{C}_2)$, as shown with (A.2.6) in Appendix A.2.1. Substituting the equation for $\mathbf{F}$ in (14.14), it is rewritten as:

$$
\begin{aligned}
f(\mathbf{F}, \mathbf{C}_1, \mathbf{C}_2) &= ||\tfrac{1}{2}(\mathbf{X}_1\mathbf{C}_1 + \mathbf{X}_2\mathbf{C}_2) - \mathbf{X}_1\mathbf{C}_1||^2 + ||\tfrac{1}{2}(\mathbf{X}_1\mathbf{C}_1 + \mathbf{X}_2\mathbf{C}_2) - \mathbf{X}_2\mathbf{C}_2||^2 \\
&= ||\tfrac{1}{2}\mathbf{X}_2\mathbf{C}_2 - \tfrac{1}{2}\mathbf{X}_1\mathbf{C}_1||^2 + ||\tfrac{1}{2}\mathbf{X}_1\mathbf{C}_1 - \tfrac{1}{2}\mathbf{X}_2\mathbf{C}_2||^2 = \tfrac{1}{2}||\mathbf{X}_1\mathbf{C}_1 - \mathbf{X}_2\mathbf{C}_2||^2,
\end{aligned}
\tag{14.15}
$$

which equals half of (14.8).

Generalized canonical correlation analysis (GCCA) can be formulated through the extension of (14.14) to the cases for $\mathbf{X} = [\mathbf{X}_1, \ldots, \mathbf{X}_J]$ with $J \geq 2$. That is, the loss function of GCCA is expressed as:

$$\eta(\mathbf{F}, \mathbf{C}) = \sum_{j=1}^{J} \left\| \mathbf{F} - \mathbf{X}_j \mathbf{C}_j \right\|^2, \tag{14.16}$$

with $\mathbf{C} = \begin{bmatrix} \mathbf{C}_1 \\ \vdots \\ \mathbf{C}_J \end{bmatrix}$. In GCCA, $\mathbf{F}$ rather than $\mathbf{X}_j \mathbf{C}_j$ is constrained as:

$$\frac{1}{n} \mathbf{F}'\mathbf{F} = \mathbf{I}_m, \tag{14.17}$$

with $m \leq r = \mathrm{rank}(\mathbf{X})$. That is, GCCA can be formulated as minimizing (14.16) over $\mathbf{F}$ and $\mathbf{C}$ subject to (14.17). The implication of this minimization is shown in Fig. 14.1, where a *single* $\mathbf{F}$ and *multiple* $\mathbf{X}_j \mathbf{C}_j$ $(j = 1, \ldots, J)$ are depicted. The double-headed arrows in the figure express the deviations of $\mathbf{X}_j \mathbf{C}_j$ from $\mathbf{F}$. They are summated as in (14.16), which is minimized so that $\mathbf{X}_j \mathbf{C}_j$ are well *matched* with $\mathbf{F}$. As a result, $\mathbf{X}_j \mathbf{C}_j$ $(j = 1, \ldots, J)$ becomes *similar* across different j, and $\mathbf{X}_j \mathbf{C}_j$ is *summarized* into a single matrix $\mathbf{F}$.

As explained later, the matrix $\mathbf{X}\mathbf{D}_{\mathbf{X}}^{-1/2}$ plays an important role in GCCA with

$$\mathbf{D}_{\mathbf{X}} = \begin{bmatrix} \mathbf{X}_1'\mathbf{X}_1 & & & \\ & \mathbf{X}_2'\mathbf{X}_2 & & \\ & & \ddots & \\ & & & \mathbf{X}_J'\mathbf{X}_J \end{bmatrix} \tag{14.18}$$

a $p \times p$ *block diagonal matrix* in which the blank cells are filled with zeros. We explain the term block diagonal matrix and the superscript $-1/2$ in $\mathbf{D}_{\mathbf{X}}^{-1/2}$ in the following two notes.

Fig. 14.1 Illustration of generalized canonical correlation analysis

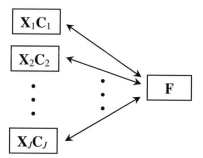

Note 14.4. Block Diagonal Matrices

A matrix $\mathbf{B}$ whose (i,j) block is a zero matrix for $i \neq j$, i.e.,

$$\mathbf{B} = \begin{bmatrix} \mathbf{B}_1 & & & \\ & \mathbf{B}_2 & & \\ & & \ddots & \\ & & & \mathbf{B}_I \end{bmatrix} \tag{14.19}$$

is called a *block diagonal matrix* and $\mathbf{B}_i$ $(i = 1, \ldots, I)$ is called the ith *diagonal block* of $\mathbf{B}$.

The products of block matrices are given as:

$$\begin{bmatrix} \mathbf{B}_1 & & & \\ & \mathbf{B}_2 & & \\ & & \ddots & \\ & & & \mathbf{B}_I \end{bmatrix} \begin{bmatrix} \mathbf{H}_1 & & & \\ & \mathbf{H}_2 & & \\ & & \ddots & \\ & & & \mathbf{H}_I \end{bmatrix}$$

$$= \begin{bmatrix} \mathbf{B}_1\mathbf{H}_1 & & & \\ & \mathbf{B}_2\mathbf{H}_2 & & \\ & & \ddots & \\ & & & \mathbf{B}_I\mathbf{H}_I \end{bmatrix}, \tag{14.20}$$

$$[\mathbf{X}_1, \mathbf{X}_2, \ldots, \mathbf{X}_J] \begin{bmatrix} \mathbf{C}_1 & & & \\ & \mathbf{C}_2 & & \\ & & \ddots & \\ & & & \mathbf{C}_J \end{bmatrix} = [\mathbf{X}_1\mathbf{C}_1, \mathbf{X}_2\mathbf{C}_2, \ldots, \mathbf{X}_J\mathbf{C}_J],$$

$$\tag{14.21}$$

$$\begin{bmatrix} \mathbf{B}_1 & & & \\ & \mathbf{B}_2 & & \\ & & \ddots & \\ & & & \mathbf{B}_I \end{bmatrix} \begin{bmatrix} \mathbf{Q}_1 \\ \mathbf{Q}_2 \\ \vdots \\ \mathbf{Q}_I \end{bmatrix} = \begin{bmatrix} \mathbf{B}_1\mathbf{Q}_1 \\ \mathbf{B}_2\mathbf{Q}_2 \\ \vdots \\ \mathbf{B}_I\mathbf{Q}_I \end{bmatrix}. \tag{14.22}$$

Here, we have supposed that the products of the blocks are defined. If (14.19) and $\mathbf{B}_1, \ldots, \mathbf{B}_I$ are square and nonsingular, the inverse matrix of (14.19) is expressed as:

$$\mathbf{B}^{-1} = \begin{bmatrix} \mathbf{B}_1^{-1} & & & \\ & \mathbf{B}_2^{-1} & & \\ & & \ddots & \\ & & & \mathbf{B}_I^{-1} \end{bmatrix}. \tag{14.23}$$

Note 14.5. Square and Square Root of a Matrix
The *square* of an $n \times n$ matrix $\mathbf{V}$ is expressed as:

$$\mathbf{V}^2 = \mathbf{VV}. \tag{14.24}$$

The *square root* of $\mathbf{V}$, denoted as $\mathbf{V}^{1/2}$, is the matrix satisfying

$$\mathbf{V}^{1/2}\mathbf{V}^{1/2} = \mathbf{V} \tag{14.25}$$

and the inverse matrix of $\mathbf{V}^{1/2}$, denoted as $\mathbf{V}^{-1/2}$, satisfies

$$\mathbf{V}^{-1/2}\mathbf{V}^{-1/2} = \mathbf{V}^{-1}. \tag{14.26}$$

Thus, $\mathbf{D}_X^{-1/2}$ is the matrix satisfying $\mathbf{D}_X^{-1/2}\mathbf{D}_X^{-1/2} = \mathbf{D}_X^{-1}$. Comparing (14.18) with (14.23) and (14.26), we find

$$\mathbf{D}_X^{-1/2} = \begin{bmatrix} (\mathbf{X}_1'\mathbf{X}_1)^{-1/2} & & & \\ & (\mathbf{X}_2'\mathbf{X}_2)^{-1/2} & & \\ & & \ddots & \\ & & & (\mathbf{X}_J'\mathbf{X}_J)^{-1/2} \end{bmatrix}, \tag{14.27}$$

and use (14.21) to get

$$\mathbf{XD}_X^{-1/2} = [\mathbf{X}_1(\mathbf{X}_1\mathbf{X}_1)^{-1/2}, \ldots, \mathbf{X}_J(\mathbf{X}_J\mathbf{X}_J)^{-1/2}]. \tag{14.28}$$

How to obtain $(\mathbf{X}_j'\mathbf{X}_j)^{-1/2}$ is described in Appendix A.4.6.

As described in Theorem A.4.6 (Appendix A.4.4), the GCCA problem, i.e., the minimization of (14.16) subject to (14.17), is equivalent to minimizing

$$f(\mathbf{F}, \mathbf{C}) = ||\mathbf{XD}_X^{-1/2} - \frac{1}{n}\mathbf{FC}'\mathbf{D}_X^{1/2}||^2 \tag{14.29}$$

over $\mathbf{F}$ and $\mathbf{C}$ subject to (14.17), which can be viewed as the reduced rank approximation of $\mathbf{XD}_X^{-1/2}$. The solution of $\mathbf{F}$ and $\mathbf{C}$ is given by

$$\mathbf{F} = \sqrt{n}\mathbf{N}_m\mathbf{T}, \tag{14.30}$$

$$\mathbf{C} = \sqrt{n}\mathbf{D}_X^{-1/2}\mathbf{M}_m\mathbf{\Phi}_m\mathbf{T}, \tag{14.31}$$

as found in Theorem A.4.6. Here, $\mathbf{T}$ is an $m \times m$ orthonormal matrix, and $\mathbf{N}_m$, $\mathbf{M}_m$, and $\mathbf{\Phi}_m$ are obtained through the SVD of $\mathbf{XD}_{XX}^{-1/2}$ defined as:

$$\mathbf{XD_{XX}^{-1/2}} = \mathbf{N\Phi M'},\qquad(14.32)$$

with $\mathbf{N'N} = \mathbf{M'M} = \mathbf{I}_r$ and $\mathbf{\Phi}$ a diagonal matrix whose diagonal elements are ordered in descending order; $\mathbf{N}_m$ and $\mathbf{M}_m$ contain the first m columns of $\mathbf{N}$ and those of $\mathbf{M}$, respectively, with $\mathbf{\Phi}_m$ the first $m \times m$ diagonal block. The matrix $\mathbf{T}$ appearing in (14.30) and (14.31) implies that the solution can be rotated as in EFA.

The importance of GCCA may not be its usefulness in real data analysis, but rather that it leads to multiple correspondence analysis for the categorical data described in the next sections.

14.4 Multivariate Categorical Data

An example of *multivariate categorical data* is given by a 10-individuals × 3-variables matrix $\mathbf{Y} = (y_{ij})$ in Table 14.2(A), where the variables are:

[V1] Faculty to which each individual belongs,
[V2] Subject at which she/he is best, and
[V3] Sciences, basic or applied, to which he/she is oriented.

We should note that the elements of $\mathbf{Y}$ are *not quantitative* scores, but the *code* numbers referring to *categories*. For example, those for [V1] are coded as

Table 14.2 Artificial example describing the faculties (FC) of students (ST), the subjects (SJ) at which they are best, and their orientation (OT), which is found in Adachi and Murakami (2011)

ST	(A) Data matrix **Y**			(B) Data matrix **Y**			(C) Indicator matrix $\mathbf{G} = [\mathbf{G}_1, \mathbf{G}_2, \mathbf{G}_3]$								
	Code number			Category[a]			$\mathbf{G}_1$ (FC)			$\mathbf{G}_2$ (SJ)				$\mathbf{G}_3$ (OT)	
	FC	SJ	OT	FC	SJ	OT	Sci	Med	Eng	Math	Bio	Phy	Chemo	Bs	Ap
1	3	4	2	Eng	Che	Ap	0	0	1	0	0	0	1	0	1
2	1	2	1	Sci	Bio	Bs	1	0	0	0	1	0	0	1	0
3	2	3	2	Med	Phy	Ap	0	1	0	0	0	1	0	0	1
4	1	1	1	Sci	Mat	Bs	1	0	0	1	0	0	0	1	0
5	2	2	1	Med	Bio	Bs	0	1	0	0	1	0	0	1	0
6	3	3	2	Eng	Phy	Ap	0	0	1	0	0	1	0	0	1
7	2	2	2	Med	Bio	Ap	0	1	0	0	1	0	0	0	1
8	1	3	1	Sci	Phy	Bs	1	0	0	0	0	1	0	1	0
9	2	4	2	Med	Che	Ap	0	1	0	0	0	0	1	0	1
10	3	1	1	Eng	Mat	Bs	0	0	1	1	0	0	0	1	0

[a]The names of categories are abbreviated as follows: Eng = engineering, Sci = sciences, Med = medicine; Che = chemistry, Bio = biology, Phy = physics, Mat = mathematics; Ap = applications, Bs = basis

1 = Sciences, 2 = Engineering, and 3 = Medicine. In Table 14.2(B), the elements of **Y** are presented as category names.

Each column of the data matrix in (A) or (B) can also be expressed as the *individuals × categories indicator* matrices

$$
\mathbf{G}_j = \begin{bmatrix} \mathbf{g}'_{1j} \\ \vdots \\ \mathbf{g}'_{ij} \\ \vdots \\ \mathbf{g}'_{nj} \end{bmatrix} \quad (j = 1, 2, 3), \tag{14.33}
$$

as in Table 14.2(C). Here, the jth variable in (A) or (B) corresponds to $\mathbf{G}_j$, and the kth element g_{ijk} in the ith row $\mathbf{g}'_{ij}$ is defined as:

$$
g_{ijk} = \begin{cases} 1 & \text{if } k = y_{ij} \\ 0 & \text{otherwise} \end{cases}. \tag{14.34}
$$

For example, $\mathbf{g}'_{82} = [0, 0, 1, 0]$, since individual-8 shows 3 (=Physics) for variable 2. The indicator matrix $\mathbf{G}_j$ in (14.33) can also be called a *membership* matrix, as described in Sect. 7.1, as $\mathbf{G}_j$ stands for the membership of individuals to categories.

Let the number of columns of $\mathbf{G}_j$ be $p_j, j = 1, \ldots, J$, and $p = \sum_{j=1}^{J} p_j$, as those for the data set in the last section. We further define an $n \times p$ block matrix as

$$
\mathbf{G} = \begin{bmatrix} \mathbf{G}_1, \ldots, \mathbf{G}_j, \ldots, \mathbf{G}_J \end{bmatrix}. \tag{14.35}
$$

In the next sections, we refer to **G** rather than $\mathbf{G}_j$ as an *indicator* matrix.

14.5   Multiple Correspondence Analysis

The loss function for *multiple correspondence analysis* (MCA) is given by replacing $\mathbf{X}_j$ by $\mathbf{G}_j$ in the GCCA function (14.16). That is, MCA is formulated as minimizing

$$
\eta(\mathbf{F}, \mathbf{C}) = \sum_{j=1}^{J} \left\| \mathbf{F} - \mathbf{G}_j \mathbf{C}_j \right\|^2 \tag{14.36}
$$

subject to (14.17) and an additional constraint,

$$\mathbf{1}'_n\mathbf{F} = \mathbf{0}'_m, \text{ or equivalently, } \mathbf{F} = \mathbf{J}\mathbf{F}, \tag{14.37}$$

with $m \leq \text{rank}(\mathbf{JG})$. The equivalence in (14.37) has been proved in Note 3.1 (Chap. 3). The K_j-*categories* × m-*dimensions* matrix $\mathbf{C}_j$ to be obtained is called a *category score* matrix, as its kth row stands for the vector of scores which is suitable to be given to category k, as explained in the next section. There, we also explain why we refer to the columns of $\mathbf{C}_j$ as dimensions. For the same reason, an n-individuals × m-dimensions matrix $\mathbf{F}$ is called an *individual score* matrix. Why constraint (14.37) is added is explained next:

Note 14.6. Avoiding Trivial Solutions
Let $m = 1$ for the sake of simplicity. Then, $\mathbf{F}$ and $\mathbf{C}_j$ in (14.36) are the column vectors. Without (14.37), the MCA loss function (14.36) attains the lower limit zero for

$$\mathbf{F} = \mathbf{1}_n \text{ and } \mathbf{C}_j = \mathbf{1}_{K_j}, \tag{14.38}$$

because (14.34) implies $\mathbf{G}_j\mathbf{1}_{K_j} = \mathbf{1}_n$. The solution in (14.38) is *trivial*, since it implies the same score of "one" given to all individuals and categories. This trivial solution does not satisfy (14.37); by adding it, the trivial one can be excluded from the solution.

As the minimization of (14.16) is equivalent to that of (14.29) in GCCA, the MCA problem, i.e., the minimization of (14.36) subject to (14.17) and (14.37), is equivalent to minimizing

$$h(\mathbf{F}, \mathbf{C}) = ||\mathbf{JGD}_{\mathrm{G}}^{-1/2} - \frac{1}{n}\mathbf{FC}'\mathbf{D}_{\mathrm{G}}^{1/2}||^2 \tag{14.39}$$

over $\mathbf{F}$ and $\mathbf{C}$ under the same constraints, which is detailed in Theorem A.4.7 (Appendix A.4.4). Further, the theorem shows that the MCA solution is given by

$$\mathbf{F} = \sqrt{n}\,\mathbf{S}_m\mathbf{T}, \tag{14.40}$$

$$\mathbf{C} = \sqrt{n}\mathbf{D}_{\mathrm{G}}^{-1/2}\mathbf{P}_m\Theta_m\mathbf{T}. \tag{14.41}$$

Here,

$$\mathbf{D}_{\mathrm{G}} = \begin{bmatrix} \mathbf{G}'_1\mathbf{G}_1 & & & \\ & \mathbf{G}'_2\mathbf{G}_2 & & \\ & & \ddots & \\ & & & \mathbf{G}'_J\mathbf{G}_J \end{bmatrix} \tag{14.42}$$

is the matrix in (14.27) with $\mathbf{X}_j$ replaced by $\mathbf{G}_j$, $\mathbf{T}$ is an $m \times m$ orthonormal matrix, and $\mathbf{S}_m$, $\mathbf{P}_m$, and $\mathbf{\Theta}_m$ are obtained through the SVD of $\mathbf{GD}_G^{-1/2}$ defined as:

$$\mathbf{JGD}_G^{-1/2} = \mathbf{S\Theta P'}. \tag{14.43}$$

Here, $\mathbf{S'S} = \mathbf{P'P} = \mathbf{I}_q$ with $q = \text{rank}(\mathbf{JG})$ and $\mathbf{\Theta}$ is a diagonal matrix whose diagonal elements are arranged in descending order. That is, $\mathbf{S}_m$ and $\mathbf{P}_m$ contain the first m columns of $\mathbf{S}$ and $\mathbf{P}$, respectively, with $\mathbf{\Theta}_m$ the first $m \times m$ diagonal block of $\mathbf{\Theta}$. In this chapter, we do not use a rotation technique by setting $\mathbf{T}$ in (14.40) and (14.41) at $\mathbf{I}_m$, as explained with (A.4.33) in Appendix A.4.4.

We must mention that the block diagonal matrix (14.42) is simply a *diagonal* one. This can be verified by the fact that the $\mathbf{G}_1$ in Table 14.2(C) implies

$\mathbf{G}_1'\mathbf{G}_1 = \begin{bmatrix} 3 & & \\ & 4 & \\ & & 3 \end{bmatrix}$. Thus, $(\mathbf{G}_1'\mathbf{G}_1)^{-1/2} = \begin{bmatrix} 1/\sqrt{3} & & \\ & 1/\sqrt{4} & \\ & & 1/\sqrt{3} \end{bmatrix}$. In general,

$\mathbf{G}_j'\mathbf{G}_j$ and $(\mathbf{G}_j'\mathbf{G}_j)^{-1/2}$ ($j = 1, \dots, J$) are diagonal matrices, which imply that $\mathbf{D}_G$ and $\mathbf{D}_G^{-1/2}$ are also diagonal.

14.6   Homogeneity Assumption

Table 14.3 presents the MCA solution of $\mathbf{F}$ and $\mathbf{C} = [\mathbf{C}_1', \dots, \mathbf{C}_J']'$ for the data set in Table 14.2 with $m = 2$. In Table 14.3, $\mathbf{f}_i'$ denotes the ith row of $\mathbf{F}$, which corresponds to the ith individual in Table 14.2, and $\mathbf{c}_{jk}'$ denotes the kth row of $\mathbf{C}_j$, which is associated with category k in variable j; for example, $\mathbf{c}_{23}'$ contains the scores for Phy (physics). The solution in Table 14.3 can be graphically represented as in Fig. 14.2, where individual i ($=1, \dots, n$) is plotted as the point with its coordinate $\mathbf{f}_i$, and category k in variable j is plotted with its coordinate $\mathbf{c}_{jk}$. We can interpret the plot by noting inter-point *distances*. The rationale for this distance-based interpretation of MCA solutions is explained in the following paragraph.

Table 14.3 MCA solution for the data in Table 14.2

F			C				
$\mathbf{f}_1'$	1.20	1.20	$\mathbf{C}_1$	$\mathbf{c}_{11}'$	Sci	−1.19	−0.10
$\mathbf{f}_2'$	−1.12	−0.94		$\mathbf{c}_{12}'$	Med	0.63	−0.84
$\mathbf{f}_3'$	0.83	−0.38		$\mathbf{c}_{13}'$	Eng	0.36	1.23
$\mathbf{f}_4'$	−1.56	0.59	$\mathbf{C}_2$	$\mathbf{c}_{21}'$	Math	−1.19	1.03
$\mathbf{f}_5'$	−0.27	−1.44		$\mathbf{c}_{22}'$	Bio	−0.26	−1.25
$\mathbf{f}_6'$	0.71	1.01		$\mathbf{c}_{23}'$	Phy	0.21	0.23
$\mathbf{f}_7'$	0.61	−1.37		$\mathbf{c}_{24}'$	Che	1.26	0.50
$\mathbf{f}_8'$	−0.90	0.05	$\mathbf{C}_3$	$\mathbf{c}_{31}'$	Bs	−0.93	−0.05
$\mathbf{f}_9'$	1.32	−0.19		$\mathbf{c}_{32}'$	Ap	0.93	0.05
$\mathbf{f}_{10}'$	−0.83	1.48					

MCA can be reformulated with the *homogeneity assumption*:

The scores for an *individual* should be *homogeneous* to

the scores for the *categories* to which the *individual belongs*.   (14.44)

Here, the underlined scores are expressed as the vector $\mathbf{c}_{jy_{ij}}$, which is the *category score* vector $\mathbf{c}'_{jk}$ with k set to the category y_{ij} (the *category number* that individual i shows for variable j). Assumption (14.44) requires $||\mathbf{f}'_i - \mathbf{c}'_{jy_{ij}}||^2$ to be small, and its sum over i and j can be expressed as:

$$\sum_{j=1}^{J}\sum_{i=1}^{n}\left\|\mathbf{f}'_i - \mathbf{c}'_{jy_{ij}}\right\|^2 = \sum_{j=1}^{J}\sum_{i=1}^{n}\left\|\mathbf{f}'_i - \mathbf{g}'_{ij}\mathbf{C}_j\right\|^2 = \sum_{j=1}^{J}\left\|\begin{bmatrix}\mathbf{f}'_1\\ \vdots\\ \mathbf{f}'_n\end{bmatrix} - \begin{bmatrix}\mathbf{g}'_{1j}\\ \vdots\\ \mathbf{g}'_{nj}\end{bmatrix}\mathbf{C}_j\right\|^2.$$

(14.45)

Here, we have used

$$\mathbf{g}'_{ij}\mathbf{C}_j = \mathbf{g}'_{ij}[\mathbf{c}_{j1}, \ldots, \mathbf{c}_{jK_j}]' = \sum_{k=1}^{K_j}g_{ijk}\mathbf{c}_{jk} = \mathbf{c}'_{jy_{ij}},$$ (14.46)

because of (14.34). We can find the equivalence of (14.45) to (14.36) by noting (14.33) and $\mathbf{F} = [\mathbf{f}_1, \ldots, \mathbf{f}_n]'$.

The inter-point distances in Fig. 14.2 allow us to capture the relationships among categories, among individuals, and between categories and individuals; we can consider the entities near one another to share similar features. For example, [1] the point for sciences is found to be close to that for basis, which shows that the students in the Department of Science tend to regard basic sciences as important; [2] individuals 1 and 6 are similar students; [3] individual 3 is involved with medicine and applied sciences.

Fig. 14.2 Scatter plot of categories and individuals

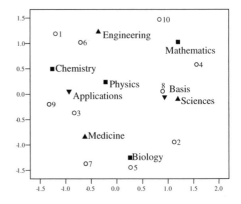

14.7   Bibliographical Notes

CCA is intelligibly introduced in Lattin et al. (2003) with real data examples and detailed in Izenman (2008) and Kock (2014). The formulations of GCCA and MCA in this chapter are detailed in Gifi (1990). MCA is also intelligibly treated in Greenacre (2007). The analysis procedure called "correspondence analysis", with the "multiple" deleted from MCA, is treated only in the next exercises. The relationships between correspondence analysis and MCA are detailed in Greenacre (1984, 2007).

We must mention that various terms have been used for referring to MCA and related procedures. For example, the term *homogeneity analysis* has been used in Gifi (1990). Other terms can be found in Hayashi (1952), Nishisato (1980), and Young (1981).

Exercises

14.1. Show that (14.16) can be rewritten as $\|\mathbf{1}_J \otimes \mathbf{F} - \mathbf{D}_X \mathbf{C}\|^2$, where $\mathbf{D}_X =$

$$\begin{bmatrix} \mathbf{X}_1 & & \\ & \ddots & \\ & & \mathbf{X}_J \end{bmatrix}$$ is the $nJ \times p$ block diagonal matrix, where blank cells

indicate zero elements and $\otimes$ denotes the *Kronecker product* defined as follows: For $\mathbf{A} = (a_{ij})$ $(n \times m)$ and $\mathbf{B} = (b_{ij})$ $(p \times q)$, $np \times mq$, matrices are

obtained   as $\mathbf{A} \otimes \mathbf{B} = \begin{bmatrix} a_{11}\mathbf{B} & \cdots & a_{1m}\mathbf{B} \\ & \vdots & \\ a_{n1}\mathbf{B} & \cdots & a_{nm}\mathbf{B} \end{bmatrix}$ and $\mathbf{B} \otimes \mathbf{A} =$

$\begin{bmatrix} b_{11}\mathbf{A} & \cdots & b_{1q}\mathbf{A} \\ & \vdots & \\ b_{p1}\mathbf{A} & \cdots & b_{pq}\mathbf{A} \end{bmatrix}$ (e.g., Harville 1997, pp. 333–339).

14.2. Discuss the similarities and differences between GCCA and the generalized orthogonal rotation in Exercise 13.10.

14.3. Let $\mathbf{Z} = [\mathbf{z}_1, \ldots, \mathbf{z}_p]$ contain standard scores with $\mathbf{1}'_n \mathbf{z}_j = 0$ and $n^{-1}\mathbf{z}'_j\mathbf{z}_j = 1$ $(j = 1, \ldots, p)$. We can substitute $\mathbf{z}_j$ for $\mathbf{X}_j$ in (14.16) to rewrite it as $\eta(\mathbf{F}, \mathbf{A}) = \sum_{j=1}^p \left\| \mathbf{F} - \mathbf{z}_j \mathbf{a}'_j \right\|^2$, with $\mathbf{C}$ in (14.16) replaced by $\mathbf{A} = [\mathbf{a}_1, \ldots, \mathbf{a}_p]'$. Note the equivalence between (14.16) and (14.29) to show that the minimization of $\eta(\mathbf{F}, \mathbf{A})$ subject to (14.17) is equivalent to the principal component analysis for $\mathbf{Z}$, i.e., minimizing $\|\mathbf{Z} - \mathbf{FA}'\|^2$ under (14.17).

14.4. Show that the function (14.29) multiplied by n can be rewritten as:

$$nf(\mathbf{F}, \mathbf{C}) = \|\mathbf{X}\mathbf{D}_V^{-1/2} - \mathbf{F}\mathbf{C}'\mathbf{D}_V^{1/2}\|^2$$

with $\mathbf{D}_V = \begin{bmatrix} \mathbf{V}_1 & & \\ & \ddots & \\ & & \mathbf{V}_J \end{bmatrix}$ the block diagonal matrix, whose jth block $\mathbf{V}_j$

is defined as $\mathbf{V}_j = n^{-1}\mathbf{X}_j'\mathbf{X}_j$ and is the covariance matrix for $\mathbf{X}_j$ if it is centered.

14.5 Let us constrain $\mathbf{C}_j$ in (14.36) to be $\mathbf{C}_j = \mathbf{q}_j\,\mathbf{a}_j'$, with $\mathbf{q}_j$ and $\mathbf{a}_j$ being $K_j \times 1$ and $m \times 1$ vectors, respectively. Then, (14.36) is rewritten as $\eta(\mathbf{F}, \mathbf{q}_j, \mathbf{a}_j) = \sum_{j=1}^{J} \|\mathbf{F} - \mathbf{G}_j\mathbf{q}_j\mathbf{a}_j'\|^2$. Show that its minimization over $\mathbf{F}$, $\mathbf{q}_1$, $\dots$, $\mathbf{q}_J$, and $\mathbf{A} = [\mathbf{a}_1, \dots, \mathbf{a}_J]$ subject to (14.37) and $\mathbf{G}_j\mathbf{q}_j$ being standardized, is equivalent to minimizing $\|\mathbf{G}_{(\mathbf{q})} - \mathbf{F}\mathbf{A}'\|^2$ under the same constraints with $\mathbf{G}_{(\mathbf{q})} = [\mathbf{G}_1\mathbf{q}_1, \mathbf{G}_2\mathbf{q}_2, \dots, \mathbf{G}_J\mathbf{q}_J]$ an $n \times J$ matrix (Gifi, 1990).

14.6. Argue that the assignment of quantitative scores to categories and PCA are simultaneously performed in the procedure considered in Exercise 14.5.

14.7. Show that $\mathbf{N} = (n_{kl}) = \mathbf{G}_1'\mathbf{G}_2$ represents the $p_1 \times p_2$ *contingency table* whose element n_{kl} expresses the number of individuals classified into category k for variable 1 and category l for variable 2.

14.8. Show that $\mathbf{G}_1'\mathbf{J}\mathbf{G}_2 = \mathbf{N} - n^{-1}\mathbf{D}_1\mathbf{1}_{p_1}\mathbf{1}_{p_2}'\mathbf{D}_2$, with $\mathbf{N}$ defined in Exercise 14.7, $\mathbf{J} = \mathbf{I}_n - n^{-1}\mathbf{1}_n\mathbf{1}_n'$, and $\mathbf{D}_j$ the $p_j \times p_j$ diagonal matrix whose kth diagonal element is the number of the individuals classified into category k for variable j ($= 1, 2$).

14.9. The procedure called "*correspondence analysis*" that involves removing "multiple" from "multiple correspondence analysis" is used for the contingency table $\mathbf{N}$ defined in Exercise 14.7 (Greenacre 1984). The loss function of correspondence analysis is expressed as:

$$f(\mathbf{C}_1, \mathbf{C}_2) = \|\tilde{\mathbf{N}} - \frac{1}{n}\mathbf{D}_1^{1/2}\mathbf{C}_1\mathbf{C}_2'\mathbf{D}_2^{1/2}\|^2, \tag{14.48}$$

which is minimized over $\mathbf{C}_1$ and $\mathbf{C}_2$, with $\tilde{\mathbf{N}} = \mathbf{D}_1^{-1/2}(\mathbf{N} - n^{-1}\mathbf{D}_1\mathbf{1}_{p_1}\mathbf{1}_{p_2}'\mathbf{D}_2)\mathbf{D}_2^{-1/2}$ $= \mathbf{D}_1^{-1/2}\mathbf{G}_1'\mathbf{J}\mathbf{G}_2\mathbf{D}_2^{-1/2}$. Show that (14.48) is minimized for

$$\mathbf{C}_1 = \sqrt{n}\,\mathbf{D}_1^{-1/2}\mathbf{U}_m\boldsymbol{\Delta}_m^{1/2} \quad \text{and} \quad \mathbf{C}_2 = \sqrt{n}\,\mathbf{D}_2^{-1/2}\mathbf{V}_m\boldsymbol{\Delta}_m^{1/2} \tag{14.49}$$

subject to $\mathbf{C}_1'\mathbf{D}_1\mathbf{C}_1 = \mathbf{C}_2'\mathbf{D}_2\mathbf{C}_2$ being a diagonal matrix. Here, $\mathbf{U}_m$ and $\mathbf{V}_m$ contain the first m columns of $\mathbf{U}$ and $\mathbf{V}$, respectively, while $\boldsymbol{\Delta}_m$ is the first $m \times m$ diagonal block of $\boldsymbol{\Delta}$, with $\mathbf{U}$, $\mathbf{V}$, and $\boldsymbol{\Delta}$ obtained from the SVD $\tilde{\mathbf{N}} = \mathbf{U}\boldsymbol{\Delta}\mathbf{V}'$.

14.10. The solution of MCA for $\mathbf{G}$ with $K = 2$, i.e., $\mathbf{G} = [\mathbf{G}_1, \mathbf{G}_2]$, is given through the SVD (14.43) with $K = 2$, which is rewritten as:

$$\mathbf{J}[\mathbf{G}_1, \mathbf{G}_2]\begin{bmatrix} \mathbf{D}_1^{-1/2} & \\ & \mathbf{D}_2^{-1/2} \end{bmatrix} = \mathbf{S}\boldsymbol{\Theta}[\mathbf{P}_1', \mathbf{P}_2'], \tag{14.50}$$

with $\mathbf{P}_1$ $(p_1 \times r)$ and $\mathbf{P}_2$ $(p_2 \times r)$ the blocks of $\mathbf{P} = \begin{bmatrix} \mathbf{P}_1 \\ \mathbf{P}_2 \end{bmatrix}$. Show that (14.50) leads to

$$\begin{bmatrix} \mathbf{D}_1^{-1/2} & \\ & \mathbf{D}_2^{-1/2} \end{bmatrix} \begin{bmatrix} \mathbf{G}_1' \\ \mathbf{G}_2' \end{bmatrix} \mathbf{J}[\mathbf{G}_1, \mathbf{G}_2] \begin{bmatrix} \mathbf{D}_1^{-1/2} & \\ & \mathbf{D}_2^{-1/2} \end{bmatrix} = \begin{bmatrix} \mathbf{P}_1 \\ \mathbf{P}_2 \end{bmatrix} \Theta[\mathbf{P}_1', \mathbf{P}_2']$$

and that its left-hand side can be rewritten as $\begin{bmatrix} \mathbf{M}_1 & \tilde{\mathbf{N}} \\ \mathbf{N}' & \mathbf{M}_2 \end{bmatrix}$, which imply the equivalence of the correspondence analysis to the MCA for $[\mathbf{G}_1, \mathbf{G}_2]$ with the constraint of $\mathbf{C}_1' \mathbf{D}_1 \mathbf{C}_1 = \mathbf{C}_2' \mathbf{D}_2 \mathbf{C}_2$ being a diagonal matrix. Here, the symbols have been the ones defined in Exercises 14.8 and 14.9, with $\mathbf{M}_j = \mathbf{I}_{p_j} - n^{-1} \mathbf{D}_j^{1/2} \mathbf{1}_{p_j} \mathbf{1}'_{p_j} \mathbf{D}_j^{1/2}$.

14.11. If the columns of $\mathbf{X}_j$, $j = 1, \ldots, J$, in (14.2) are occupied by the same set of K entities, show that (14.2) can be rewritten as *three-way data array* $\{x_{ijk};$ $i = 1, \ldots, n, j = 1, \ldots, J, k = 1, \ldots, K\}$.

14.12. For the three-way array in Exercise 14.11, a modified PCA procedure called *Tucker3* is modeled as $x_{ijk} = \sum_{p=1}^{P} \sum_{q=1}^{Q} \sum_{r=1}^{R} a_{ip} b_{jq} c_{kr} z_{pqr} + e_{ijk}$, with e_{ijk} an error, $P \leq n$, $Q \leq J$, and $R \leq K$ (Tucker 1966). Show that the Tucker3 model is rewritten as $\mathbf{X} = \mathbf{A} \mathbf{Z} (\mathbf{C} \otimes \mathbf{B})' + \mathbf{E}$. Here, $\otimes$ denotes the Kronecker product introduced in Exercise 14.1, $\mathbf{A} = (a_{ip})$ $(n \times P)$, $\mathbf{B} = (b_{jq})$ $(J \times Q)$, $\mathbf{C} = (c_{kr})$ $(K \times R)$, $\mathbf{E}$ contains errors, and $\mathbf{Z} = [\mathbf{Z}_1, \ldots, \mathbf{Z}_R]$ is the $P \times RQ$ block matrix whose rth block $\mathbf{Z}_r$ is the $P \times Q$ matrix with its (p, q) element z_{pqr}.

14.13. The solutions of the Tucker3 model in Exercise 14.12 are given by minimizing $f = \|\mathbf{X} - \mathbf{A} \mathbf{Z} (\mathbf{C} \otimes \mathbf{B})'\|^2$ over $\mathbf{A}$, $\mathbf{B}$, $\mathbf{C}$, and $\mathbf{Z}$. Use $\mathbf{V}_A = (\mathbf{A}' \mathbf{A})^{-1} \mathbf{A}'$, $\mathbf{H} = \mathbf{C} \otimes \mathbf{B}$, and $\mathbf{V}_H = (\mathbf{H}' \mathbf{H})^{-1} \mathbf{H}'$ to show that f can be rewritten as $\|\mathbf{X} - \mathbf{A} \mathbf{Z} \mathbf{H}'\|^2 = \|\mathbf{X} - \mathbf{A} \mathbf{V}_A \mathbf{X} \mathbf{V}_H' \mathbf{H}'\|^2 + \|\mathbf{A} \mathbf{V}_A \mathbf{X} \mathbf{V}_H' \mathbf{H}' - \mathbf{A} \mathbf{Z} \mathbf{H}'\|^2$, which implies the solution of $\mathbf{Z}$ satisfying $\mathbf{Z} = \mathbf{V}_A \mathbf{X} \mathbf{V}_H'$.

14.14. For the three-way array in Exercise 14.11, a modified PCA procedure called *Parafac* is modeled as $x_{ijk} = \sum_{p=1}^{P} a_{ip} b_{jp} c_{kp} + e_{ijk}$, with e_{ijk} an error and $P \leq \min(n, J, K)$ (Harshman 1970; Hitchcock 1927). Show that the Parafac model can be rewritten as $\mathbf{X} = \mathbf{A} (\mathbf{C} \cdot \mathbf{B})' + \mathbf{E}$. Here, $\mathbf{A} = (a_{ip}) = [\mathbf{a}_1, \ldots, \mathbf{a}_P]$ $(n \times P)$, $\mathbf{B} = (b_{jq}) = [\mathbf{b}_1, \ldots, \mathbf{b}_P]$ $(J \times P)$, $\mathbf{C} = (c_{kr}) = [\mathbf{c}_1, \ldots, \mathbf{c}_P]$ $(K \times P)$, and $\cdot$ denotes the *Khatri-Rao product* defined as $\mathbf{C} \cdot \mathbf{B} = [\mathbf{c}_1, \ldots, \mathbf{c}_P] \cdot [\mathbf{b}_1, \ldots, \mathbf{b}_P] = [\mathbf{c}_1 \otimes \mathbf{b}_1, \ldots, \mathbf{c}_P \otimes \mathbf{b}_P]$ (Rao and Mira 1971).

14.15. Parafac is formulated as minimizing

$$\begin{aligned} f_P(\mathbf{A}, \mathbf{B}, \mathbf{C}) &= \|\mathbf{X} - \mathbf{A} (\mathbf{C} \cdot \mathbf{B})'\|^2 = \|\mathbf{X}^\# - \mathbf{B} (\mathbf{C} \cdot \mathbf{A})'\|^2 \\ &= \|\mathbf{X}^* - \mathbf{C} (\mathbf{B} \cdot \mathbf{A})'\|^2 \end{aligned} \tag{14.54}$$

over $\mathbf{A}$, $\mathbf{B}$, and $\mathbf{C}$. Here, $\mathbf{X}^{\#} = [\mathbf{X}'_1, \ldots, \mathbf{X}'_K]$ and $\mathbf{X}^* = [\text{vec}(\mathbf{X}_1), \ldots, \text{vec}$

$(\mathbf{X}_K)]'$ with vec($\mathbf{M}$) denoting the vec operator defined as vec $(\mathbf{M}) =$

$\begin{bmatrix} \mathbf{m}_1 \\ \vdots \\ \mathbf{m}_p \end{bmatrix}$ $(pn \times 1)$ for $n \times p$ matrix $\mathbf{M} = [\mathbf{m}_1, \ldots, \mathbf{m}_p]$ (e.g., Harville 1997,

pp. 339–343). Show the equality among the three functions in (14.54).

14.16. Let $\mathbf{W}_{(\mathbf{C},\mathbf{B})}$ be the function of $\mathbf{B}$ and $\mathbf{C}$ yielding $\mathbf{W}_{(\mathbf{C},\mathbf{B})} = (\mathbf{C} \cdot \mathbf{B})\{(\mathbf{C} \cdot \mathbf{B})'(\mathbf{C} \cdot \mathbf{B})\}^{-1}$. Show that the algorithm for minimizing (15.54) can be formed by the following steps:

Step 1. Initialize $\mathbf{B}$ and $\mathbf{C}$.
Step 2. Update $\mathbf{A} = \mathbf{X}\mathbf{W}_{(\mathbf{C},\mathbf{B})}$.
Step 3. Update $\mathbf{B} = \mathbf{X}^{\#}\mathbf{W}_{(\mathbf{C},\mathbf{A})}$.
Step 4. Update $\mathbf{C} = \mathbf{X}^* \mathbf{W}_{(\mathbf{B}, \mathbf{A})}$.
Step 5. Finish if convergence is reached; otherwise, go back to Step 2.

Chapter 15
Discriminant Analysis

Discriminant analysis refers to a group of statistical procedures for analyzing a data set with individuals *classified* into certain *groups*, where the results of the analysis are used for *finding* the *group* of a *new individual* that is not included in the above data set. The sections in this chapter can be classified into two parts: [1] Sects. 15.1–15.3 concern an approach *without using probabilities* and [2] the remaining sections concern *probabilistic approaches*. In [1], a *canonical discriminant analysis* (*CDA*) procedure is introduced by modifying the multiple correspondence analysis in the last chapter. In [2], we introduce two probabilistic procedures using multivariate normal distributions. One of them is *linear discriminant analysis* (*LDA*), which is rooted in Fisher (1936). The other is a *generalization* of LDA.

15.1 Modification of Multiple Correspondence Analysis

The *multiple correspondence analysis* (*MCA*) in the last chapter is performed for the n-individuals $\times$ K-categories membership matrix (14.35). Here, let us consider a case where $J = 1$, i.e., $\mathbf{G} = \mathbf{G}_1$, and an n-individuals $\times$ p-variables quantitative data matrix $\mathbf{X}$ is given, with $\mathbf{1}'_n\mathbf{X} = \mathbf{0}'_p$. That is, the data set is expressed as an $n \times (K + p)$ block matrix $[\mathbf{G}, \mathbf{X}]$. An example of $[\mathbf{G}, \mathbf{X}]$ is shown in Table 15.1 (Fisher 1936), in which individuals are irises whose categories are indicated by $\mathbf{G}$ and the individuals' features are described by $\mathbf{X}$. In this chapter, the column entities of $\mathbf{G}$ are called *groups* rather than categories.

For the above $\mathbf{G}$, the MCA loss function (14.36) is simplified into $\|\mathbf{F} - \mathbf{GC}\|^2$ without the symbol of summation and the subscript for $\mathbf{C}$. Here, let the individual score matrix $\mathbf{F}$ be constrained as:

© Springer Nature Singapore Pte Ltd. 2016
K. Adachi, *Matrix-Based Introduction to Multivariate Data Analysis*,
DOI 10.1007/978-981-10-2341-5_15

Table 15.1 Membership of irises for groups 1, 2, and 3 (**G**) and standardized scores for features of the irises (**X**)

Iris	G			X			
	1	2	3	SL*	SW*	PL*	PW*
1	1	0	0	−0.90	1.02	−1.34	−1.31
2	1	0	0	−1.14	−0.13	−1.34	−1.31
.	.	.	.	.	.	.	.
.	.	.	.	.	.	.	.
.	.	.	.	.	.	.	.
50	1	0	0	−1.02	0.56	−1.34	−1.31
51	0	1	0	1.40	0.33	0.53	0.26
52	0	1	0	0.67	0.33	0.42	0.39
.	.	.	.	.	.	.	.
.	.	.	.	.	.	.	.
.	.	.	.	.	.	.	.
100	0	1	0	−0.17	−0.59	0.19	0.13
101	0	0	1	0.55	0.56	1.27	1.71
102	0	0	1	−0.05	−0.82	0.76	0.92
.	.	.	.	.	.	.	.
.	.	.	.	.	.	.	.
.	.	.	.	.	.	.	.
150	0	0	1	0.07	−0.13	0.76	0.79

The original data are available at http://astro.temple.edu/∼alan/ MMST/datasets.htm (Izenman 2008)

*SL sepal length, SW sepal width, PL petal length, PW petal width

$$\mathbf{F} = \mathbf{XB}, \tag{15.1}$$

with **B** a $p \times m$ coefficient matrix. Using (15.1) in $||\mathbf{F} - \mathbf{GC}||^2$, it is rewritten as:

$$\eta(\mathbf{B}, \mathbf{C}) = ||\mathbf{XB} - \mathbf{GC}||^2. \tag{15.2}$$

Further, the substitution of (15.1) into constraint (14.17) leads to

$$\frac{1}{n}\mathbf{B}'\mathbf{X}'\mathbf{XB} = \mathbf{I}_m. \tag{15.3}$$

Minimizing (15.2) over **B** and **C** subject to (15.3) is called *canonical discriminant analysis* (*CDA*), whose solution is detailed in the following section.

Note 15.1. Comparison to Cluster Analysis

Let us compare (15.2) with the loss function (7.4) in *k*-means clustering (Chap. 7). Deleting **B** from (15.2) leads to (7.4). Further, **G** is *known* in (15.2) (discriminant analysis), while **G** is *unknown* and to be obtained in (7.4) (cluster analysis).

15.2 Canonical Discriminant Analysis

As shown in Appendix A.2.2, (15.2) is minimized for

$$\mathbf{C} = (\mathbf{G}'\mathbf{G})^{-1}\mathbf{G}'\mathbf{F} = \mathbf{D}_G^{-1}\mathbf{G}'\mathbf{XB}, \tag{15.4}$$

given **G**, with $\mathbf{D}_G = \mathbf{G}'\mathbf{G}$, a $K \times K$ diagonal matrix. We can substitute (15.4) in (15.2) to rewrite it as:

$$\begin{aligned}
\eta(\mathbf{B}) &= \|\mathbf{XB} - \mathbf{GD}_G^{-1}\mathbf{G}'\mathbf{XB}\|^2 \\
&= \operatorname{tr} \mathbf{B}'\mathbf{X}'\mathbf{XB} - \operatorname{tr} \mathbf{B}'\mathbf{X}'\mathbf{GD}_G^{-1}\mathbf{G}'\mathbf{XB} = nm - \operatorname{tr} \mathbf{B}'\mathbf{X}'\mathbf{GD}_G^{-1}\mathbf{G}'\mathbf{XB},
\end{aligned} \tag{15.5}$$

where we have used (15.3). The minimization of (15.5) under (15.3) is equivalent to maximizing $\operatorname{tr} \mathbf{B}'\mathbf{X}'\mathbf{GD}_G^{-1}\mathbf{G}'\mathbf{XB}$ subject to (15.3), whose solution is given as in Theorem A.4.9 (Appendix A.4.5). There, by setting $\mathbf{M} = \mathbf{UU}'$ and **V** in (A.4.41) to $\mathbf{X}'\mathbf{GD}_G^{-1}\mathbf{G}'\mathbf{X}$ and $\mathbf{V} = n^{-1}\mathbf{X}'\mathbf{X}$, respectively, we have the solution for **B**, as in (A.4.43).

> **Note 15.2. Another Formulation of CDA**
> As found above, CDA can be formulated as maximizing $\rho(\mathbf{B}) = \operatorname{tr}\mathbf{B}'\mathbf{SB}$ over **B** subject to (15.3) with $\mathbf{S} = \mathbf{X}'\mathbf{GD}_G^{-1}\mathbf{G}'\mathbf{X}$. In a more popular introduction of CDA, 15.3 is replaced by $\mathbf{B}'\mathbf{WB} = \mathbf{I}_m$ with $\mathbf{W} = n^{-1}(\mathbf{X}'\mathbf{X}-\mathbf{S})$: CDA is formulated as maximizing $\rho(\mathbf{B})$ under $\mathbf{B}'\mathbf{WB} = \mathbf{I}_m$. A reason for using (15.3) in this book is relate CDA to MCA and use (15.2) as the CDA loss function.

Let us express the ith row of **X** as $\mathbf{x}_i' = [x_{i1}, \ldots, x_{ip}]$ and the lth column of **B** as $\mathbf{b}_l = [b_{1l}, \ldots, b_{pl}']$. Then, the (i, l) element of (15.1) is expressed as:

$$f_{il} = \mathbf{x}_i'\mathbf{b}_l = b_{1l}x_{i1} + \cdots + b_{pl}x_{ip}, \tag{15.6}$$

the weighted sum of the p variables in $\mathbf{x}_i$. Sum (15.6) is called the lth *discriminant score* for individual i and the weights $b_{1l}, \ldots, b_{pl}$ are called the lth *discriminant coefficients*, with $l = 1, \ldots, m$. The other parameter matrix in CDA is **C**. Its rows of $\mathbf{C} = [\mathbf{c}_1, \ldots, \mathbf{c}_K]'$ are associated with groups and the kth row $\mathbf{c}_k'$ ($1 \times m$) can be called the kth *group score vector*, as it stands for the features of the group.

Let us consider performing CDA for the iris data in Table 15.1, setting $m = 2$. This gives us **F**, whose ith row is expressed as $\mathbf{f}_i' = [f_{i1}, f_{i2}] = [\mathbf{x}_i'\mathbf{b}_1, \mathbf{x}_i'\mathbf{b}_2] = \mathbf{x}_i'\mathbf{B}$,

Fig. 15.1 Plots of
individuals' discriminant
scores and group scores

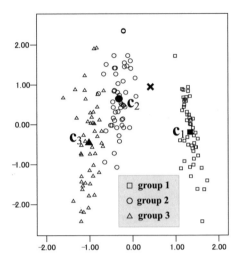

i.e., two discriminant scores for each individual. The resulting scores for the data set
are expressed as:

$$\mathbf{x}_i'\mathbf{b}_1 = 0.12 \times \text{SL} + 0.12 \times \text{SW} - 0.68 \times \text{PL} - 0.38 \times \text{PW}, \tag{15.7}$$

$$\mathbf{x}_i'\mathbf{b}_2 = -0.02 \times \text{SL} - 0.84 \times \text{SW} + 1.47 \times \text{PL} - 1.94 \times \text{PW}, \tag{15.8}$$

where the names of the variables in Table 15.1 and the solutions of the coefficients
are substituted into x_{i1}, …, x_{ip} and b_{1l}, … b_{pl} in (15.6), respectively. For example,
the elements of the data vector $\mathbf{x}_1'$ (= [SL, SW, PL, PW]) = [−0.90, 1.02, −1.34,
−1.31] for individual 1 can be substituted into the variables in (15.7) and (15.8), so
that the discriminant score vector for individual 1 is given as $\mathbf{f}_1' = \mathbf{x}_1'\mathbf{B} = [\mathbf{x}_1'\mathbf{b}_1,$
$\mathbf{x}_1'\mathbf{b}_2] = [1.42, -0.28]$. In Fig. 15.1, the vectors for all individuals, $\mathbf{f}_i' = \mathbf{x}_i'\mathbf{B}$ ($i = 1,$
…, 150), are plotted, with squares, circles, and triangles used for the individuals in
the group 1, 2, and 3, respectively.

The CDA for the data in Table 15.1 also give the solution of the group scores as:

$$\mathbf{C} = \begin{bmatrix} \mathbf{c}_1' \\ \mathbf{c}_2' \\ \mathbf{c}_3' \end{bmatrix} = \begin{bmatrix} 1.33 & -0.19 \\ -0.31 & 0.65 \\ -1.01 & -0.46 \end{bmatrix}. \tag{15.9}$$

In Fig. 15.1, $\mathbf{c}_1$, $\mathbf{c}_2$, and $\mathbf{c}_3$ are represented as a filled square, circle, and triangle,
respectively. There, we can find that the discriminant scores for the individuals in
the same group are distributed mutually close, with their *center* being the *group
score* vector. This can be mathematically shown in the next section.

15.3   Minimum Distance Classification

Equation (15.4) for $\mathbf{C} = [\mathbf{c}_1, \ldots, \mathbf{c}_K]'$ implies that its kth row $\mathbf{c}'_k$ is the *centroid*, i.e., the averaged vector of the discriminant score vectors $\mathbf{f}'_i$ for the individuals belonging to the group k:

$$\mathbf{c}'_k = \frac{1}{n_k}\sum_{i\in g_k}\mathbf{f}'_i = \frac{1}{n_k}\sum_{i\in g_k}\mathbf{x}'_i\mathbf{B}. \tag{15.10}$$

Here, g_k expresses the set of the individuals in the group k with their number denoted by n_k, and $\sum_{i\in g_k}\mathbf{f}'_i$ stands for the summation of $\mathbf{f}'_i$ over i in the group k. The rows of $\mathbf{C}$ being averages, can be verified by the following example: (15.4) is expressed as:

$$\mathbf{C} = \mathbf{D}_G^{-1}\mathbf{G}'\begin{bmatrix}\mathbf{f}'_1\\\mathbf{f}'_2\\\mathbf{f}'_3\\\mathbf{f}'_4\\\mathbf{f}'_5\end{bmatrix} = \begin{bmatrix}\frac{1}{3}(\mathbf{f}'_2+\mathbf{f}'_3+\mathbf{f}'_5)\\\frac{1}{2}(\mathbf{f}'_1+\mathbf{f}'_4)\end{bmatrix}, \text{ when } \mathbf{G} = \begin{bmatrix}0&1\\1&0\\1&0\\0&1\\1&0\end{bmatrix}, \text{ thus } \mathbf{D}_G = \begin{bmatrix}3&0\\0&2\end{bmatrix}.$$

Further, we can ascertain the closeness of the vectors $\mathbf{f}'_i = \mathbf{x}'_i\mathbf{B}$ in the group k to $\mathbf{c}'_k$ from the fact that (15.2) is rewritten as:

$$\eta(\mathbf{B},\mathbf{C}) = \sum_{i=1}^{n}\left\|\mathbf{x}'_i\mathbf{B} - \mathbf{g}'_i\mathbf{C}\right\|^2 = \sum_{i=1}^{n}\left\|\mathbf{x}'_i\mathbf{B} - \mathbf{c}'_{y_i}\right\|^2. \tag{15.11}$$

Here, $\mathbf{g}'_i$ is the ith row of $\mathbf{G}$, y_i is the index number of the group to which individual i belongs, and we have used $\mathbf{g}'_i\mathbf{C} = \mathbf{c}'_{y_i}$. This implies that CDA is also based on the *homogeneity assumption*:

> the scores for an *individual* should be *homogeneous* to
> the scores for the *group* to which the *individual belongs*, $\tag{15.12}$

which is the same as (14.44) except the term "categories" has been replaced by "group." Minimizing (15.11) allows $\mathbf{x}'_i\mathbf{B}$ to be close to $\mathbf{c}'_{y_i}$, with $\mathbf{c}'_{y_i}$ being the score of the group including individual i, which is also the centroid of the individual scores in that group, as shown in (15.10).

Let $\mathbf{x}'_?$ be a $1 \times p$ vector which is *not included* in $\mathbf{X}$, so that it is *unknown* to what *group* $\mathbf{x}'_?$ belongs. That is, our task is to *classify* $\mathbf{x}'_?$ into one of the groups $k = 1, \ldots, K$, in other words, to find the group to which $\mathbf{x}'_?$ should belong. Assumption (15.12) leads to the following *minimum distance classification*:

$$\mathbf{x} \text{ is classified into group } k^* \text{ with}\left\|\mathbf{x}'\mathbf{B} - \mathbf{c}'_{k*}\right\| = \min_{1\le k\le K}\left\|\mathbf{x}'\mathbf{B} - \mathbf{c}'_k\right\|. \tag{15.13}$$

Here, $\mathbf{x}'$ generally expresses a $1 \times p$ vector whose elements are associated with the p variables in $\mathbf{X}$. We illustrate the classification rule (15.13) with $\mathbf{x}$ equaling

$\mathbf{x}_?' = [1.8, 0.4, 0.1, -0.6]$. This is substituted into (15.6) to provide $\mathbf{x}_?'\mathbf{B} = [0.42, 0.94]$, with the elements of $\mathbf{B}$ give as in (15.7) and (15.8). The location of $\mathbf{x}_?'\mathbf{B}$ is shown by "×" in Fig. 15.1. By comparing its *distances* to $\mathbf{c}_1$, $\mathbf{c}_2$, and $\mathbf{c}_3$, we can find that $\mathbf{x}_?'\mathbf{B}$ is closest to $\mathbf{c}_2$; thus, $\mathbf{x}_?$ is reasonably classified into the group 2.

15.4   Maximum Probability Classification

Beginning with this section, discriminant analysis will be formulated in a different manner: we start with a classification rule, in which the *distances* and "min" in (15.13) are replaced by *probabilities* and "max," respectively. The rule is stated as follows:

$$\mathbf{x} \text{ is classified into group } k^* \text{ with } P(g_{k^*}|\mathbf{x}) = \max_{1 \le k \le m} P(g_k|\mathbf{x}). \qquad (15.14)$$

Here, $P(g_k|\mathbf{x})$ stands for the probability that the individual showing $\mathbf{x}$ belongs to the group k. This particular probability is called a *posterior probability*, as it is related to considering the group from which $\mathbf{x}$ arises a posteriori, after $\mathbf{x}$ was observed. Interchanging g_k and $\mathbf{x}$ in $P(g_k|\mathbf{x})$ gives the symbol $P(\mathbf{x}|g_k)$, which is called a *group-conditional density* and stands for the probability density of an individual in the group k showing $\mathbf{x}$. Between $P(g_k|\mathbf{x})$ and $P(\mathbf{x}|g_k)$, the following equation is known to hold:

$$P(g_k|\mathbf{x}) = \frac{P(g_k)P(\mathbf{x}|g_k)}{\sum_{l=1}^{m} P(g_l)P(\mathbf{x}|g_l)}. \qquad (15.15)$$

Here, $P(g_k)$ is a probability of a randomly chosen individual belonging to the group k and called a *prior probability*, as it is given a priori, before $\mathbf{x}$ is observed. Equation (15.15) is known as the *Bayes theorem*, as it was found by English clergyman, Thomas Bayes, early in the eighteenth century. Thus, (15.14) is called the *Bayes classification rule*.

As found in (15.15), we can obtain the posterior probability $P(g_k|\mathbf{x})$ necessary for classifying $\mathbf{x}$ with (15.14), if group-conditional densities $P(\mathbf{x}|g_k)$ and prior probabilities $P(g_k)$ ($k = 1, \ldots, K$) are estimated. This estimation is made using the data set $[\mathbf{G}, \mathbf{X}]$. The facts described in $[\mathbf{G}, \mathbf{X}]$ can also be expressed *without* using $\mathbf{G}$, by means of *rearranging* the individuals in $\mathbf{X}$ so that the ones belonging to the same group are collected in the same block. The rearrangement gives an n-individuals $\times$ p-variables block matrix

$$\mathbf{X} = \begin{bmatrix} \mathbf{X}_1 \\ \vdots \\ \mathbf{X}_k \\ \vdots \\ \mathbf{X}_K \end{bmatrix} \quad \text{with} \quad \mathbf{X}_k = \begin{bmatrix} \mathbf{x}_{k1}' \\ \vdots \\ \mathbf{x}_{ki}' \\ \vdots \\ \mathbf{x}_{kn_k}' \end{bmatrix}. \qquad (15.16)$$

Here, $n = n_1 + \cdots + n_K$, and $\mathbf{x}_{ki}$ is the $p \times 1$ data vector for the ith one of the individuals belonging to the group k. In the remaining sections, (15.16) is used for a data matrix with the memberships of individuals to groups known. Further, $P(\mathbf{x}|g_k)$ is supposed to be the probability density of a *multivariate normal (MVN) distribution*:

$$\mathbf{x} \sim N_p(\boldsymbol{\mu}_k, \Sigma_k) \quad \text{for} \quad \mathbf{x} \in g_k, \tag{15.17}$$

with $\mathbf{x} \in g_k$ representing the fact that the individual showing $\mathbf{x}$ belongs to the group k. That is, the group-conditional density for the group k is given as:

$$P(\mathbf{x}|g_k) = P(\mathbf{x}|\boldsymbol{\mu}_k, \Sigma_k) = \frac{1}{(2\pi)^{p/2}|\Sigma_k|^{1/2}} \exp\left\{ -\frac{1}{2}(\mathbf{x} - \boldsymbol{\mu}_k)'\Sigma_k^{-1}(\mathbf{x} - \boldsymbol{\mu}_k) \right\}$$

$$\tag{15.18}$$

by adding the subscript k to (8.9).

15.5 Normal Discrimination for Two Groups

In this section, the number of *groups* is restricted to *two* ($K = 2$), and the *covariance* matrix in (15.18) is supposed to be *homogeneous* between two groups:

$$\Sigma_1 = \Sigma_2 = \Sigma. \tag{15.19}$$

Then, (15.14) is rewritten as follows: $\mathbf{x}$ is classified into g_1 if $P(g_1)P(\mathbf{x}|g_1) \geq P(g_2)P(\mathbf{x}|g_2)$ or, equivalently,

$$\frac{p(\mathbf{x}|g_1)}{p(\mathbf{x}|g_2)} \geq \frac{p(g_2)}{p(g_1)}; \tag{15.20}$$

otherwise, $\mathbf{x}$ is classified into g_2. By changing both sides of (15.20) into their logarithm, we can rewrite it as:

$$f(\mathbf{x}) = \log P(\mathbf{x}|g_1) - \log P(\mathbf{x}|g_2) + \log\{P(g_1)/P(g_2)\} \geq 0. \tag{15.21}$$

Further, by substituting (15.18) into (15.21) with the use of (15.19), we can rewrite the function in (15.21) as:

$$f(\mathbf{x}) = -\frac{1}{2}(\mathbf{x} - \boldsymbol{\mu}_1)'\Sigma^{-1}(\mathbf{x} - \boldsymbol{\mu}_1) + \frac{1}{2}(\mathbf{x} - \boldsymbol{\mu}_2)'\Sigma^{-1}(\mathbf{x} - \boldsymbol{\mu}_2) + \log\{P(g_1)/P(g_2)\}$$

$$= (\boldsymbol{\mu}_1 - \boldsymbol{\mu}_2)'\Sigma^{-1}\mathbf{x} + \frac{1}{2}(\boldsymbol{\mu}_2'\Sigma^{-1}\boldsymbol{\mu}_2 - \boldsymbol{\mu}_1'\Sigma^{-1}\boldsymbol{\mu}_1) + \log\{P(g_1)/P(g_2)\}$$

$$= \mathbf{b}'\mathbf{x} + c = b_1 x_1 + \cdots + b_p x_p + c, \tag{15.22}$$

with

$$\mathbf{b} = \left[b_1, \ldots, b_p\right]' = \Sigma^{-1}(\mathbf{\mu}_1 - \mathbf{\mu}_2), \tag{15.23}$$

$$c = \frac{1}{2}(\mathbf{\mu}_2'\Sigma^{-1}\mathbf{\mu}_2 - \mathbf{\mu}_1'\Sigma^{-1}\mathbf{\mu}_1) + \log\{P(\mathrm{g}_1)/P(\mathrm{g}_2)\}. \tag{15.24}$$

Rule (15.14) is thus simplified as:

$\mathbf{x}$ is classified into group 1 if $f(\mathbf{x}) > 0$; otherwise, into group 2. $\tag{15.25}$

As (15.22) is a linear function of $\mathbf{x}$, which is the weighted composite of variables, (15.22) is called a *linear discriminant function (LDF)* and the procedure for obtaining (15.22) is called *linear discriminant analysis (LDA)*. As described in Appendix A.5.2, the maximum likelihood estimates of $\mathbf{\mu}_1$, $\mathbf{\mu}_2$, and Σ needed for obtaining (15.22) are given by

$$\hat{\mathbf{\mu}}_k = \bar{\mathbf{x}}_k = \frac{1}{n}\mathbf{1}_k'\mathbf{X}_k(k = 1, 2), \tag{15.26}$$

$$\hat{\Sigma} = \frac{1}{n}\left\{\sum_{i=1}^{n_1}(\mathbf{x}_{1i} - \bar{\mathbf{x}}_1)(\mathbf{x}_{1i} - \bar{\mathbf{x}}_1)' + \sum_{i=1}^{n_2}(\mathbf{x}_{2i} - \bar{\mathbf{x}}_2)(\mathbf{x}_{2i} - \bar{\mathbf{x}}_2)'\right\}. \tag{15.27}$$

These are substituted into $\mathbf{\mu}_1$, $\mathbf{\mu}_2$, and Σ in (15.23) and (15.24) for providing $\mathbf{b}$ and c, though $P(\mathrm{g}_1)/P(\mathrm{g}_2)$ must also be estimated for obtaining c.

For example, we consider a case of $p = 2$, where

$$P(\mathrm{g}_1) = P(\mathrm{g}_2) \tag{15.28}$$

is supposed, and $\hat{\mathbf{\mu}}_1 = [76.20, 61.42]'$, $\hat{\mathbf{\mu}}_2 = [66.93, 72.16]'$, and $\hat{\Sigma} = \begin{bmatrix} 120.77 & 60.05 \\ 60.05 & 146.98 \end{bmatrix}$. By substituting these into (15.23) and (15.24), we have $\mathbf{b} = [0.14, -0.13]$ and $c = -1.40$. They lead to the LDF

$$f(\mathbf{x}) = 0.14x_1 - 0.13x_2 - 1.40. \tag{15.29}$$

The classification in which (15.29) is used for (15.25) can be graphically illustrated as in Fig. 15.2, where the bird's-eye view of the group-conditional densities for the two groups are depicted. As found there, the LDF (15.29) value of $\mathbf{x} = [x_1, x_2]$ is the coordinate on the axis called a *discriminant axis*. For example, let the triangle in the figure indicate $\mathbf{x}_? = [58, 62]'$, i.e., a new observation to be classified. This leads to

$$f([58, 62]') = 0.14 \times 58 - 0.13 \times 62 - 1.40 = -1.34, \tag{15.30}$$

which is a coordinate on the discriminant axis. The LDF value in (15.30) is called a *discriminant score*. Since (15.30) is negative, (15.25) shows that $\mathbf{x}_? = [58, 62]'$ is to

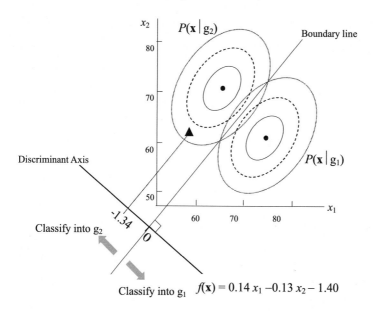

Fig. 15.2 Illustration of linear discriminant analysis

be classified into the group 2. Let us note the *boundary line* in Fig. 15.2. It defines the regions for two groups: the observations **x** located to the right of/below the line are classified into g_1 and those on the other side are classified into g_2.

15.6 Interpreting Solutions

For illustrating the interpretation of LDA solutions, we use the 27 (employees) × 4 (personality traits) data matrix $\mathbf{X} = [\mathbf{X}_1', \mathbf{X}_2']'$ in Table 15.2(A), where the 27 employees are supposed to be classified into groups according to their departments and competency for jobs in the departments to which they belong. The LDA for the data set provides the LDF as:

$$f(\mathbf{x}) = 0.719x_1 + 0.139x_2 - 0.462x_3 - 0.084x_4 - 2.069. \tag{15.31}$$

Here, we have used $P(g_1) = 15/27$ and $P(g_2) = 12/27$, i.e., the proportions of the members in groups 1 and 2 in Table 15.2.

Let us consider assessing how correctly/incorrectly individuals are classified by the LDF in (15.31). An easy way to do so is to substitute each row vector of **X** into (15.31) and examine whether the resulting discriminant score shows the *correct classification or not*. For example, the substitution of the first and second row vectors yields

Table 15.2 Artificial example of the data for LDF (Adachi 2006) and results of classification

Department		Social	Cooperative	Diligent	Creative
(A) Data (artificial)					
g1	1	15	14	15	14
	2	11	13	17	17
	3	16	14	17	26
	4	19	21	18	15
	5	18	26	21	15
	6	15	28	18	12
	7	17	19	12	10
	8	12	15	18	12
	9	13	22	16	10
	10	14	26	18	6
	11	16	20	18	18
	12	11	15	20	15
	13	20	21	17	20
	14	15	20	19	12
	15	13	13	17	16
g2	16	11	15	18	17
	17	10	13	16	9
	18	11	14	24	16
	19	10	10	13	12
	20	10	14	22	18
	21	13	19	23	24
	22	11	10	20	28
	23	15	20	20	16
	24	12	22	23	13
	25	10	11	18	10
	26	12	10	19	27
	27	10	14	21	19
Obs		Score		Result	
(B) Classification					
1		2.783		g1	
2		−1.407		g2*	
3		1.568		g1	
4		5.165		g1	
5		3.758		g1	
6		3.518		g1	
7		6.639		g1	
8		−0.450		g2*	
9		2.336		g1	
10		3.026		g1	

(continued)

Table 15.2 (continued)

Obs	Score	Result
11	2.616	g1
12	−2.344	g2*
13	5.924	g1
14	1.942	g1
15	0.115	g1
16	−1.590	g2
17	−0.990	g2
18	−4.414	g2
19	−0.276	g2
20	−4.379	g2
21	−2.493	g2
22	−4.137	g2
23	1.143	g1*
24	−1.866	g2
25	−2.276	g2
26	−2.872	g2
27	−4.001	g2

*Misclassification

$$f(\mathbf{x}_{11} = [15, 14, 15, 14]')$$
$$= 0.719 \times 15 + 0.139 \times 14 - 0.462 \times 15 - 0.084 \times 14 - 2.069 = 1.24,$$
$$(15.32)$$

$$f(\mathbf{x}_{12} = [11, 13, 17, 17]')$$
$$= 0.719 \times 11 + 0.139 \times 13 - 0.462 \times 17 - 0.084 \times 17 - 2.069 = -0.62,$$
$$(15.33,)$$

respectively. Here, (15.32) implies correct classification, since it gives a positive value, showing that $\mathbf{x}_{11}$ is to be classified into *the* group 1 and, in reality, $\mathbf{x}_{11}$ belongs to the group 1. On the other hand, (15.33) implies *misclassification*, since (15.33) is negative and shows that $\mathbf{x}_{12}$ is to be classified into the group 2. The scores obtained as above are shown in Table 15.2(B), with the asterisks indicating misclassification.

By counting those asterisks, we can assess *misclassification rates*; the rate is 3/15 in the group 1, while it is 1/12 for the group 2, and the total rate is $(3 + 1)/27 = 0.15$. This assessment is known to *underestimate* the misclassification rate, since the classification is made for the data vectors from which LDFs are obtained. This differs from a usual setting, in which a *new* data vector $\mathbf{x}_?$ is *not included* in the

Table 15.3 Standardized discriminant coefficients for the data in Table 15.2(A)

Social	Cooperative	Diligent	Creative
2.079	0.704	−1.289	−0.459

data set **X**. However, procedures for more accurately assessing the rate are out of the scope of this book.

LDA is used not only for classification, but also for finding the variables that characterize groups. For this purpose, the *standardized discriminant coefficients* are to be used that are obtained with LDA for standardized data. Table 15.3 presents the coefficients for the standard scores transformed from the data in Table 15.2(A). There, we can find the following:

[1]   The persons to be classified into the group 1 (or the ones regarded as competent for the corresponding department) are social and cooperative, but not diligent and creative, with particularly important characteristics being social and less diligent.

[2]   The persons to be classified into the group 2 are diligent and creative, but not social and cooperative, with important characteristics being diligent and less social.

The *CDA* in earlier sections can also be performed for the data set in Table 15.2 (A) with $m = 1$. For the data, CDA provides $\mathbf{B} = \mathbf{b} = [0.226, 0.044, -0.145, -0.26]'$, every element of which equals the corresponding coefficient in (15.31) divided by 3.1. Indeed, it is known that the coefficients for CDA are proportional to those of LDA, and the classifications made by CDA are *equivalent* to those by LDA when $P(g_1) = P(g_2)$, though its proof is omitted here. The discriminant analyses differing from LDA and CDA are described in the following section.

15.7   Generalized Normal Discrimination

In this section, the classification by (15.14) is illustrated for the cases with the number of groups $K \geq 2$ and $\mathbf{\Sigma}_k$ *heterogeneous* among groups. We consider the data matrix (15.16) with $n = 150$, $p = 2$, $K = 3$, and the 150 individuals randomly sampled. Let the statistics obtained from $\mathbf{X}_1$, $\mathbf{X}_2$, and $\mathbf{X}_3$ be summarized as in Fig. 15.3a; for example, the average vector $40^{-1}\mathbf{1}'_{40}\mathbf{X}_2$ for the group 2 is [25.9, 74.8] and the covariance matrix $44^{-1}\mathbf{X}'_3\mathbf{J}\mathbf{X}_3$ for the group 3 is $\begin{bmatrix} 435.1 & 212.6 \\ 212.6 & 168.4 \end{bmatrix}$ with $\mathbf{J}$ the centering matrix defined as (2.10).

Group	g_1		g_2		g_3	
n_k	66		40		44	
Variable	x_1	x_2	x_1	x_2	x_1	x_2
Average	52.1	43.3	25.9	74.8	71.0	23.4
Covar-	355.7	203.8	180.4	-198.8	435.1	212.6
iances	203.8	252.5	-198.8	369.4	212.6	168.4

(**a**) Statistics for each group

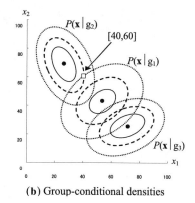

(**b**) Group-conditional densities

Fig. 15.3 Statistics and probability densities or generalized normal discrimination

Prior probabilities can be estimated as $P(g_k) = n_k/n$:

$$P(g_1) = \frac{66}{150}, \quad P(g_2) = \frac{40}{150}, \quad \text{and} \quad P(g_3) = \frac{44}{150} \tag{15.34}$$

for the data set in Fig. 15.3a. The *group-conditional density* is given as (15.18), whose parameters $\boldsymbol{\mu}_k$ and $\boldsymbol{\Sigma}_k$ can be estimated by the maximum likelihood method as shown in Sect. 8.6 and detailed in Appendix A.5.1. The MLE of $\boldsymbol{\mu}_k$ and $\boldsymbol{\Sigma}_k$ are given by Eqs. (8.21) and (8.22) with the subscript k added as:

$$\hat{\boldsymbol{\mu}}_k = \bar{\mathbf{x}}_k = \frac{1}{n_k} \sum_{i=1}^{n_k} \mathbf{x}_{ki}, \tag{15.35}$$

$$\hat{\boldsymbol{\Sigma}}_k = \frac{1}{n_k} \sum_{i=1}^{n_k} (\mathbf{x}_{ki} - \bar{\mathbf{x}}_k)(\mathbf{x}_{ki} - \bar{\mathbf{x}}_k)'. \tag{15.36}$$

Let us substitute the statistics in Fig. 15.3a into the corresponding parts of (15.35) and (15.36). Using its results in (15.18), we have the group-conditional densities

$$P(\mathbf{x}|g_1) = (2\pi)^{-p/2} \left| \begin{bmatrix} 355.7 & 203.8 \\ 203.8 & 252.5 \end{bmatrix} \right|^{-1/2} \exp\left\{ -\frac{1}{2}\left(\mathbf{x} - \begin{bmatrix} 52.1 \\ 43.3 \end{bmatrix}\right)' \begin{bmatrix} 355.7 & 203.8 \\ 203.8 & 252.5 \end{bmatrix}^{-1} \left(\mathbf{x} - \begin{bmatrix} 52.1 \\ 43.3 \end{bmatrix}\right) \right\},$$
$$\tag{15.37}$$

$$P(\mathbf{x}|g_2) = (2\pi)^{-p/2} \left| \begin{bmatrix} 180.4 & -198.8 \\ -198.8 & 369.4 \end{bmatrix} \right|^{-1/2} \exp\left\{ -\frac{1}{2}\left(\mathbf{x} - \begin{bmatrix} 25.9 \\ 74.8 \end{bmatrix}\right)' \begin{bmatrix} 180.4 & -198.8 \\ -198.8 & 369.4 \end{bmatrix}^{-1} \left(\mathbf{x} - \begin{bmatrix} 25.9 \\ 74.8 \end{bmatrix}\right) \right\},$$
$$\tag{15.38}$$

$$P(\mathbf{x}|g_3) = (2\pi)^{-p/2} \left| \begin{bmatrix} 435.1 & 212.6 \\ 212.6 & 166.4 \end{bmatrix} \right|^{-1/2} \exp\left\{ -\frac{1}{2}\left(\mathbf{x} - \begin{bmatrix} 71.0 \\ 23.4 \end{bmatrix}\right)' \begin{bmatrix} 435.1 & 212.6 \\ 212.6 & 166.4 \end{bmatrix}^{-1} \left(\mathbf{x} - \begin{bmatrix} 71.0 \\ 23.4 \end{bmatrix}\right)\right\},$$

$$(15.39)$$

with $\pi = 3.1416 \dots$ the circle ratio. In Fig. 15.3b, a bird's-eye view of (15.37) to (15.39) is drawn as in Fig. 8.4b. We may consider the figure as a map depicting three mountains whose tops are indicated by filled circles, and counters are expressed by ellipses.

Let $\mathbf{x}_? = [40, 60]'$ indicated by a blank square in Fig. 15.3b be a new data vector for the individual whose membership to a group is *unknown*; our task is to classify $\mathbf{x}_?$ into one of the groups 1, 2, and 3. This can be achieved by performing the calculus in the *Bayes theorem* (15.15) and by using the *classification rule* (15.14).

By substituting $\mathbf{x}_? = [40, 60]'$ into (15.37), (15.38), and (15.39), we have the values of the group-conditional densities as: $P([40, 60]'|g_1) = 7.534 \times 10^{-5}$, $P([40, 60]'|g_2) = 5.560 \times 10^{-4}$, and $P([40, 60]'|g_3) = 1.600 \times 10^{-13}$, respectively. Using these with (15.34), the numerator in the right-hand side of (15.15) is obtained as:

$$P(g_1)P\big([40,60]'|g_1\big) = \frac{66}{150} \times 7.534 \times 10^{-5} = 3.315 \times 10^{-5}, \qquad (15.40)$$

$$P(g_2)P\big([40,60]'|g_2\big) = \frac{40}{150} \times 5.560 \times 10^{-4} = 1.483 \times 10^{-4}, \qquad (15.41)$$

$$P(g_3)P\big([40,60]'|g_3\big) = \frac{44}{150} \times 1.600 \times 10^{-13} = 4.693 \times 10^{-14} \qquad (15.42)$$

for each group.

It should be noted that the denominator in the right-hand side of (15.15) is equivalent among different groups; we may *only compare* its *numerator* between groups for classification. This implies that (15.14) may be simplified to

$$\mathbf{x} \text{ is classified into group } k^* \text{ with } P(g_{k^*})P(g_{k^*}|\mathbf{x}) = \max_{1 \le k \le m} P(g_k)P(g_k|\mathbf{x}). \quad (15.43)$$

By this rule, we can compare (15.40), (15.41), and (15.42) to classify $\mathbf{x}_? = [40, 60]'$ into the group 2, since (15.41) is the *highest* of the three values. However, if we wish to perform not only the classification, but also obtain the *posterior probability* of $\mathbf{x}_?$ belonging to the group, the denominator in the right-side hand of (15.15) must be obtained, which is the sum of $P(g_k)P(g_k|\mathbf{x})$ over k. The sum of (15.40) to (15.42) is given by

$$\sum_{l=1}^{m} P(g_l)P(\mathbf{x}|g_l) = 3.315 \times 10^{-5} + 1.483 \times 10^{-4} + 4.693 \times 10^{-14}$$

$$= 1.815 \times 10^{-4}. \qquad (15.44)$$

The use of the above value and (15.41) in (15.15) leads to the posterior probability

$$P(g_2|[40, 60]') = \frac{1.483 \times 10^{-4}}{1.815 \times 10^{-4}} = 0.82. \tag{15.45}$$

Thus, the probability of $\mathbf{x}_?$ belonging to the group 2 is 0.82. This value can be regarded as expressing the *confidence* with which we classify $\mathbf{x}_?$ into g_2. In a parallel manner, the probability of $\mathbf{x}_?$ belonging to the group 1 can be obtained as: $P(g_1|[40, 60]') = \frac{3.315 \times 10^{-5}}{1.815 \times 10^{-4}} = 0.18$, and $P(g_3|[40, 60]') = 1 - P(g_1|[40, 60]') - P(g_2|[40, 60]')$ can be found to be almost zero.

15.8 Bibliographical Notes

A variety of discriminant analysis procedures are described in McLachlan (1992) and Hand (1997). Some new procedures in discriminant analysis are detailed in Hastie et al. (2009). An introduction to CDA as a modification of MCA is found in Adachi (2004).

Exercises

15.1. Matrices $\mathbf{V}_B = \frac{1}{n} \sum_{k=1}^{K} \sum_{i=1}^{n_k} (\bar{\mathbf{x}}_k - \bar{\mathbf{x}})(\bar{\mathbf{x}}_k - \bar{\mathbf{x}})' = \frac{1}{n} \sum_{k=1}^{K} n_k (\bar{\mathbf{x}}_k - \bar{\mathbf{x}})(\bar{\mathbf{x}}_k - \bar{\mathbf{x}})'$,
$\mathbf{V}_W = \frac{1}{n} \sum_{k=1}^{K} \sum_{i=1}^{n_k} (\mathbf{x}_{ki} - \bar{\mathbf{x}}_k)(\mathbf{x}_{ki} - \bar{\mathbf{x}}_k)'$, and $\mathbf{V}_T = \frac{1}{n} \sum_{k=1}^{K} \sum_{i=1}^{n_k} (\mathbf{x}_{ki} - \bar{\mathbf{x}})(\mathbf{x}_{ki} - \bar{\mathbf{x}})'$
are called *between-group*, *within-group*, and *total* covariance matrices, respectively, with $\mathbf{x}_{ki}$ the $p \times 1$ data vector for the ith individual in the group k, $n = \sum_{i=1}^{K} n_k$, $\bar{\mathbf{x}}_k = n_k^{-1} \sum_{i=1}^{n_k} \mathbf{x}_{ki}$, and $\bar{\mathbf{x}} = n^{-1} \sum_{k=1}^{K} \sum_{i=1}^{n_k} \mathbf{x}_{ki}$. Show $\mathbf{V}_T = \mathbf{V}_B + \mathbf{V}_W$.

15.2. Let $\mathbf{x}_l'$ be the lth row vector of n-individuals $\times$ p-variables data matrix $\mathbf{X} = [\mathbf{x}_1, \ldots, \mathbf{x}_n]'$ in (15.16) and $\mathbf{X}_{[l]}$ be the $(n - 1) \times p$ matrix obtained by removing $\mathbf{x}_l'$ from $\mathbf{X}$. In a *leaving-one-out* procedure, the following assessment is replicated over $l = 1, \ldots, n$: (15.23) and (15.24) are estimated with $\mathbf{X}_{[l]}$ and classification (15.25) with $\mathbf{x} = \mathbf{x}_l$ performed in order to assess whether the resulting classification is correct or not. It is known that misclassification rates are estimated better in the leaving-one-out procedure than in that illustrated in Sect. 15.6. Discuss the reason for this.

15.3. In *logistic discriminant analysis* for two groups, the posterior probability for the group 1 is expressed as: $P(g_1|\mathbf{x}) = \frac{1}{1 + \exp(-\mathbf{b}'\mathbf{x} + c)}$, with $\mathbf{x}$ the vector containing observed variables, $\mathbf{b}$ the vector of coefficients, and $P(g_2|\mathbf{x}) = 1 - P(g_1|\mathbf{x})$. Discuss the similarities and differences between the logistic and linear discriminant analyses.

15.4. The *Mahalanobis distance* of $\mathbf{x}$ to the group k is defined as: $(\mathbf{x} - \boldsymbol{\mu}_k)'\boldsymbol{\Sigma}_k^{-1}(\mathbf{x} - \boldsymbol{\mu}_k)$, with $\boldsymbol{\mu}_k$ and $\boldsymbol{\Sigma}_k$ the mean vector and covariance matrix for the group k, respectively. Show that classification (15.13), with its distances replaced by the Mahalanobis distances, is equivalent to the classification in Sect. 15.7 with $P(g_k)$ and $|\boldsymbol{\Sigma}_k|$ constrained to be homogeneous among the groups.

15.5. Let us consider a case in which the vector $\mathbf{x}$ in (15.15) consists of ones and zeros, and its jth element x_j take one with probability θ_{jk} for $\mathbf{x}$ being included in the group k. If the elements of $\mathbf{x}$ are observed mutually independently, show that classification (15.14) is feasible using $P(\mathbf{x}|g_k) = \prod_{j=1}^{p} \theta_{jk}^{x_j}(1 - \theta_{jk})^{1-x_j}$ in (15.15).

15.6. Let us consider a version of CDA with $\mathbf{G}$ unknown. This version is formulated as minimizing $\|\mathbf{XB} - \mathbf{GC}\|^2$ over $\mathbf{G} = (g_{ik})$, $\mathbf{C} = [\mathbf{c}_1, \ldots, \mathbf{c}_K]'$ and $\mathbf{B}$ subject to (7.1), (7.2), and (15.3). Show that the minimization can be attained by the following algorithm:

Step 1. Initialize $\mathbf{G}$ and obtain $\mathbf{V} = \mathbf{V}^{1/2}\mathbf{V}^{1/2}$ with $\mathbf{V} = n^{-1}\mathbf{X}'\mathbf{X}$.
Step 2. Obtain the SVD $\mathbf{X}'\mathbf{G}\mathbf{D}_G^{-1}\mathbf{G}'\mathbf{X} = \mathbf{Q}\boldsymbol{\Theta}^2\mathbf{Q}'$ to set $\mathbf{B} = \mathbf{V}^{-1/2}\mathbf{Q}_m$ with $\mathbf{D}_G = \mathbf{G}'\mathbf{G}$.
Step 3. Obtain $\mathbf{C}$ by (15.4).
Step 4. Set $g_{ik} = 1$ if $\|\mathbf{x}_i'\mathbf{B} - \mathbf{c}_k\|^2 = \min_{1 \leq l \leq K} \|\mathbf{x}_i'\mathbf{B} - \mathbf{c}_l\|^2$ and $g_{ik} = 0$ otherwise, for $i = 1, \ldots, n$; $k = 1, \ldots, K$.
Step 5. Finish if convergence is reached; otherwise, go back to Step 2.

In Vichi and Kiers' (2001) *factorial K-means analysis (FKM)*, (15.3) is replaced by $\mathbf{B}'\mathbf{B} = \mathbf{I}_m$.

15.7. There exists a *Bayesian method* for estimating parameters besides the least squares and maximum likelihood methods. In the Bayesian method, the fact is used that Bayes' theorem (15.15) can be generalized as:

$$P(\boldsymbol{\theta}|\mathbf{X}) = \frac{P(\boldsymbol{\theta})P(\mathbf{X}|\boldsymbol{\theta})}{P(\mathbf{X})}. \tag{15.46}$$

Here, $\boldsymbol{\theta}$ is the vector containing parameters, while $\mathbf{X}$ is a data matrix, with $P(\boldsymbol{\theta})$ denoting the probability density function (PDF) of $\boldsymbol{\theta}$, $P(\mathbf{X})$ the PDF of $\mathbf{X}$ observed, $P(\mathbf{X}|\boldsymbol{\theta})$ the PDF of $\mathbf{X}$ for given $\boldsymbol{\theta}$, and $P(\boldsymbol{\theta}|\mathbf{X})$ the PDF of $\boldsymbol{\theta}$ for given $\mathbf{X}$. As found in (15.46), the parameters are also viewed as being randomly distributed in the Bayesian method. This method is formulated as the maximization of (15.46) over $\boldsymbol{\theta}$, or equivalently, maximizing $P(\boldsymbol{\theta})P(\mathbf{X}|\boldsymbol{\theta})$. Argue that $P(\boldsymbol{\theta})P(\mathbf{X}|\boldsymbol{\theta})$ is the product of the prior information for parameters and their likelihood.

15.8. A *penalized least squares method* for $n \times p$ data matrix $\mathbf{X}$ can be formulated as minimizing $\|\mathbf{X} - \mathbf{H}(\boldsymbol{\theta})\|^2 + \tau g(\boldsymbol{\theta})$ over parameter vector $\boldsymbol{\theta}$, with τ a specified value and $\mathbf{H}(\boldsymbol{\theta})$ a function of $\boldsymbol{\theta}$ providing an $n \times p$ matrix. An example of the method is found in Exercise 4.11. Show that the Bayesian estimation method in Exercise 15.7 is equivalent to the penalized least squares one, if $P(\mathbf{X}|\boldsymbol{\theta})$ in (15.46) takes the form of $P(\mathbf{X}|\boldsymbol{\theta}) = a \times \exp\{-b\|\mathbf{X} - \mathbf{H}(\boldsymbol{\theta})\|^2\}$ and τ is set to a certain value.

Chapter 16
Multidimensional Scaling

The key words for describing *multidimensional scaling* (*MDS*) are the *coordinates* of objects, the *distances* between objects, and the corresponding *quasi-distances* given as *data*. For example, let us suppose that the objects are cities such as London, Paris, and Amsterdam. Then, their *coordinates* are the locations of those cities on a map, which *define* the inter-city *distances*. We further suppose that the flight times between those cities are given as data, which are regarded as *quasi-distance data*, since they are approximately proportional to distances but are *not equivalent* to them. The purpose of MDS is to *estimate the coordinates of objects*, i.e., their locations, *from quasi-distance data*; the coordinates are obtained so that their defined distances approximate quasi-distance data.

The origin of MDS can be found in Torgerson (1952). His approach is called *classical scaling*, which is equivalent to Gower's (1966) *principal coordinate analysis*. Those procedures are formulated with inter-object inner-products rather than distances. Also, though they are not treated, their squares are considered in Takane et al. (1977) procedure known as *alternating least squares scaling* (ALSCAL). In this chapter, only an MDS procedure is introduced in which *distances themselves* are considered, and a computational technique called a *majorization algorithm* is used. This technique for MDS is rooted in de Leeuw (1977) and has been developed by Groenen (1993), Heiser (1991), and others.

16.1 Linking Coordinates to Quasi-Distances

Let us use q_{ij} for the *observed quasi-distance* between objects i and j. Then, the data set of quasi-distances among n-objects can be expressed as an $n \times n$ matrix:

© Springer Nature Singapore Pte Ltd. 2016
K. Adachi, *Matrix-Based Introduction to Multivariate Data Analysis*,
DOI 10.1007/978-981-10-2341-5_16

Table 16.1 Rated dissimilarities between sports

	Baseball	Volleyball	Football	Tennis	Ping-pong	Basketball	Rugby	Softball
Baseball	0.0	5.6	5.0	4.6	4.4	5.4	6.0	1.2
Volleyball	5.6	0.0	5.4	4.4	4.2	3.0	5.4	5.4
Football	5.0	5.4	0.0	5.6	6.2	4.0	2.8	4.8
Tennis	4.6	4.4	5.6	0.0	2.0	5.8	6.4	4.2
Ping-pong	4.4	4.2	6.2	2.0	0.0	5.2	6.4	4.8
Basketball	5.4	3.0	4.0	5.8	5.2	0.0	4.6	5.2
Rugby	6.0	5.4	2.8	6.4	6.4	4.6	0.0	5.6
Softball	1.2	5.4	4.8	4.2	4.8	5.2	5.6	0.0

$$
\mathbf{Q} = \begin{bmatrix}
 & q_{12} & q_{13} & \cdots & q_{1n} \\
 & & q_{23} & \cdots & q_{2n} \\
 & & & \ddots & \vdots \\
 & & & & q_{n-1,n} \\
 & & & &
\end{bmatrix}. \tag{16.1}
$$

Here, the lower left elements, i.e., the parts for q_{ij} with $i \geq j$, are blank, as it is supposed that $\mathbf{Q}$ is *symmetric* with $q_{ij} = q_{ji}$, and q_{ii} (the quasi-distance between an object and itself) is *not observed*. We also suppose $q_{ij} \geq 0$ in this chapter. One feature of data matrix $\mathbf{Q}$ is that the *same set of entities* occupies the *rows* and *columns*, which differs from the n-observations $\times$ p-variables data matrices that have been treated in other chapters. Table 16.1 presents an example of $\mathbf{Q}$, which describes the perceived dissimilarities between objects (sports). They are quasi-distance data in that a pair of objects perceived *similar/dissimilar* corresponds to their being *close/distant*, but the dissimilarities *differ from genuine inter-object distances*.

The purpose of MDS is to obtain an n-objects $\times$ m-dimensions matrix of the objects' *coordinates* from (16.1):

$$
\mathbf{A} = (a_{ik}) = \begin{bmatrix} \mathbf{a}'_1 \\ \vdots \\ \mathbf{a}'_n \end{bmatrix}, \tag{16.2}
$$

Here, $\mathbf{a}_i$ ($m \times 1$) is the coordinate vector indicating the location of object i, with the kth element a_{ik} the coordinate of i on dimension k. The *distance* between $\mathbf{a}_i$ and $\mathbf{a}_j$ is expressed as:

$$
\|\mathbf{a}_i - \mathbf{a}_j\| = \sqrt{\sum_{k=1}^{m} (a_{ik} - a_{jk})^2}. \tag{16.3}
$$

This particular distance is called *Euclidean distance*, from the ancient Greek mathematician Euclid, for distinguishing it from the other special definitions of distances.

Distance (16.3) can be linked with its *quasi*-version q_{ij} in (16.1) as:

$$q_{ij} = \|\mathbf{a}_i - \mathbf{a}_j\| + e_{ij}, \tag{16.4}$$

with e_{ij} an error. Thus, MDS is formulated as minimizing the sum of squared errors, i.e.,

$$h(\mathbf{A}) = \sum_{i<j} (q_{ij} - \|\mathbf{a}_i - \mathbf{a}_j\|)^2, \tag{16.5}$$

is minimized over $\mathbf{A}$.

Note 16.1. Summation for $i < j$

The symbol $\sum_{i<j} x_{ij}$ stands for the summation of a set of x_{ij} that satisfies $i < j$. For example, let $\mathbf{X} = (x_{ij})$ be a 4×4 matrix, then $\sum_{i<j} x_{ij} = x_{12} + x_{13} + x_{14} + x_{23} + x_{24} + x_{34}$.

As found in (16.3), the distance is the *squared root* of $\|\mathbf{a}_i - \mathbf{a}_j\|^2$, which is far more *difficult to handle* than $\|\mathbf{a}_i - \mathbf{a}_j\|^2$. For dealing with that difficulty, some MDS procedures are formulated as fitting $\|\mathbf{a}_i - \mathbf{a}_j\|^2 = \|\mathbf{a}_i\|^2 + \|\mathbf{a}_j\|^2 - 2\mathbf{a}'_i\mathbf{a}_j$ to *squared* q_{ij} (Takane et al. 1977) or fitting *inner-product* $\mathbf{a}'_i\mathbf{a}_j$ to the corresponding counterpart transformed from q_{ij} (Togerson 1952; Gower 1966), rather than minimizing (16.5). But, we will directly treat it in this chapter.

16.2 Illustration of an MDS Solution

For matrix $\mathbf{Q}$ in Table 16.1, MDS loss function (16.5) is minimized for coordinate matrix $\mathbf{A}$ in Fig. 16.1a. This solution is graphically represented as in Fig. 16.1b, where the objects (sports) are plotted according to their *coordinates* in Fig. 16.1a. We can see the plot as a usual *map*; the close/distant objects in the plot are similar/dissimilar in their features. For example, baseball and softball are closely located, which implies both are perceived to be similar, while rugby and ping-pong are distant, implying that they are dissimilar. This illustrates that we can *visually capture inter-objects relationships* in MDS solutions.

The solution in Fig. 16.1 cannot explicitly be given. The iterative algorithm that provides the solution is described in the remaining sections.

Dimension	1	2
Baseball	1.7	2.7
Volleyball	−2.8	−1.4
Football	2.6	−1.7
Tennis	−2.4	1.9
Ping-pong	−2.2	2.7
Basketball	−1.1	−2.7
Rugby	2.2	−3.4
Softball	1.9	2.0

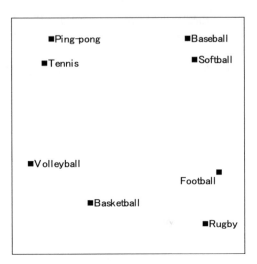

(**a**) Coordinates   (**b**) Configuration

Fig. 16.1 MDS solution for the data in Table 16.1

16.3 Iterative Algorithm

Loss function (16.5) is expanded as $h(\mathbf{A}) = \Sigma_{i<j} q_{ij}^2 + \Sigma_{i<j} \|\mathbf{a}_i - \mathbf{a}_j\|^2 - 2\Sigma_{i<j} q_{ij} \|\mathbf{a}_i - \mathbf{a}_j\|$. Here, $\Sigma_{i<j} q_{ij}^2$ is a constant irrelevant to $\mathbf{A}$. Thus, the minimization of (16.5) is equivalent to minimizing:

$$f(\mathbf{A}) = \sum_{i<j} \|\mathbf{a}_i - \mathbf{a}_j\|^2 - 2\sum_{i<j} q_{ij} \|\mathbf{a}_i - \mathbf{a}_j\|. \tag{16.6}$$

We will consider the latter.

Using $\mathbf{A}_{[t]}$ for the coordinate matrix $\mathbf{A}$ obtained at the tth iteration, the outline of the iterative algorithm for minimizing (16.6) can be listed as follows:

Step 1. Initialize $\mathbf{A}_{[t]}$ with $t = 0$.
Step 2. Update $\mathbf{A}_{[t]}$ to $\mathbf{A}_{[t+1]}$ so that $f(\mathbf{A}_{[t]}) \geq f(\mathbf{A}_{[t+1]})$.
Step 3. Finish if convergence is reached; otherwise, increase t by one and return to Step 2.

In Step 3, the convergence can be defined as $f(\mathbf{A}_{[t]}) - f(\mathbf{A}_{[t+1]})$ is small enough to be ignored.

The update formula in Step 2 is given by:

$$\mathbf{A}_{[t+1]} = \frac{1}{n} \mathbf{Q}_{(\mathbf{A}_{[t]})} \mathbf{A}_{[t]}. \tag{16.7}$$

Here, $\mathbf{Q}_{(\mathbf{A}_{[t]})}$ is the $n \times n$ matrix which is a function of $\mathbf{A}_{[t]}$ and is expressed as:

$$
\mathbf{Q}_{(\mathbf{A}_{[t]})} = \begin{bmatrix} \sum_{i=1}^{n} q_{i1}^{(\mathbf{A}_{[t]})} & & \\ & \ddots & \\ & & \sum_{i=1}^{n} q_{in}^{(\mathbf{A}_{[t]})} \end{bmatrix} - \begin{bmatrix} q_{11}^{(\mathbf{A}_{[t]})} & \cdots & q_{1n}^{(\mathbf{A}_{[t]})} \\ \vdots & \cdots & \vdots \\ q_{n1}^{(\mathbf{A}_{[t]})} & \cdots & q_{nn}^{(\mathbf{A}_{[t]})} \end{bmatrix}, \tag{16.8}
$$

with the blanks standing for zero elements and $q_{ij}^{(\mathbf{A}_{[t]})}$ defined, using $\mathbf{a}_i^{[t]\prime}$ ($1 \times m$) for the ith row of $\mathbf{A}_{[t]}$, as:

$$
q_{ij}^{(\mathbf{A}_{[t]})} = \begin{cases} 0 & \text{if } \mathbf{a}_i^{[t]} = \mathbf{a}_j^{[t]} \\ \dfrac{q_{ij}}{\|\mathbf{a}_i^{[t]} - \mathbf{a}_j^{[t]}\|} & \text{otherwise} \end{cases}. \tag{16.9}
$$

Why does (16.7) guarantee $f(\mathbf{A}_{[t]}) \geq f(\mathbf{A}_{[t+1]})$? In order to explain this, we need a *long story* continuing over the next three sections. There, the following tasks are attained in turn:

[1] $\Sigma_{i<j}\|\mathbf{a}_i - \mathbf{a}_j\|^2$ in (16.6) is expressed in matrix form (Sect. 16.4).
[2] An inequality for $\Sigma_{i<j}q_{ij}\|\mathbf{a}_i - \mathbf{a}_j\|$ in (16.6) is derived (Sect. 16.5).
[3] We use the results of [1] and [2] to derive an algorithm which leads to (16.7) (Sect. 16.6).

16.4 Matrix Expression for Squared Distances

In order to express *squared distance* $\|\mathbf{a}_i - \mathbf{a}_j\|^2$ in matrix form using $\mathbf{A}$, we introduce the elementary vectors in the following note:

Note 16.2. Elementary Vectors
Let $\mathbf{e}_i$ denote the $n \times 1$ vector filled with zeros, with only the ith element taking one. Such a vector is called an *elementary vector*. For example, $\mathbf{e}_2 = [0, 1, 0]'$ for $n = 3$. We can easily find that $\mathbf{e}_i'\mathbf{A} = \mathbf{a}_i'$ with $\mathbf{A}$ defined as (16.2); $\mathbf{e}_i'$ serves for selecting the ith row of a matrix.

Let $\mathbf{B} = \begin{bmatrix} \mathbf{b}_1' \\ \vdots \\ \mathbf{b}_n' \end{bmatrix}$ be an $n \times m$ matrix, like $\mathbf{A}$. Then, we have:

$$(\mathbf{a}_i - \mathbf{a}_j)'(\mathbf{b}_i - \mathbf{b}_j) = (\mathbf{e}_i'\mathbf{A} - \mathbf{e}_j'\mathbf{A})(\mathbf{e}_i'\mathbf{B} - \mathbf{e}_j'\mathbf{B})' = (\mathbf{e}_i - \mathbf{e}_j)'\mathbf{A}\mathbf{B}'(\mathbf{e}_i - \mathbf{e}_j)$$
$$= \operatorname{tr}(\mathbf{e}_i - \mathbf{e}_j)'\mathbf{A}\mathbf{B}'(\mathbf{e}_i - \mathbf{e}_j) = \operatorname{tr}\mathbf{B}'(\mathbf{e}_i - \mathbf{e}_j)(\mathbf{e}_i - \mathbf{e}_j)'\mathbf{A}$$
$$= \operatorname{tr}\mathbf{A}'\mathbf{H}_{ij}\mathbf{B},$$

$$(16.10)$$

with

$$\mathbf{H}_{ij} = (\mathbf{e}_i - \mathbf{e}_j)(\mathbf{e}_i - \mathbf{e}_j)'. \tag{16.11}$$

For example, when $n = 3$,

$$\mathbf{H}_{12} = \begin{bmatrix} 1 & -1 & 0 \\ -1 & 1 & 0 \\ 0 & 0 & 0 \end{bmatrix}, \mathbf{H}_{13} = \begin{bmatrix} 1 & 0 & -1 \\ 0 & 0 & 0 \\ -1 & 0 & 1 \end{bmatrix}, \mathbf{H}_{23} = \begin{bmatrix} 0 & 0 & 0 \\ 0 & 1 & -1 \\ 0 & -1 & 1 \end{bmatrix}.$$

$$(16.12)$$

If $\mathbf{B}$ is set to $\mathbf{A}$ in (16.10), we have the *squared distance:*

$$||\mathbf{a}_i - \mathbf{a}_j||^2 = \operatorname{tr}\mathbf{A}'\mathbf{H}_{ij}\mathbf{A}. \tag{16.13}$$

For rewriting the summation of (16.10) over $i < j$, the matrix in the following note is used:

Note 16.3. Use of Centering Matrix
It can be found that:

$$\sum_{i<j}\mathbf{H}_{ij} = n\mathbf{I}_n - \mathbf{1}_n\mathbf{1}_n' = n\mathbf{J}, \tag{16.14}$$

with $\mathbf{J} = \mathbf{I}_n - n^{-1}\mathbf{1}_n\mathbf{1}_n'$, the *centering matrix* defined in (2.10). Using (16.12), we can verify (16.14) as:

$$\sum_{i<j}\mathbf{H}_{ij} = \mathbf{H}_{12} + \mathbf{H}_{13} + \mathbf{H}_{23} = \begin{bmatrix} 2 & -1 & -1 \\ -1 & 2 & -1 \\ -1 & -1 & 2 \end{bmatrix}$$

$$= 3\begin{bmatrix} 1 & 0 & 0 \\ 0 & 1 & 0 \\ 0 & 0 & 1 \end{bmatrix} - \begin{bmatrix} 1 & 1 & 1 \\ 1 & 1 & 1 \\ 1 & 1 & 1 \end{bmatrix}. \tag{16.15}$$

Using (16.13) and (16.14), we can rewrite $\Sigma_{i<j}\|\mathbf{a}_i - \mathbf{a}_j\|^2$ in (16.6) as:

$$\sum_{i<j}\left\|\mathbf{a}_i - \mathbf{a}_j\right\|^2 = \sum_{i<j}\operatorname{tr}\mathbf{A}'\mathbf{H}_{ij}\mathbf{A} = \operatorname{tr}\mathbf{A}'\sum_{i<j}\mathbf{H}_{ij}\mathbf{A} = n\operatorname{tr}\mathbf{A}'\mathbf{J}\mathbf{A}. \qquad (16.16)$$

16.5   Inequality for Distances

This section concerns the term $\Sigma_{i<j}q_{ij}\|\mathbf{a}_i - \mathbf{a}_j\|$ in (16.6). The *distance* $\|\mathbf{a}_i - \mathbf{a}_j\|$ in that term is more difficult to handle than $\|\mathbf{a}_i - \mathbf{a}_j\|^2$. This difficulty can be dealt with by finding an *inequality* for $\Sigma_{i<j}q_{ij}\|\mathbf{a}_i - \mathbf{a}_j\|$ and $\Sigma_{i<j}q_{ij}(\mathbf{a}_i - \mathbf{a}_j)'(\mathbf{b}_i - \mathbf{b}_j)$, with $\mathbf{b}_i'$ being a row vector of $\mathbf{B}$ defined in Note 16.2. The first step for that task is using the following famous theorem:

Note 16.4. The Cauchy–Schwarz Inequality

$$\mathbf{x}'\mathbf{y} \leq \|\mathbf{x}\| \times \|\mathbf{y}\|. \qquad (16.17)$$

Setting $\mathbf{x} = \mathbf{a}_i - \mathbf{a}_j$ and $\mathbf{y} = \mathbf{b}_i - \mathbf{b}_j$ in (16.17) and using $q_{ij} \geq 0$, we have:

$$q_{ij}(\mathbf{a}_i - \mathbf{a}_j)'(\mathbf{b}_i - \mathbf{b}_j) \leq q_{ij}\|\mathbf{a}_i - \mathbf{a}_j\| \times \|\mathbf{b}_i - \mathbf{b}_j\|, \qquad (16.18)$$

which leads to:

$$q_{ij}\|\mathbf{a}_i - \mathbf{a}_j\| \geq q_{ij}^{(\mathbf{B})}(\mathbf{a}_i - \mathbf{a}_j)'(\mathbf{b}_i - \mathbf{b}_j), \qquad (16.19)$$

with

$$q_{ij}^{(\mathbf{B})} = \begin{cases} 0 & \text{if } \mathbf{b}_i = \mathbf{b}_j \\ \frac{q_{ij}}{\|\mathbf{b}_i - \mathbf{b}_j\|} & \text{otherwise} \end{cases}. \qquad (16.20)$$

Here, it has been taken into consideration that the division by $\|\mathbf{b}_i - \mathbf{b}_j\| = 0$ cannot be defined, and $q_{ij}^{(\mathbf{B})}$ has the superscript "$^{(\mathbf{B})}$" because $q_{ij}^{(\mathbf{B})}$ is a function of the rows of $\mathbf{B}$.

We can use (16.10) to rewrite the right-hand side of (16.19) as:

$$q_{ij}^{(\mathbf{B})}(\mathbf{a}_i - \mathbf{a}_j)'(\mathbf{b}_i - \mathbf{b}_j) = \operatorname{tr}\mathbf{A}'(q_{ij}^{(\mathbf{B})}\mathbf{H}_{ij})\mathbf{B}. \qquad (16.21)$$

The left-hand side of (16.19) can be rewritten as:

$$q_{ij}\|\mathbf{a}_i - \mathbf{a}_j\| = \begin{cases} 0 & \text{if } \mathbf{a}_i = \mathbf{a}_j \\ \frac{q_{ij}}{\|\mathbf{a}_i - \mathbf{a}_j\|}(\mathbf{a}_i - \mathbf{a}_j)'(\mathbf{a}_i - \mathbf{a}_j) & \text{otherwise} \end{cases} \cdot \tag{16.22}$$

Its comparison with (16.20) allows us to find that (16.22) is further rewritten as:

$$q_{ij}\|\mathbf{a}_i - \mathbf{a}_j\| = q_{ij}^{(A)}(\mathbf{a}_i - \mathbf{a}_j)'(\mathbf{a}_i - \mathbf{a}_j) = \operatorname{tr} \mathbf{A}'(q_{ij}^{(A)} \mathbf{H}_{ij})\mathbf{A}, \tag{16.23}$$

where $q_{ij}^{(A)}$ is defined by replacing $\mathbf{b}_i$ by $\mathbf{a}_i$ in (16.20), and we have also used (16.10).

The summation of both sides of (16.19) leads to:

$$\sum_{i<j} q_{ij}\|\mathbf{a}_i - \mathbf{a}_j\| \geq \sum_{i<j} q_{ij}^{(B)}(\mathbf{a}_i - \mathbf{a}_i)'(\mathbf{b}_i - \mathbf{b}_j). \tag{16.24}$$

Here, we can use (16.21) and (16.23) to rewrite the left- and right-hand sides of (16.24) as:

$$\sum_{i<j} q_{ij}\|\mathbf{a}_i - \mathbf{a}_j\| = \operatorname{tr} \mathbf{A}'\mathbf{Q}_{(A)}\mathbf{A}, \tag{16.25}$$

$$\sum_{i<j} q_{ij}^{(B)}(\mathbf{a}_i - \mathbf{a}_i)'(\mathbf{b}_i - \mathbf{b}_j) = \operatorname{tr} \mathbf{A}'\mathbf{Q}_{(B)}\mathbf{B}, \tag{16.26}$$

respectively, with

$$\mathbf{Q}_{(A)} = \sum_{i<j} q_{ij}^{(A)}\mathbf{H}_{ij} \quad \text{and} \quad \mathbf{Q}_{(B)} = \sum_{i<j} q_{ij}^{(B)}\mathbf{H}_{ij}. \tag{16.27}$$

Thus, (16.24) is rewritten as:

$$\operatorname{tr} \mathbf{A}'\mathbf{Q}_{(A)}\mathbf{A} \geq \operatorname{tr} \mathbf{A}'\mathbf{Q}_{(B)}\mathbf{B}, \tag{16.28}$$

which allows us to form the MDS algorithm described in the following section.

16.6 Majorization Algorithm

Using (16.16) and (16.25), MDS loss function (16.6) is rewritten as:

$$f(\mathbf{A}) = n \operatorname{tr} \mathbf{A}'\mathbf{JA} - 2 \operatorname{tr} \mathbf{A}'\mathbf{Q}_{(\mathbf{A})}\mathbf{A}. \tag{16.29}$$

We also consider another function in which tr $\mathbf{A}'\mathbf{Q}_{(\mathbf{A})}\mathbf{A}$ in (16.29) is replaced by (16.26):

$$g(\mathbf{A}, \mathbf{B}) = n \operatorname{tr} \mathbf{A}'\mathbf{JA} - 2 \operatorname{tr} \mathbf{A}'\mathbf{Q}_{(\mathbf{B})}\mathbf{B}. \tag{16.30}$$

By comparing (16.29) and (16.30) with (16.28), we can find:

$$g(\mathbf{A}, \mathbf{B}) \geq f(\mathbf{A}). \tag{16.31}$$

Also, it should be noted that the replacement of $\mathbf{A}$ by $\mathbf{B}$ in (16.29) and (16.30) gives:

$$g(\mathbf{B}, \mathbf{B}) = f(\mathbf{B}). \tag{16.32}$$

Inequality (16.31) and equality (16.32) lead to:

$$f(\mathbf{B}) = g(\mathbf{B}, \mathbf{B}) \geq g(\mathbf{A}^*, \mathbf{B}) \geq f(\mathbf{A}^*), \tag{16.33}$$

where $\mathbf{A}^*$ is the matrix $\mathbf{A}$ that minimizes $g(\mathbf{A}, \mathbf{B})$ for a given $\mathbf{B}$.

For finding $\mathbf{A}^*$, we use the fact that $\mathbf{Q}_{(\mathbf{B})}$ in (16.27) satisfies:

$$\mathbf{JQ}_{(\mathbf{B})} = \mathbf{Q}_{(\mathbf{B})} \quad \text{or} \quad \sum_{i<j} q_{ij}^{(\mathbf{B})} \mathbf{JH}_{ij} = \sum_{i<j} q_{ij}^{(\mathbf{B})} \mathbf{H}_{ij}. \tag{16.34}$$

This follows from the fact that (3.21) shows the equivalence of $\mathbf{JH}_{ij} = \mathbf{H}_{ij}$ to $\mathbf{1}'_n\mathbf{H}_{ij} = \mathbf{0}'_n$, and it can be verified by the premultiplication of matrices (16.15) by [1, 1, 1]. Using (16.34), we can rewrite (16.30) as:

$$g(\mathbf{A}, \mathbf{B}) = n \operatorname{tr} \mathbf{A}'\mathbf{JA} - 2 \operatorname{tr} \mathbf{A}'\mathbf{JQ}_{(\mathbf{B})}\mathbf{B}$$
$$= \left\| n^{1/2}\mathbf{JA} - n^{-1/2}\mathbf{Q}_{(\mathbf{B})}\mathbf{B} \right\|^2 - n^{-1} \operatorname{tr} \mathbf{B}'\mathbf{Q}'_{(\mathbf{B})}\mathbf{Q}_{(\mathbf{B})}\mathbf{B}, \tag{16.35}$$

because of (2.11) and (2.12). Given $\mathbf{B}$, (16.35) is minimized over $\mathbf{A}$ for:

$$n^{1/2}\mathbf{JA} = \frac{1}{\sqrt{n}}\mathbf{Q}_{(\mathbf{B})}\mathbf{B}. \tag{16.36}$$

Here, we can suppose $\mathbf{A} = \mathbf{JA}$; equivalently, $n^{-1}\mathbf{1}'_n\mathbf{A} = \mathbf{0}'_n$, as the *center of coordinates* $n^{-1}\mathbf{1}'_n\mathbf{A}$ *may be anywhere*; thus, we can set it to the *origin*. This allows (16.36) to be rewritten as $\mathbf{A} = n^{-1}\mathbf{Q}_{(\mathbf{B})}\mathbf{B}$. That is, when

$$\mathbf{A}^* = \mathbf{J}\mathbf{A}^* = \frac{1}{n}\mathbf{Q}_{(\mathbf{B})}\mathbf{B}, \tag{16.37}$$

(16.33) holds true. By setting $\mathbf{A}^* = \mathbf{A}_{[t+1]}$ and $\mathbf{B} = \mathbf{A}_{[t]}$ in (16.33) and (16.37), respectively, we have $f(\mathbf{A}_{[t]}) = g(\mathbf{A}_{[t]}, \mathbf{A}_{[t]}) \geq g(\mathbf{A}_{[t+1]}, \mathbf{A}_{[t]}) \geq f(\mathbf{A}_{[t+1]})$, i.e., $f(\mathbf{A}_{[t]}) \geq f(\mathbf{A}_{[t+1]})$, and the update formula (16.7) for the coordinate matrix $\mathbf{A}$ to be obtained in MDS.

One feature of the derived algorithm is using an *auxiliary* function $g(\mathbf{A}, \mathbf{B})$ beside $f(\mathbf{A})$. The auxiliary function $g(\mathbf{A}, \mathbf{B})$ is called a *majorizing function*, as it majorizes $f(\mathbf{A})$ with (16.31). Algorithms with such auxiliary functions are called *majorization algorithms*. Though quasi-distance q_{ij} is restricted to a nonnegative value in this chapter, Heiser (1991) generalized the algorithm so that it is feasible for q_{ij} being negative.

16.7 Bibliographical Notes

Multidimensional scaling is detailed in Borg and Groenen (2005) and Cox and Cox (2000). A book-length description of majorization algorithms is found in Groenen (1993). Applications of MDS are intelligibly illustrated in Borg et al. (2014).

Exercises

16.1. Let $\mathbf{D}^{(2)}$ be the $n \times n$ matrix whose (i, j) element is the squared distance $\|\mathbf{a}_i - \mathbf{a}_j\|^2$ between the ith and jth rows of $\mathbf{A} = [\mathbf{a}_1, \ldots, \mathbf{a}_n]'$. Show that $\mathbf{D}^{(2)} = \mathbf{1}_n\mathbf{1}'_n \operatorname{diag}(\mathbf{A}\mathbf{A}') - 2\mathbf{A}\mathbf{A}' + \operatorname{diag}(\mathbf{A}\mathbf{A}')\mathbf{1}_n\mathbf{1}'_n$.

16.2. Show that $-2^{-1}\mathbf{J}\mathbf{D}^{(2)}\mathbf{J} = \mathbf{A}\mathbf{A}'$, subject to $\mathbf{A} = \mathbf{J}\mathbf{A}$, with $\mathbf{J} = \mathbf{I}_n - n^{-1}\mathbf{1}_n\mathbf{1}'_n$ and $\mathbf{D}^{(2)}$ defined in Exercise 16.1, and discuss the rationale of minimizing $\|-2^{-1}\mathbf{J}\mathbf{Q}\mathbf{J} - \mathbf{A}\mathbf{A}'\|^2$ over $\mathbf{A}$ (Gower1966; Torgerson 1952).

16.3. It is known that the differentiation of (16.3) with respect to a_{ij} is proportional to $\|\mathbf{a}_i - \mathbf{a}_j\|^{-1}$, which implies that an algorithm using the differentiation for MDS fails when $\mathbf{a}_i = \mathbf{a}_j$ arises. Show that the majorization algorithm in this chapter does not fail for $\mathbf{a}_i = \mathbf{a}_j$.

16.4. Show that the MDS solution minimizing (16.6) can be rotated.

16.5. Let $\mathbf{Q}_s = (q_{sij})$ be an $n \times n$ quasi-distance data matrix obtained from source $s = 1, \ldots, S$. In an extended version of MDS for $\mathbf{Q}_1, \ldots, \mathbf{Q}_S$, the loss function is defined as $\sum_{s=1}^{S} \sum_{i<j} (q_{sij} - d_{sij})^2$, with

$$d_{sij} = \sqrt{\sum_{k=1}^{m} w_{sk}^2 (a_{ik} - a_{jk})^2} \tag{16.38}$$

the *weighted Euclidean distance*; the loss function is minimized over $\mathbf{A}$ and w_{sk} ($s = 1, \ldots, S; k = 1, \ldots, m$) is subject to a certain constraint on $\mathbf{A}$. Here, $\mathbf{A} = (a_{ik})$ does not have subscript s, while w_{sk} does, implying that w_{sk} serves to explain the differences of $\mathbf{Q}_s$ across sources $s = 1, \ldots, S$. Discuss how w_{sk} explains those differences.

16.6. Show that (16.38) is rewritten as $\{(\mathbf{a}_i - \mathbf{a}_j)'\mathbf{W}_s^2(\mathbf{a}_i - \mathbf{a}_j)\}^{1/2} = \|\mathbf{W}_s\mathbf{a}_i - \mathbf{W}_s\mathbf{a}_j\|$

$= \|\mathbf{W}_s(\mathbf{a}_i - \mathbf{a}_j)\|$ with $\mathbf{W}_s = \begin{bmatrix} w_{s1} & & \\ & \ddots & \\ & & w_{sm} \end{bmatrix}$ an $m \times m$ diagonal matrix.

16.7. Show that $\mathbf{A}$ cannot be rotated in the extended MDS considered in Exercise 16.5, except for special cases.

16.8. Distance (16.38) can be rewritten as $d_{sij} = \sqrt{\sum_{k=1}^m \frac{w_{sk}^2}{c_k^2}(c_k a_{ik} - c_k a_{jk})^2}$. Show that the solution is not unique without a constraint on $\mathbf{A}$ and the solution can be determined uniquely by constraining each column of $\mathbf{A}$ to be standardized.

16.9. Show that:

$$\mathbf{D}_{(\mathbf{F},\mathbf{C})}^{(2)} = \text{diag}\,(\mathbf{FF}')\mathbf{1}_n\mathbf{1}_p' - 2\mathbf{FC}' + \mathbf{1}_n\mathbf{1}_p'\,\text{diag}\,(\mathbf{CC}') \tag{16.39}$$

expresses the $n \times p$ matrix whose (i, j) element is the squared distance between the ith row of $\mathbf{F}$ $(n \times m)$ and the jth row of $\mathbf{C}$ $(p \times m)$.

16.10. Let us consider approximating $n \times p$ data matrix $\mathbf{X} = (x_{ij})$ by $\mathbf{D}_{(\mathbf{F},\,\mathbf{C})}$, whose elements are the square roots of the corresponding ones in (16.39), i.e., minimizing $\|\mathbf{X} - \mathbf{D}_{(\mathbf{F},\mathbf{C})}\|^2$ over $\mathbf{F}$ and $\mathbf{C}$, with $m \leq \min(n, p)$. Discuss for what types of $\mathbf{X}$ the above minimization is useful.

Appendix

The fundamentals of matrix algebra and computations for multivariate data analysis, which had not been treated in the main chapters of this book, are described in Appendices A.1 to A.4. That is followed by supplements for Chaps. 8 and 15 in Appendix A.5. Finally, iterative algorithms are summarized and a gradient method for them is illustrated in Appendix A.6.

A.1 Geometric Understanding of Matrices and Vectors

A.1.1 Angles Between Vectors

Vectors can be depicted as lines (with arrows) as in Fig. A.1. There, we find the triangle formed by $\mathbf{a}$, $\mathbf{b}$, and $\mathbf{a} - \mathbf{b}$ with θ the *angle* between $\mathbf{a}$ and $\mathbf{b}$. For this triangle, the *cosine theorem*

$$||\mathbf{a} - \mathbf{b}||^2 = ||\mathbf{a}||^2 + ||\mathbf{b}||^2 - 2||\mathbf{a}||\,||\mathbf{b}||\cos\theta \qquad (A.1.1)$$

holds, which readers should have learned in high school. Its left-hand side can be expanded as $||\mathbf{a}||^2 + ||\mathbf{b}||^2 - 2\mathbf{a}'\mathbf{b}$, which implies

$$\mathbf{a}'\mathbf{b} = ||\mathbf{a}||\,||\mathbf{b}||\cos\theta. \qquad (A.1.2)$$

The inner product of two vectors is the multiplication of their lengths and the cosine of their angle. Equation (A.1.2) is rewritten as

$$\cos\theta = \frac{\mathbf{a}'\mathbf{b}}{||\mathbf{a}||\,||\mathbf{b}||}. \qquad (A.1.3)$$

The *cosine of the angle* between two vectors equals the *division* of their *inner product* by their *lengths*.

© Springer Nature Singapore Pte Ltd. 2016
K. Adachi, *Matrix-Based Introduction to Multivariate Data Analysis*,
DOI 10.1007/978-981-10-2341-5

Fig. A.1 Geometric illustration of vectors **a** and **b** with **0** = [0, …, 0]′

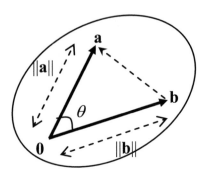

Let the angle between vectors **s** and **t** be 90°, with their lengths not being zero. Then, **s** and **t** satisfies

$$\mathbf{s}'\mathbf{t} = 0, \tag{A.1.4}$$

because of (A.1.2) and cos 90° = 0. The two vectors in (A.1.4) are said to be mutually *orthogonal*.

A.1.2 Orthonormal Matrix

The $p \times m$ matrix $\mathbf{W} = [\mathbf{w}_1, \ldots, \mathbf{w}_m]$ satisfying

$$\mathbf{W}'\mathbf{W} = \mathbf{I}_m \tag{A.1.5}$$

is said to be *column-orthonormal*, as (A.1.5) implies that the column vectors are mutually orthogonal with $\mathbf{w}_j'\mathbf{w}_k = 0$ for $j \neq k$ and of unit-length $\|\mathbf{w}_j\| = 1$. The term "orthonormal" is a composite of "orthogonal" and "normal", with the latter adjective standing for $\|\mathbf{w}_j\| = 1$.

Let a matrix **T** be a column-orthonormal and *square* of $p \times p$. It implies $\mathbf{T}' = \mathbf{T}^{-1}$ (the inverse matrix of **T**):

$$\mathbf{T}'\mathbf{T} = \mathbf{T}\mathbf{T}' = \mathbf{I}_p. \tag{A.1.6}$$

Such a **T** is simply said to be *orthonormal*. For $p \times 1$ vectors **a** and **b**,

$$\|\mathbf{Ta}\|^2 = \mathbf{a}'\mathbf{T}'\mathbf{Ta} = \mathbf{a}'\mathbf{a} = \|\mathbf{a}\|^2, \tag{A.1.7}$$

$$\|\mathbf{Ta} - \mathbf{Tb}\|^2 = (\mathbf{a} - \mathbf{b})'\mathbf{T}'\mathbf{T}(\mathbf{a} - \mathbf{b}) = (\mathbf{a} - \mathbf{b})'(\mathbf{a} - \mathbf{b}) = \|\mathbf{a} - \mathbf{b}\|^2 : \tag{A.1.8}$$

the premultiplication of vectors by an orthonormal matrix **T** does *not change* the *length* of the vectors or the *distance* between the vectors. This implies that the pre-multiplication simply *rotates* the vectors, as illustrated in Fig. A.2.

Fig. A.2 Rotation of vectors by an orthonormal matrix

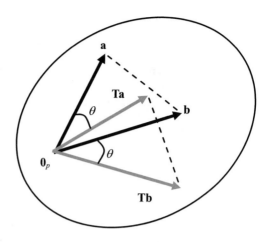

A.1.3 Vector Space

Let $\mathbf{H} = [\mathbf{h}_1, \ldots, \mathbf{h}_p]$ be a $n \times p$ matrix with $n > p$ and $\mathbf{b} = [b_1, \ldots, b_p]'$ a $p \times 1$ vector. The purpose of this section is to show what the *linear combination* of the column vectors in $\mathbf{H}$, i.e.,

$$\mathbf{h}^* = b_1\mathbf{h}_1 + \cdots + b_p\mathbf{h}_p = \mathbf{Hb}, \qquad (A.1.9)$$

geometrically represents. Here, $\mathbf{H} = [\mathbf{h}_1, \ldots, \mathbf{h}_p]$ is fixed, while each element of $\mathbf{b} = [b_1, \ldots, b_p]'$ can take any real value: $-\infty < b_j < \infty$ for $j = 1, \ldots, p$.

We start with the cases with $p = 2$, where (A.1.9) is simplified as

$$\mathbf{h}^* = b_1\mathbf{h}_1 + b_2\mathbf{h}_2. \qquad (A.1.10)$$

In Fig. A.3a, the two vectors obtained with (A.1.10) are illustrated when $[b_1, b_2] = [0.6, -1.3]$ and when $[b_1, b_2] = [1.2, 0.7]$. Since vectors $\mathbf{h}_1$, $\mathbf{h}_2$, and $\mathbf{h}^*$ are $n \times 1$, they extend in an n-dimensional space; this is depicted as an ellipse in Fig. A.3. However, $\mathbf{h}^*$ cannot extend in arbitrary directions; they are *restricted*. As illustrated

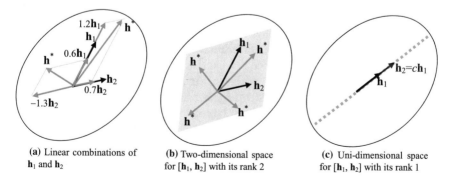

(a) Linear combinations of $\mathbf{h}_1$ and $\mathbf{h}_2$

(b) Two-dimensional space for $[\mathbf{h}_1, \mathbf{h}_2]$ with its rank 2

(c) Uni-dimensional space for $[\mathbf{h}_1, \mathbf{h}_2]$ with its rank 1

Fig. A.3 Spaces spanned by $\mathbf{h}_1$ and $\mathbf{h}_2$

in Fig. A.3b, $\mathbf{h}^*$ can only extend on the grayed plane, i.e., on a *two*-dimensional space, on which $\mathbf{h}_1$ and $\mathbf{h}_2$ extend. Here, it should be noted that $-\infty < b_1 < \infty$ and $-\infty < b_2 < \infty$, which implies the plane extends *infinitely*, though that cannot be depicted in the figure due to the limitations of the page. The plane in Fig. A.3b is called a *two-dimensional space spanned by* $\mathbf{h}_1$ and $\mathbf{h}_2$. Obviously, this space is *included* in the *n*-dimensional one for $n > p = 2$. Thus, the grayed plane is illustrated inside the ellipse in Fig. A.3b. The notions in this paragraph can be captured intuitively as follows:

Note A.1.1. Intuitive Understanding of Vector Spaces

Let us view the vectors $\mathbf{h}_1$ and $\mathbf{h}_2$ in Fig. A.3b as *pencils* before our eyes, with $n = 3$. The *pencils* extend anywhere in the three-dimensional space where we are. Then, we can verify that a *sheet* (or a thin *notebook*) can be located as the grayed plane in Fig. A.3b, i.e., so that two *pencils*, $\mathbf{h}_1$ and $\mathbf{h}_2$, extend on the *sheet*.

Further, let $\mathbf{h}^*$ be another *pencil* extending in the direction satisfying (A.1.10). Then, we can verify that *pencil* $\mathbf{h}^*$ necessarily extends in the direction of the *sheet*, i.e., it cannot extend in a direction different from the *sheet*, regardless of the values b_1 and b_2 take. Here, the world in which we, $\mathbf{h}_1$, $\mathbf{h}_2$, $\mathbf{h}^*$, and the sheet exist is a three-dimensional space, but the sheet in whose direction $\mathbf{h}_1$, $\mathbf{h}_2$, and $\mathbf{h}^*$ extend is restricted to the two-dimensional space included in the three-dimensional one.

Though we have supposed so far that $\mathbf{h}_1$ and $\mathbf{h}_2$ are linearly *independent* with rank $([\mathbf{h}_1, \mathbf{h}_2]) = 2$ in Fig. A.3a, b, the case with $\mathbf{h}_2 = c\mathbf{h}_1$ (linearly *dependent*) and rank$([\mathbf{h}_1, \mathbf{h}_2]) = 1$ is illustrated in Fig. A.3c; linear dependence and the rank of a matrix was introduced in Sects. 3.9 and 3.10. Then, the space spanned by $\mathbf{h}_1$ and $\mathbf{h}_2$ is *one-dimensional*; the space is a line when rank$([\mathbf{h}_1, \mathbf{h}_2]) = 1$. It can also be ascertained that $\mathbf{h}_2 = c\mathbf{h}_1$ allows (A.1.10) to be rewritten as $\mathbf{h}^* = b_1\mathbf{h}_1 + b_2 c\mathbf{h}_1 = (b_1 + b_2 c)\mathbf{h}_1$ for $\mathbf{h}_2 = c\mathbf{h}_1$.

Now, let us consider the cases of $p = 3$, where (A.1.9) is expressed as

$$\mathbf{h}^* = b_1\mathbf{h}_1 + b_2\mathbf{h}_2 + b_3\mathbf{h}_3. \tag{A.1.11}$$

This gives the same story as in the previous paragraph. The *three-dimensional space spanned by* $\mathbf{h}_1$, $\mathbf{h}_2$, and $\mathbf{h}_3$, which are linearly independent, is depicted as the grayed object in Fig. A.4a. Though that space (grayed object) is depicted as a "plane" in the figure, it is of three dimensions.

In Fig. A.4b, the case is illustrated in which $\mathbf{h}_1$, $\mathbf{h}_2$, and $\mathbf{h}_3$ are linearly *dependent*, with $\mathbf{h}_2 = c_1\mathbf{h}_1 + c_2\mathbf{h}_3$, but $\mathbf{h}_1$ and $\mathbf{h}_3$ are linearly independent, and rank$([\mathbf{h}_1, \mathbf{h}_2, \mathbf{h}_3]) = 2$. In this case, the space spanned by $\mathbf{h}_1$, $\mathbf{h}_2$, and $\mathbf{h}_3$ is *two*-dimensional, since (A.1.11) can be rewritten as $\mathbf{h}^* = b_1\mathbf{h}_1 + b_2(c_1\mathbf{h}_1 + c_2\mathbf{h}_3) + b_3\mathbf{h}_3 = (b_1 + b_2 c_1)\mathbf{h}_1 + (b_2 c_2 + b_3)\mathbf{h}_3$, which implies that the space spanned by $\mathbf{h}_1$, $\mathbf{h}_2$, and $\mathbf{h}_3$ is equivalent to the two-dimensional space spanned by $\mathbf{h}_1$ and $\mathbf{h}_3$.

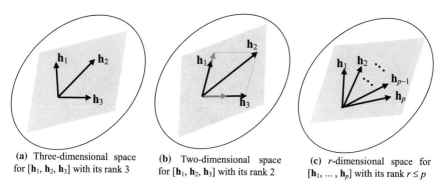

(a) Three-dimensional space for $[\mathbf{h}_1, \mathbf{h}_2, \mathbf{h}_3]$ with its rank 3

(b) Two-dimensional space for $[\mathbf{h}_1, \mathbf{h}_2, \mathbf{h}_3]$ with its rank 2

(c) r-dimensional space for $[\mathbf{h}_1, \ldots, \mathbf{h}_p]$ with its rank $r \leq p$

Fig. A.4 Spaces spanned by $\mathbf{h}_1, \ldots, \mathbf{h}_p$ for $p = 3$ and for $p > 3$

The space spanned by $\mathbf{h}_1, \ldots, \mathbf{h}_p$ can be defined for $p > 3$ in the same manner as for $p = 2, 3$. This is illustrated in Fig. A.4c. That space is called the *column space* of $\mathbf{H} = [\mathbf{h}_1, \ldots, \mathbf{h}_p]$ and is formally expressed as

$$\Xi(\mathbf{H}) = \{\mathbf{h}^* : \mathbf{h}^* = \mathbf{H}\mathbf{b} = b_1\mathbf{h}_1 + \cdots + b_p\mathbf{h}_p; -\infty < b_j < \infty, j = 1, \ldots, p\}.$$
$$(A.1.12)$$

The *dimensionality of the space* is equal to $r = \text{rank}(\mathbf{H})$. As $n > p$, this space is included in the n-dimensional space depicted as the ellipse in Fig. A.4c. Thus, the r-dimensional space spanned by $\mathbf{h}_1, \ldots, \mathbf{h}_p$, i.e., the column space of $\mathbf{H}$, is a *subspace* of n-dimensional space, since a space included in another space is called a subspace of the latter.

A.1.4 Projection onto a Subspace

Let us consider a two-dimensional subspace (i.e., plane), which is included in a p-dimensional space and spanned by the $p \times 1$ vectors $\mathbf{w}_1$ and $\mathbf{w}_2$. Here, they are of *unit length* and mutually *orthogonal* with $\|\mathbf{w}_1\| = \|\mathbf{w}_2\| = 1$ and $\mathbf{w}_1'\mathbf{w}_2 = 0$. Those equations are summarized into

$$\begin{bmatrix} \mathbf{w}_1' \\ \mathbf{w}_2' \end{bmatrix} [\mathbf{w}_1, \mathbf{w}_2] = \mathbf{W}'\mathbf{W} = \mathbf{I}_m. \qquad (A.1.13)$$

with $m = 2$ and $\mathbf{W} = [\mathbf{w}_1, \mathbf{w}_2]$ ($p \times 2$). This implies that $\mathbf{w}_1$ and $\mathbf{w}_2$ define the *orthogonal axes on the subspace*, as illustrated in Fig. A.5. Using $\mathbf{s} = [s_1, s_2]'$, whose elements can take arbitrary real values, any point on the subspace is expressed as

$$\mathbf{g} = \mathbf{W}\mathbf{s} = s_1\mathbf{w}_1 + s_2\mathbf{w}_2. \qquad (A.1.14)$$

Fig. A.5 Projection of a data
vector on a plane

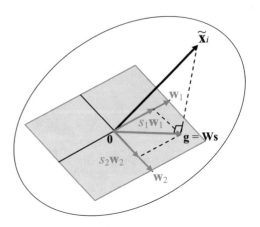

Now, we consider what values the elements of $\mathbf{s} = [s_1, s_2]$ should take, subject to the condition that $\mathbf{g}$ is the *projection* of $\tilde{\mathbf{x}}_i$ onto the subspace (plane) spanned by $\mathbf{w}_1$ and $\mathbf{w}_2$. This condition is restated as the difference vector $(\tilde{\mathbf{x}}_i - \mathbf{g})$ being *orthogonal* to the subspace, which is equivalent to $(\tilde{\mathbf{x}}_i - \mathbf{g})$ being orthogonal to $\mathbf{w}_1$ and $\mathbf{w}_2$ with $(\tilde{\mathbf{x}}_i - \mathbf{g})'\mathbf{w}_1 = 0$ and $(\tilde{\mathbf{x}}_i - \mathbf{g})'\mathbf{w}_2 = 0$. These two equations are summarized into

$$(\tilde{\mathbf{x}}_i - \mathbf{g})'\mathbf{W} = \mathbf{0}_2'. \tag{A.1.15}$$

Substituting (A.1.14) into (A.1.15), we have $(\tilde{\mathbf{x}}_i - \mathbf{W}\mathbf{s})'\mathbf{W} = \mathbf{0}_2'$, which is rewritten as $\tilde{\mathbf{x}}_i'\mathbf{W} = \mathbf{s}'\mathbf{W}'\mathbf{W}$. In this equation, we can use (A.1.13) to get

$$\mathbf{s}' = \tilde{\mathbf{x}}_i'\mathbf{W} \quad \text{or} \quad \mathbf{s} = \mathbf{W}'\tilde{\mathbf{x}}_i. \tag{A.1.16}$$

The above discussions can be generalized to the cases with $m \geq 2$. That is, (A.1.16) expresses the *coordinates* of the *projection* of $\tilde{\mathbf{x}}_i$ onto the *subspace spanned by the columns* of $\mathbf{W} = [\mathbf{w}_1, \dots, \mathbf{w}_m]$ under the condition $\mathbf{W}'\mathbf{W} = \mathbf{I}_m$ in (A.1.13).

A.2 Decomposition of Sums of Squares

As shown in (1.31), the squared norm $\|\mathbf{A}\|^2 = \text{tr}\mathbf{A}'\mathbf{A}$ expresses the sum of the squared elements in $\mathbf{A}$. Thus, $\|\mathbf{A}\|^2$ is called a *sum of squares*. It can often be rewritten as the sum of *other* sums of squares as $\|\mathbf{A}\|^2 = \|\mathbf{B}\|^2 + \|\mathbf{C}\|^2$. Such an equality is generally called the *decomposition of the sum of squares*. The decomposition is utilized in the *least squares method* in which the parameter values are found that minimize a sum of squares.

A.2.1 Decomposition Using Averages

Let us consider the sum of squares

$$f(c) = ||\mathbf{h} - c\mathbf{1}_n||^2, \tag{A.2.1}$$

with $\mathbf{h}$ an $n \times 1$ vector and c a scalar. We can find that (A.2.1) is minimized when c equals the *average* of the elements in $\mathbf{h}$:

$$\hat{c} = \frac{1}{n}\mathbf{1}_n'\mathbf{h}. \tag{A.2.2}$$

This result follows from the fact that (A.2.1) is decomposed as

$$||\mathbf{h} - c\mathbf{1}_n||^2 = ||\mathbf{h} - \frac{1}{n}\mathbf{1}_n\mathbf{1}_n'\mathbf{h}||^2 + ||\frac{1}{n}\mathbf{1}_n\mathbf{1}_n'\mathbf{h} - c\mathbf{1}_n||^2 : \tag{A.2.3}$$

only the term $g(c) = ||n^{-1}\mathbf{1}_n\mathbf{1}_n'\mathbf{h} - c\mathbf{1}_n||^2$ is relevant to c in the right-hand side of (A.2.3), and (A.2.2) allows $g(c)$ to attain its lower limit as $g(\hat{c}) = ||n^{-1}\mathbf{1}_n\mathbf{1}_n'\mathbf{h} - \mathbf{1}_n \times \hat{c}||^2 = ||n^{-1}\mathbf{1}_n\mathbf{1}_n'\mathbf{h} - n^{-1}\mathbf{1}_n\mathbf{1}_n'\mathbf{h}||^2 = 0$. The decomposition (A.2.3) is derived as follows: (A.2.1) can be rewritten as

$$\begin{aligned} ||\mathbf{h} - c\mathbf{1}_n||^2 &= ||\mathbf{h} - \hat{c}\mathbf{1}_n + \hat{c}\mathbf{1}_n - c\mathbf{1}_n||^2 \\ &= ||\mathbf{h} - \hat{c}\mathbf{1}_n||^2 + ||\hat{c}\mathbf{1}_n - c\mathbf{1}_n||^2 + 2v, \end{aligned} \tag{A.2.4}$$

with $v = (\mathbf{h} - \hat{c}\mathbf{1}_n)'(\hat{c}\mathbf{1}_n - c\mathbf{1}_n) = \hat{c}\mathbf{h}'\mathbf{1}_n - c\mathbf{h}'\mathbf{1}_n - \hat{c}^2n + \hat{c}cn = \hat{c}^2n - \hat{c}cn - \hat{c}^2n + \hat{c}cn = 0$ following from (A.2.2) or, equivalently, $\mathbf{1}_n'\mathbf{h} = n\hat{c}$.

Next, let us consider the *sum* of the sums of squares

$$h(\mathbf{F}) = \sum_{j=1}^{J} ||\mathbf{F} - \mathbf{Z}_j||^2, \tag{A.2.5}$$

with $\mathbf{F}$ and $\mathbf{Z}_j$ $n \times m$ matrices. We can find that (A.2.5) is minimized when $\mathbf{F}$ equals

$$\hat{\mathbf{F}} = \bar{\mathbf{Z}} = \frac{1}{J}\sum_{j=1}^{J}\mathbf{Z}_j, \tag{A.2.6}$$

using the fact that (A.2.5) is decomposed as

$$\sum_{j=1}^{J} ||\mathbf{F} - \mathbf{Z}_j||^2 = J||\mathbf{F} - \bar{\mathbf{Z}}|| + \sum_{j=1}^{J} ||\mathbf{Z} - \bar{\mathbf{Z}}_j||^2. \tag{A.2.7}$$

In the right-hand side, only the term $J\|\mathbf{F} - \bar{\mathbf{Z}}\|^2$ is relevant to $\mathbf{F}$ and that term attains zero for (A.2.6). Decomposition (A.2.7) is derived as follows: (A.2.5) can be rewritten as

$$\sum_{j=1}^{J} \|\mathbf{F} - \mathbf{Z}_j\|^2 = \sum_{j=1}^{J} \|\mathbf{F} - \bar{\mathbf{Z}} + \bar{\mathbf{Z}} - \mathbf{Z}_j\|^2 = J\|\mathbf{F} - \bar{\mathbf{Z}}\| + \sum_{j=1}^{J} \|\bar{\mathbf{Z}} - \mathbf{Z}_j\|^2 + 2\mathrm{tr}\mathbf{S},$$

(A.2.8)

with

$$\mathbf{S} = \sum_{j=1}^{J} (\mathbf{F} - \bar{\mathbf{Z}})' (\bar{\mathbf{Z}} - \mathbf{Z}_j)$$

$$= \sum_{j=1}^{J} \mathbf{F}'\bar{\mathbf{Z}} - \sum_{j=1}^{J} \mathbf{F}'\mathbf{Z}_j - \sum_{j=1}^{J} \bar{\mathbf{Z}}'\bar{\mathbf{Z}} + \sum_{j=1}^{J} \bar{\mathbf{Z}}'\mathbf{Z}_j$$

$$= J\mathbf{F}'\bar{\mathbf{Z}} - \mathbf{F}' \sum_{j=1}^{J} \mathbf{Z}_j - J\bar{\mathbf{Z}}'\bar{\mathbf{Z}} + \bar{\mathbf{Z}}' \sum_{j=1}^{J} \mathbf{Z}_j$$

$$= J\mathbf{F}'\bar{\mathbf{Z}} - J\mathbf{F}'\bar{\mathbf{Z}} - J\bar{\mathbf{Z}}'\bar{\mathbf{Z}} + J\bar{\mathbf{Z}}'\bar{\mathbf{Z}} = {}_m\mathbf{O}_m,$$

(A.2.9)

where we have used the fact that (A.2.6) implies $\sum_{j=1}^{J} \mathbf{Z}_j = J\bar{\mathbf{Z}}$.

A.2.2 Decomposition Using a Projector Matrix

The $n \times n$ matrix

$$\mathbf{P}_{\mathbf{X}} = \mathbf{X}(\mathbf{X}'\mathbf{X})^{-1}\mathbf{X}'$$

(A.2.10)

is called a *projection matrix* for $\mathbf{X}$ ($n \times p$). Though the use of (A.2.10) allows us to generalize the discussions in A.1.4 (e.g., Banerjee and Roy 2014; Yanai et al. 2011), that is beyond the scope of this book. Here, we focus only on the decomposition of sums of squares using (A.2.10).

Let us consider the sum of squares

$$f(\mathbf{B}) = \|\mathbf{Y} - \mathbf{X}\mathbf{B}\|^2,$$

(A.2.11)

with $\mathbf{Y}$ and $\mathbf{B}$ being $n \times q$ and $p \times q$ matrices, respectively, and $\mathbf{X}'\mathbf{X}$ nonsingular. We find that (A.2.11) is minimized when

$$\mathbf{XB} = \mathbf{P_X Y}, \text{ i.e., } \mathbf{B} = (\mathbf{X'X})^{-1}\mathbf{X'Y}, \tag{A.2.12}$$

using the fact that (A.2.11) is decomposed as

$$||\mathbf{Y} - \mathbf{XB}||^2 = ||\mathbf{Y} - \mathbf{P_X Y}||^2 + ||\mathbf{P_X Y} - \mathbf{XB}||^2. \tag{A.2.13}$$

On the right-hand side, only the term $||\mathbf{P_X Y} - \mathbf{XB}||^2$ is relevant to $\mathbf{B}$ and that term attains zero for (A.2.12). Decomposition (A.2.13) is derived as follows: (A.2.13) can be rewritten as

$$\begin{aligned}
||\mathbf{Y} - \mathbf{XB}||^2 &= ||\mathbf{Y} - \mathbf{P_X Y} + \mathbf{P_X Y} - \mathbf{XB}||^2 \\
&= ||\mathbf{Y} - \mathbf{P_X Y}||^2 + ||\mathbf{P_X Y} - \mathbf{XB}||^2 + 2\mathrm{tr}\mathbf{C}, \tag{A.2.14}
\end{aligned}$$

with

$$\begin{aligned}
\mathbf{C} &= (\mathbf{Y} - \mathbf{P_X Y})'(\mathbf{P_X Y} - \mathbf{XB}) \\
&= \mathbf{Y'P_X Y} - \mathbf{Y'XB} - \mathbf{Y'P_X^2 Y} + \mathbf{Y'P_X'XB} = {}_q\mathbf{O}_q, \tag{A.2.15}
\end{aligned}$$

where we have used $\mathbf{P_X'} = \mathbf{P_X}$, $\mathbf{P_X^2} = \mathbf{P_X}$, and $\mathbf{P_X X} = \mathbf{X}$.

Solution (4.12) in Chap. 4 is obtained by setting $q = 1$ and substituting $\mathbf{JX}$ and $\mathbf{y}$ for $\mathbf{X}$ and $\mathbf{Y}$ in (A.2.12):

$$\hat{\mathbf{b}} = (\mathbf{X'J'JX})^{-1}\mathbf{X'Jy} = (\mathbf{X'JX})^{-1}\mathbf{X'Jy}, \tag{A.2.16}$$

where $\mathbf{B}$ in (A.2.12) is replaced by $\hat{\mathbf{b}}$ (a $p \times 1$ vector).

We should note that $\hat{c}\mathbf{1}_n = n^{-1}\mathbf{1}_n'\mathbf{1}_n$ in (A.2.3) is also a projection matrix, since substituting $\mathbf{1}_n$ for $\mathbf{X}$ in (A.2.10) leads to $\mathbf{P_{1_n}} = \mathbf{1}_n(\mathbf{1}_n'\mathbf{1}_n)^{-1}\mathbf{1}_n' = n^{-1}\mathbf{1}_n\mathbf{1}_n'$.

A.3   Singular Value Decomposition

The author believes that *singular value decomposition* (*SVD*) is the *most important* tool in matrix algebra, as SVD can be defined for any matrix, a number of facts can be easily derived from SVD, and it plays important roles in matrix computations as found in Appendix A.4.

A.3.1   SVD: Extended Version

Please, *learn* this theorem (SVD) by *heart* as *absolute truth*!

Theorem A.3.1. SVD (extended version)
Any $n \times p$ matrix $\mathbf{X}$ with $n \geq p$ can be decomposed as

$$\mathbf{X} = \tilde{\mathbf{K}}\tilde{\mathbf{\Lambda}}\tilde{\mathbf{L}}'$$
(A.3.1)

Here, $\tilde{\mathbf{K}}(n \times p)$ is an $n \times p$ *column-orthonormal* matrix and $\tilde{\mathbf{L}}$ $(p \times p)$ is a $p \times p$ *orthonormal* matrix:

$$\tilde{\mathbf{K}}'\tilde{\mathbf{K}} = \tilde{\mathbf{L}}'\tilde{\mathbf{L}} = \tilde{\mathbf{L}}\tilde{\mathbf{L}}' = \mathbf{I}_p.$$
(A.3.2)

$\tilde{\mathbf{\Lambda}}$ is a $p \times p$ *diagonal* matrix

$$\tilde{\mathbf{\Lambda}} = \begin{bmatrix} \lambda_1 & & & & & \\ & \ddots & & & & \\ & & \lambda_r & & & \\ & & & 0 & & \\ & & & & \ddots & \\ & & & & & 0 \end{bmatrix}$$
(A.3.3)

with its diagonal elements arranged in *decreasing* order

$$\lambda_1 \geq \cdots \geq \lambda_r > 0,$$
(A.3.4)

the *number of the positive diagonal* elements being the *rank* of $\mathbf{X}$:

$$r = \text{rank}(\mathbf{X}),$$
(A.3.5)

and the blank cells standing for zero elements.

Theorem A.3.1 concerns the SVD of a matrix with the number of rows greater than or equal to that of columns. The SVD of a matrix with more columns than rows can be defined simply by *transposing* both sides of (A.3.1): any matrix $\mathbf{X}'(p \times n)$ with $p \leq n$ can be decomposed as

$$\mathbf{X}' = \tilde{\mathbf{L}}\tilde{\mathbf{\Lambda}}\tilde{\mathbf{K}}',$$
(A.3.6)

with (A.3.2)–(A.3.5).

Theorem A.3.1 shows that we can easily find rank($\mathbf{X}$) by *counting the number of nonzero diagonal elements* in $\tilde{\mathbf{\Lambda}}$, if the SVD of $\mathbf{X}$ is given. Further, SVD leads to the following fact: for an $n \times p$ matrix $\mathbf{X}$,

$$\text{rank}(\mathbf{X}) = \text{rank}(\mathbf{X}') = \text{rank}(\mathbf{XX}') = \text{rank}(\mathbf{X'X}). \qquad (A.3.7)$$

Here, the first equality directly follows from (A.3.6), and the rank$(\mathbf{X}) = \text{rank}(\mathbf{XX}') =$ rank$(\mathbf{X'X})$ follows from the fact that Theorem A.3.1 implies $\mathbf{X'X} = \tilde{\mathbf{L}}\tilde{\mathbf{\Lambda}}\tilde{\mathbf{K}}'\tilde{\mathbf{K}}\tilde{\mathbf{\Lambda}}\tilde{\mathbf{L}}' = \tilde{\mathbf{L}}\tilde{\mathbf{\Lambda}}^2\tilde{\mathbf{L}}'$ and $\mathbf{XX}' = \tilde{\mathbf{K}}\tilde{\mathbf{\Lambda}}^2\tilde{\mathbf{K}}'$; the SVD of $\mathbf{X'X}$ and that of $\mathbf{XX}'$ are $\tilde{\mathbf{L}}\tilde{\mathbf{\Lambda}}^2\tilde{\mathbf{L}}'$ and $\tilde{\mathbf{K}}\tilde{\mathbf{\Lambda}}^2\tilde{\mathbf{K}}'$, respectively.

A.3.2   SVD: Compact Version

Let us consider the same matrices as in Theorem A.3.1, and let $\mathbf{K}$ and $\mathbf{L}$ be the matrices containing the first r columns of $\tilde{\mathbf{K}}$ and $\tilde{\mathbf{L}}$, respectively, with $\tilde{\mathbf{K}} = [\mathbf{K}, \mathbf{K}_{(p-r)}]$ and $\tilde{\mathbf{L}} = [\mathbf{L}, \mathbf{L}_{(p-r)}]$ being block matrices (whose introduction is found in Sect. 14.1). Here, $\mathbf{K}_{(p-r)}$ and $\mathbf{L}_{(p-r)}$ contain the last $p - r$ columns of $\tilde{\mathbf{K}}$ and $\tilde{\mathbf{L}}$, respectively. Further, let $\mathbf{\Lambda}$ be the $r \times r$ diagonal matrix whose diagonal elements are $\lambda_1 \geq \cdots \geq \lambda_r$. Then, the right-hand side of (A.3.1) is rewritten as

$$\tilde{\mathbf{K}}\tilde{\mathbf{\Lambda}}\tilde{\mathbf{L}}' = [\mathbf{K}, \mathbf{K}_{(p-r)}]\begin{bmatrix} \mathbf{\Lambda} & \\ & {}_{p-r}\mathbf{O}_{p-r} \end{bmatrix}\begin{bmatrix} \mathbf{L}' \\ \mathbf{L}'_{(p-r)} \end{bmatrix} = \mathbf{K}\mathbf{\Lambda}\mathbf{L}'. \qquad (A.3.8)$$

Thus, we have the *compact* version of Theorem A.3.1.

Theorem A.3.2   SVD (compact version)

   Any $n \times p$ matrix $\mathbf{X}$ with rank$(\mathbf{X}) = r$ can be decomposed as

$$\mathbf{X} = \mathbf{K}\mathbf{\Lambda}\mathbf{L}'. \qquad (A.3.9)$$

Here, $\mathbf{K}$ $(n \times r)$ and $\mathbf{L}$ $(p \times r)$ are *column-orthonormal* matrices with

$$\mathbf{K}'\mathbf{K} = \mathbf{L}'\mathbf{L} = \mathbf{I}_r \qquad (A.3.10)$$

and $\mathbf{\Lambda}$ is the $r \times r$ *diagonal* matrix

$$\mathbf{\Lambda} = \begin{bmatrix} \lambda_1 & & \\ & \cdots & \\ & & \lambda_r \end{bmatrix}, \qquad (A.3.11)$$

whose diagonal elements are *positive* and arranged in *decreasing* order with

$$\lambda_1 \geq \cdots \geq \lambda_r > 0. \qquad (A.3.12)$$

The lth diagonal element (λ_l) of Λ is called the lth largest *singular value* of $\mathbf{X}$. The lth columns of $\mathbf{K}$ and $\mathbf{L}$ are called the left and right *singular vectors* of $\mathbf{X}$ corresponding to λ_l, respectively. Obviously, the SVD of $\mathbf{X}'$ is defined as $\mathbf{X}' = \mathbf{L}\Lambda\mathbf{K}'$ with (A.3.10), (A.3.11), and (A.3.12).

Theorem A.3.2 shows that the SVD of $\mathbf{XX}'$ and $\mathbf{X}'\mathbf{X}$ are defined as

$$\mathbf{XX}' = \mathbf{K}\Lambda^2\mathbf{K}', \tag{A.3.13}$$

$$\mathbf{X}'\mathbf{X} = \mathbf{L}\Lambda^2\mathbf{L}', \tag{A.3.14}$$

respectively, which have already been used in the proof for (A.3.7). The SVDs (A.3.13) and (A.3.14) lead to the *sum of squares elements* in $\mathbf{X}$ equaling the *sum of its squared singular values*:

$$||\mathbf{X}||^2 = \mathrm{tr}\mathbf{X}'\mathbf{X} = \mathrm{tr}\mathbf{XX}' = \mathrm{tr}\Lambda^2 = \Lambda_1^2 + \cdots + \Lambda_r^2, \tag{A.3.15}$$

since $\mathrm{tr}\mathbf{X}'\mathbf{X} = \mathrm{tr}\mathbf{L}\Lambda\mathbf{K}'\mathbf{K}\Lambda\mathbf{L}' = \mathrm{tr}\mathbf{L}\Lambda\Lambda\mathbf{L}' = \mathrm{tr}\mathbf{L}\Lambda^2\mathbf{L}' = \mathrm{tr}\Lambda^2\mathbf{L}'\mathbf{L} = \mathrm{tr}\Lambda^2$. If rank($\mathbf{X}'\mathbf{X}$) = p, then it is a nonsingular square matrix and its *inverse* matrix is given by

$$(\mathbf{X}'\mathbf{X})^{-1} = \mathbf{L}\Lambda^{-2}\mathbf{L}'. \tag{A.3.16}$$

If $p = n$ and $\mathbf{X}$ is nonsingular, then

$$\mathbf{X}^{-1} = \mathbf{L}\Lambda^{-1}\mathbf{K}'. \tag{A.3.17}$$

A.3.3 *Other Expressions of SVD*

Let us express the matrices $\mathbf{K}$ and $\mathbf{L}$ in Theorem A.3.2 as $\mathbf{K} = [\mathbf{k}_1, \ldots, \mathbf{k}_m, \mathbf{k}_{m+1}, \ldots, \mathbf{k}_r] = [\mathbf{K}_m, \mathbf{K}_{[m]}]$ and $\mathbf{L} = [\mathbf{l}_1, \ldots, \mathbf{l}_m, \mathbf{l}_{m+1}, \ldots, \mathbf{l}_r] = [\mathbf{L}_m, \mathbf{L}_{[m]}]$. Here,

$$\mathbf{K}_m = [\mathbf{k}_1, \ldots, \mathbf{k}_m] \quad \text{and} \quad \mathbf{L}_m = [\mathbf{l}_1, \ldots, \mathbf{l}_m] \tag{A.3.18}$$

contain the *first m columns* of $\mathbf{K}$ and $\mathbf{L}$, respectively, while

$$\mathbf{K}_{[m]} = [\mathbf{k}_{m+1}, \ldots, \mathbf{k}_r] \quad \text{and} \quad \mathbf{L}_{[m]} = [\mathbf{l}_{m+1}, \ldots, \mathbf{l}_r] \tag{A.3.19}$$

contain the *r−m remaining columns* of $\mathbf{K}$ and $\mathbf{L}$, respectively. Then, (A.3.10) is rewritten as

$$\mathbf{k}_u'\mathbf{k}_u = \mathbf{l}_u'\mathbf{l}_u = 1 \quad \text{and} \quad \mathbf{k}_u'\mathbf{k}_v = \mathbf{l}_u'\mathbf{l}_v = 0 \quad \text{for } u \neq v, \tag{A.3.20}$$

with $u = 1, \ldots, r$ and $v = 1, \ldots, r$. Further, SVD (A.3.9) can be rewritten as $\mathbf{X} = \lambda_1 \mathbf{k}_1 \mathbf{l}_1' + \cdots + \lambda_m \mathbf{k}_m \mathbf{l}_m' + \lambda_{m+1} \mathbf{k}_{m+1} \mathbf{l}_m' + \cdots + \lambda_r \mathbf{k}_r \mathbf{l}_r'$, which is expressed in matrix form as

$$\mathbf{X} = \mathbf{K}\mathbf{\Lambda}\mathbf{L}' = \mathbf{K}_m \mathbf{\Lambda}_m \mathbf{L}_m' + \mathbf{K}_{[m]} \mathbf{\Lambda}_{[m]} \mathbf{L}_{[m]}', \tag{A.3.21}$$

with

$$\mathbf{\Lambda}_m = \begin{bmatrix} \lambda_1 & & \\ & \ddots & \\ & & \lambda_m \end{bmatrix} \text{ and } \mathbf{\Lambda}_{[m]} = \begin{bmatrix} \lambda_{m+1} & & \\ & \ddots & \\ & & \lambda_r \end{bmatrix}; \text{ i.e., } \mathbf{\Lambda} = \begin{bmatrix} \mathbf{\Lambda}_m & \\ & \mathbf{\Lambda}_{[m]} \end{bmatrix}. \tag{A.3.22}$$

By noting (A.3.20), we find

$$\mathbf{K}' \mathbf{K}_m = \mathbf{L}' \mathbf{L}_m = \begin{bmatrix} 1 & & \\ & \ddots & \\ & & 1 \\ 0 & \cdots & 0 \\ \vdots & & \vdots \\ 0 & \cdots & 0 \end{bmatrix} = \begin{bmatrix} \mathbf{I}_m \\ {}_{r-m}\mathbf{O}_m \end{bmatrix} : \tag{A.3.23}$$

$\mathbf{K}'\mathbf{K}_m = \mathbf{L}'\mathbf{L}_m$ equals the $r \times m$ matrix whose first m rows are those of $\mathbf{I}_m$ and the remaining rows are filled with zeros. Post-multiplying both sides of (A.3.9) by $\mathbf{L}_m$ and using (A.3.23) leads to

$$\mathbf{X}\mathbf{L}_m = \mathbf{K}\mathbf{\Lambda}\mathbf{L}' \mathbf{L}_m = [\mathbf{K}_m, \mathbf{K}_{[m]}] \begin{bmatrix} \mathbf{\Lambda}_m & \\ & \mathbf{\Lambda}_{[m]} \end{bmatrix} \begin{bmatrix} \mathbf{I}_m \\ {}_{r-m}\mathbf{O}_m \end{bmatrix} = [\mathbf{K}_m, \mathbf{K}_{[m]}] \begin{bmatrix} \mathbf{\Lambda}_m \\ {}_{r-m}\mathbf{O}_m \end{bmatrix}$$
$$= \mathbf{K}_m \mathbf{\Lambda}_m,$$

that is,

$$\mathbf{K}_m \mathbf{\Lambda}_m = \mathbf{X}\mathbf{L}_m. \tag{A.3.24}$$

Further, post-multiplying both side by $\mathbf{L}_m'$ gives

$$\mathbf{K}_m \mathbf{\Lambda}_m \mathbf{L}_m' = \mathbf{X}\mathbf{L}_m \mathbf{L}_m'. \tag{A.3.25}$$

We can also use (A.3.23) to rewrite SVD (A.3.9) as

$$\mathbf{L}_m \mathbf{\Lambda}_m = \mathbf{X}' \mathbf{K}_m, \tag{A.3.26}$$

which follows from

$$\mathbf{X}'\mathbf{K}_m = \mathbf{L}\mathbf{\Lambda}\mathbf{K}'\mathbf{K}_m = [\mathbf{L}_m, \mathbf{L}_{[m]}]\begin{bmatrix} \mathbf{\Lambda}_m & \\ & \mathbf{\Lambda}_{[m]} \end{bmatrix}\begin{bmatrix} \mathbf{I}_m \\ {}_{r-m}\mathbf{O}_m \end{bmatrix} = [\mathbf{L}_m, \mathbf{L}_{[m]}]\begin{bmatrix} \mathbf{\Lambda}_m \\ {}_{r-m}\mathbf{O}_m \end{bmatrix}$$
$$= \mathbf{L}_m\mathbf{\Lambda}_m.$$

A.3.4 SVD and Eigenvalue Decomposition for Symmetric Matrices

Let us define $\mathbf{C} = \mathbf{X}'\mathbf{X}$ with $\mathbf{X}$ treated in Theorem 3.2. As shown in A.3.14, the SVD of $\mathbf{C}$ is given by $\mathbf{C} = \mathbf{L}\mathbf{\Lambda}^2\mathbf{L}'$. This is also the *eigenvalue decomposition* (*EVD*) of $\mathbf{C}$ as found in Note 6.1. The SVD and EVD are equivalent for a symmetric matrix which is the product of a matrix and its transpose. However, it does *not hold* true for a symmetric matrix which is not such product, as shown next.

Let $\mathbf{S}$ be an $n \times n$ symmetric matrix with rank($\mathbf{S}$) = $r \leq n$. The EVD of $\mathbf{S}$ can be expressed as

$$\mathbf{S} = \mathbf{E}\mathbf{\Theta}\mathbf{E}'. \tag{A.3.27}$$

Here, $\mathbf{E}'\mathbf{E} = \mathbf{I}_r$, and $\mathbf{\Theta}$ is the $r \times r$ diagonal matrix with its kth diagonal element θ_k satisfying $|\theta_k| \geq |\theta_{k+1}|$. In general, θ_k, an eigenvalue of $\mathbf{S}$, can be *negative*, which implies that (A.3.27) is not the SVD of $\mathbf{S}$. Its SVD can be expressed as

$$\mathbf{S} = \mathbf{E}\mathbf{D}\mathbf{D}\mathbf{\Theta}\mathbf{E}'. \tag{A.3.28}$$

Here, $\mathbf{D}$ is the $r \times r$ diagonal matrix whose kth diagonal element is 1 if $\theta_k > 0$, but -1 otherwise. We can find that $\mathbf{D}\mathbf{\Theta}$ is the diagonal matrix with positive diagonal elements, i.e., the singular values of $\mathbf{S}$, and the corresponding singular vectors are contained in $\mathbf{E}\mathbf{D}$ and $\mathbf{E}$, with $(\mathbf{E}\mathbf{D})'\mathbf{E}\mathbf{D} = \mathbf{D}\mathbf{E}'\mathbf{E}\mathbf{D} = \mathbf{I}_r$.

A.4 Matrix Computations Using SVD

The purpose of this appendix is to present solutions for the problems of *maximizing* some *traces* of matrix products and *reduced rank approximations*. Their foundation is given by the Theorem in A.4.1.

A.4.1 ten Berge's Theorem with Suborthonormal Matrices

Definition A.4.1. Suborthonormal Matrix
A matrix is *suborthonormal* if it can be completed to be an orthonormal matrix by appending rows, columns, or both, or if it is orthonormal (ten Berge 1993, pp. 27–28).

An example of a suborthonormal matrix is $\mathbf{A} = \begin{bmatrix} 0.8 & 0 \\ 0 & 1 \end{bmatrix}$ (ten Berge 1993, p. 28),

since we can append the row $[0.6, 0]$ and the column $\begin{bmatrix} 0.6 \\ 0 \\ -0.8 \end{bmatrix}$ to $\mathbf{A}$ so that it can be

completed to be orthonormal $\tilde{\mathbf{A}} = \begin{bmatrix} 0.8 & 0 & 0.6 \\ 0 & 1 & 0 \\ 0.6 & 0 & -0.8 \end{bmatrix}$ with $\tilde{\mathbf{A}}'\tilde{\mathbf{A}} = \tilde{\mathbf{A}}\tilde{\mathbf{A}}' = \mathbf{I}_3$.

$$\text{A } p \times m \text{ column-orthonormal matrix } \mathbf{B}$$
$$\text{and } \mathbf{B}' \text{ are suborthonormal with } p > m, \qquad (A.4.1)$$

since the $p \times p$ matrices $[\mathbf{B}, \mathbf{C}]$ and $\begin{bmatrix} \mathbf{B}' \\ \mathbf{C}' \end{bmatrix}$ are orthonormal, with $\mathbf{C}$ a $p \times (p-m)$ matrix satisfying $\mathbf{B}'\mathbf{C} = {}_m\mathbf{O}_{p-m}$ and $\mathbf{C}'\mathbf{C} = \mathbf{I}_{p-m}$. A suborthonormal matrix has the following property:

$$\text{the } product \text{ of } suborthonormal \text{ matrices} = suborthonormal \text{ matrix} \qquad (A.4.2)$$

(ten Berge 1983, 1993).
   The following theorem concerning suborthonormal matrices takes an important role for the purpose of this section:

Theorem A.4.1. ten Berge's (1993) Theorem

If $\mathbf{S}$ is a $p \times p$ *suborthonormal* matrix with $\text{rank}(\mathbf{S}) = m \leq p$ and $\mathbf{D} =$

$\begin{bmatrix} d_1 & & \\ & \ddots & \\ & & d_p \end{bmatrix}$ is a $p \times p$ *diagonal* matrix with $d_1 \geq \cdots \geq d_p \geq 0$, then

$$f(\mathbf{S}) = \text{tr}\mathbf{SD} \leq \text{tr}\mathbf{D}_m = d_1 + \cdots + d_m \leq \text{tr}\mathbf{D}, \qquad (A.4.3)$$

with $\mathbf{D}_m = \begin{bmatrix} d_1 & & \\ & \ddots & \\ & & d_m \end{bmatrix}$ the first $m \times m$ diagonal block of $\mathbf{D}$.

The generalization of this theorem is found in ten Berge (1983). As $d_1 + \cdots + d_m \leq$ tr$\mathbf{D}$ obviously holds, it has been added to ten Berge's (1993, p. 28) inequality in (A.4.3).

A.4.2 Maximization of Trace Functions

In this section, we consider the *maximization* problems for three forms of *trace functions*. Here, the sentence "maximize $f(\mathbf{B})$ over $\mathbf{B}$ s.t. $g(\mathbf{B}) = c$" means "obtain the matrix $\mathbf{B}$ that maximizes $f(\mathbf{B})$ subject to the constraint $g(\mathbf{B}) = c$" with "s.t." the abbreviation for "*subject to*".

> **Theorem A.4.2**
> For an $n \times p$ matrix $\mathbf{Y}$ with rank$(\mathbf{Y}) = p$, we consider the problem:
>
> $$\text{maximize} f(\mathbf{C}) = \text{tr}\mathbf{Y}'\mathbf{C} \text{ over } \mathbf{C}(n \times p) \text{ s.t. } \mathbf{C}'\mathbf{C} = \mathbf{I}_p. \qquad (A.4.4)$$
>
> This is attained for
>
> $$\mathbf{C} = \mathbf{U}\mathbf{V}'. \qquad (A.4.5)$$
>
> Here, $\mathbf{U}$ and $\mathbf{V}$ are given by the SVD of $\mathbf{Y}$ defined as $\mathbf{Y} = \mathbf{U}\mathbf{D}\mathbf{V}'$ with $\mathbf{U}'\mathbf{U} = \mathbf{V}'\mathbf{V} = \mathbf{I}_p$ and $\mathbf{D}$ a $p \times p$ diagonal matrix whose diagonal elements are all positive.

Proof. By substituting $\mathbf{Y} = \mathbf{U}\mathbf{D}\mathbf{V}'$ in $f(\mathbf{C}) = \text{tr}\mathbf{Y}'\mathbf{C}$, it is rewritten as $f(\mathbf{C}) = \text{tr}\mathbf{V}\mathbf{D}\mathbf{U}'\mathbf{C} = \text{tr}\mathbf{U}'\mathbf{C}\mathbf{V}\mathbf{D}$. The column-orthonormality of $\mathbf{U}$, $\mathbf{V}$, and $\mathbf{C}$ implies that $\mathbf{U}'\mathbf{C}\mathbf{V}$ is sub-orthonormal, because of (A.4.1) and (A.4.2). Further, $r = \text{rank}(\mathbf{U}'\mathbf{C}\mathbf{V}) \leq p$, while $\mathbf{D}$ is a $p \times p$ diagonal matrix with all diagonal elements positive. Those facts and Theorem A.4.1 lead to $f(\mathbf{C}) = \text{tr}\mathbf{U}'\mathbf{C}\mathbf{V}\mathbf{D} \leq \text{tr}\mathbf{D}$, where the upper bound is attained for (A.4.5), since $f(\mathbf{U}\mathbf{V}') = \text{tr}\mathbf{V}\mathbf{D}\mathbf{U}'\mathbf{U}\mathbf{V}' = \text{tr}\mathbf{D}$. $\qquad \square$

Theorem A.4.3

For the $n \times p$ matrix $\mathbf{X}$ in Theorem A.3.2, we consider the following problem:

$$\text{maximize} f(\mathbf{A}, \mathbf{B}) = \text{tr}\mathbf{A}'\mathbf{X}\mathbf{B} \text{ over } \mathbf{A}(n \times m) \text{ and } \mathbf{B}(p \times m)$$
$$\text{s.t. } \mathbf{A}'\mathbf{A} = \mathbf{B}'\mathbf{B} = \mathbf{I}_m \text{ with } m \leq r = \text{rank}(\mathbf{X}). \tag{A.4.6}$$

This is attained for

$$\mathbf{A} = \mathbf{K}_m\mathbf{T} \quad \text{and} \quad \mathbf{B} = \mathbf{L}_m\mathbf{T} \tag{A.4.7}$$

with $\mathbf{K}_m$ and $\mathbf{L}_m$ defined as in (A.3.18) and $\mathbf{T}$ an $m \times m$ orthonormal matrix.

Proof. By substituting (A.3.9) (the SVD of $\mathbf{X}$) in $f(\mathbf{A}, \mathbf{B}) = \text{tr}\mathbf{A}'\mathbf{X}\mathbf{B}$, it is rewritten as $f(\mathbf{A}, \mathbf{B}) = \text{tr}\mathbf{A}'\mathbf{K}\boldsymbol{\Lambda}\mathbf{L}'\mathbf{B} = \text{tr}\mathbf{L}'\mathbf{B}\mathbf{A}'\mathbf{K}\boldsymbol{\Lambda}$. As found in (A.3.10) and (A.4.6), $\mathbf{K}$, $\mathbf{L}$, $\mathbf{A}$, and $\mathbf{B}$ are column-orthonormal, and $\mathbf{L}'\mathbf{B}\mathbf{A}'\mathbf{K}$ is suborthonormal because of (A.4.1) and (A.4.2). Further, $\text{rank}(\mathbf{L}'\mathbf{B}\mathbf{A}'\mathbf{K}) \leq m \leq r$, while $\boldsymbol{\Lambda}$ is an $r \times r$ diagonal matrix with all diagonal elements positive. Those facts and Theorem A.4.1 lead to $f(\mathbf{A}, \mathbf{B}) = \text{tr}\mathbf{L}'\mathbf{B}\mathbf{A}'\mathbf{K}\boldsymbol{\Lambda} \leq \text{tr}\boldsymbol{\Lambda}_m$ with $\boldsymbol{\Lambda}_m$ defined as (A.3.22). Here, the upper bound is attained for (A.4.7) as $f(\mathbf{K}_m\mathbf{T}, \mathbf{L}_m\mathbf{T}) = \text{tr}\mathbf{L}'\mathbf{L}_m\mathbf{T}\mathbf{T}'\mathbf{K}'_m\mathbf{K}\boldsymbol{\Lambda} = \text{tr}\mathbf{L}'\mathbf{L}_m\mathbf{K}'_m\mathbf{K}\boldsymbol{\Lambda} = \text{tr}\boldsymbol{\Lambda}_m$, with (A.4.7) satisfying the constraints in (A.4.6) as $\mathbf{T}'\mathbf{K}'_m\mathbf{K}_m\mathbf{T} = \mathbf{T}'\mathbf{L}'_m\mathbf{L}_m\mathbf{T} = \mathbf{I}_m$ because of (A.1.6) and (A.3.10). $\qquad\square$

Solution (A.4.7) shows that it is not unique; we can choose an arbitrary $m \times m$ orthonormal matrix as $\mathbf{T}$. Thus, we can choose $\mathbf{T}$ in the rotation methods described in Chap. 13, after obtaining (A.4.7) with $\mathbf{T} = \mathbf{I}_m$. Solutions that can be rotated as (A.4.7) are said to have *rotational indeterminacy*. This can be avoided by adding the following constraint to (A.4.6): $\mathbf{A}'\mathbf{X}\mathbf{B}$ is a diagonal matrix whose diagonal elements are arranged in descending order. Then, the solution is restricted to $\mathbf{A} = \mathbf{K}_m$ and $\mathbf{B} = \mathbf{L}_m$, which leads to $\mathbf{A}'\mathbf{X}\mathbf{B} = \boldsymbol{\Lambda}_m$.

Theorem A.4.4

For the $n \times p$ matrix $\mathbf{X}$ in Theorem A.3.2, we consider the following problem:

$$\text{maximize } f(\mathbf{W}) = \text{tr}\mathbf{W}'\mathbf{X}'\mathbf{X}\mathbf{W} \text{ over } \mathbf{W} \text{ } (p \times m)$$
$$\text{s.t. } \mathbf{W}'\mathbf{W} = \mathbf{I}_m \text{ with } m \leq r = \text{rank}(\mathbf{X}). \tag{A.4.8}$$

This is attained for

$$\mathbf{W} = \mathbf{L}_m\mathbf{T}, \tag{A.4.9}$$

with $\mathbf{L}_m$ defined as in (A.3.18) and $\mathbf{T}$ an $m \times m$ orthonormal matrix.

Proof. By substituting (A.3.9) in $f(\mathbf{W})$, it is rewritten as $f(\mathbf{W}) = \mathrm{tr}\mathbf{W}'\mathbf{L}\Lambda^2\mathbf{L}'\mathbf{W} = \mathrm{tr}\mathbf{L}'\mathbf{W}\mathbf{W}'\mathbf{L}\Lambda^2$. As found in (A.3.10) and (A.4.8), $\mathbf{L}$ and $\mathbf{W}$ are column-orthonormal, and $\mathbf{L}'\mathbf{W}\mathbf{W}'\mathbf{L}$ is suborthonormal because of (A.4.1) and (A.4.2). Further, rank $(\mathbf{L}'\mathbf{W}\mathbf{W}'\mathbf{L}) \le m \le r$, while Λ^2 is an $r \times r$ diagonal matrix with all diagonal elements positive. This fact and Theorem A.4.1 lead to $f(\mathbf{W}) = \mathrm{tr}\mathbf{W}'\mathbf{X}'\mathbf{X}\mathbf{W} \le \mathrm{tr}\Lambda_m^2$, with Λ_m defined as (A.3.22). Here, the upper bound is attained for (A.4.9) as $f(\mathbf{L}_m\mathbf{T}) = \mathrm{tr}\mathbf{L}'\mathbf{L}_m\mathbf{T}\mathbf{T}'\mathbf{L}'_m\mathbf{L}\Lambda^2 = \mathrm{tr}\Lambda_m^2$, with (A.4.9) satisfying the constraint in (A.4.8) as $\mathbf{T}'\mathbf{L}'_m\mathbf{L}_m\mathbf{T} = \mathbf{I}_m$.   □

Solution (A.4.9) also has rotational indeterminacy, which can be avoided by adding the following constraint to (A.4.8): $\mathbf{W}'\mathbf{X}'\mathbf{X}\mathbf{W}$ is a diagonal matrix whose diagonal elements are arranged in descending order.

A.4.3 Reduced Rank Approximation

In Chap. 3, *principal component analysis* (PCA) is introduced as a problem of obtaining the matrix product $\mathbf{F}\mathbf{A}'$ that well approximates a data matrix $\mathbf{X}$, subject to the number of the columns of $\mathbf{F}$ and that of $\mathbf{A}$ not exceeding the rank of $\mathbf{X}$. Such a problem can be restated as approximating $\mathbf{X}$ by another matrix of lower rank and is called *reduced rank approximation*. The theorem for the approximation is presented next:

Theorem A.4.5

For the $n \times p$ matrix $\mathbf{X}$ in Theorem A.3.2, we consider the following problem:

$$\text{Minimize} f(\mathbf{M}) = ||\mathbf{X} - \mathbf{M}||^2 \text{ over } \mathbf{M} \text{ s.t. rank}(\mathbf{M}) \le m \le \text{rank}(\mathbf{X}).$$
$$(A.4.10)$$

This is attained for

$$\mathbf{M} = \mathbf{K}_m\Lambda_m\mathbf{L}'_m.\qquad (A.4.11)$$

Here, it should be noted that the constraint in (A.4.10) is rank($\mathbf{M}$) equaling or being less than m, but solution (A.4.11) is restricted to rank($\mathbf{M}$) = m.

Proof. Using the extended version of SVD for $\mathbf{M}$, it is expressed as $\mathbf{M} = \mathbf{P}\Omega\mathbf{Q}'$, with $\mathbf{P}'\mathbf{P} = \mathbf{Q}'\mathbf{Q} = \mathbf{I}_m$ and Ω an $m \times m$ diagonal matrix whose elements are nonnegative. Then, $f(\mathbf{M})$ is rewritten as

$$\begin{aligned}
f(\mathbf{P}\Omega\mathbf{Q}') &= ||\mathbf{X} - \mathbf{P}\Omega\mathbf{Q}'||^2 \\
&= ||\mathbf{X} - \mathbf{X}\mathbf{Q}\mathbf{Q}' + \mathbf{X}\mathbf{Q}\mathbf{Q}' - \mathbf{P}\Omega\mathbf{Q}'||^2 \\
&= ||\mathbf{X} - \mathbf{X}\mathbf{Q}\mathbf{Q}'||^2 + ||\mathbf{X}\mathbf{Q}\mathbf{Q}' - \mathbf{P}\Omega\mathbf{Q}'||^2 + 2c. \qquad (A.4.12)
\end{aligned}$$

Here, we can use $\mathbf{Q'Q} = \mathbf{I}_m$ to get

$$
\begin{aligned}
c &= \mathrm{tr}(\mathbf{X} - \mathbf{XQQ'})'(\mathbf{XQQ'} - \mathbf{P\Omega Q'}) \\
&= \mathrm{tr}\mathbf{X'XQQ'} - \mathrm{tr}\mathbf{X'P\Omega Q'} - \mathrm{tr}\mathbf{QQ'X'XQQ'} + \mathrm{tr}\mathbf{QQ'X'P\Omega Q'} \\
&= \mathrm{tr}\mathbf{Q'X'XQ} - \mathrm{tr}\mathbf{X'P\Omega Q'} - \mathrm{tr}\mathbf{Q'X'XQ} + \mathrm{tr}\mathbf{X'P\Omega Q'} = 0
\end{aligned}
\tag{A.4.13}
$$

and

$$
||\mathbf{X} - \mathbf{XQQ'}||^2 = ||\mathbf{X}||^2 - 2\mathrm{tr}\mathbf{X'XQQ'} + \mathrm{tr}\mathbf{QQ'X'XQQ'} = ||\mathbf{X}||^2 - 2\mathrm{tr}\mathbf{Q'X'XQ}.
\tag{A.4.14}
$$

Using (A.4.13) and (A.4.14) in (A.4.12), it is further rewritten as

$$
f(\mathbf{P\Omega Q'}) = ||\mathbf{X}||^2 - 2\mathrm{tr}\mathbf{Q'X'XQ} + ||\mathbf{XQQ'} - \mathbf{P\Omega Q'}||^2.
\tag{A.4.15}
$$

It can be minimized, if $\mathbf{P}$, $\mathbf{\Omega}$, and $\mathbf{Q}$ are found that simultaneously maximize $\mathrm{tr}\mathbf{Q'X'XQ}$ and minimize $||\mathbf{XQQ'} - \mathbf{P\Omega Q'}||^2$. Such $\mathbf{P}$, $\mathbf{\Omega}$, and $\mathbf{Q}$ are given by

$$
\mathbf{P} = \mathbf{K}_m, \ \mathbf{\Omega} = \mathbf{\Lambda}_m, \quad \text{and} \quad \mathbf{Q} = \mathbf{L}_m,
\tag{A.4.16}
$$

which is shown as follows: (A.4.16) allows $||\mathbf{XQQ'} - \mathbf{P\Omega Q'}||^2$ to attain its lower limit, zero, as $||\mathbf{XL}_m\mathbf{L}'_m - \mathbf{K}_m\mathbf{\Lambda}_m\mathbf{L}'_m||^2 = 0$ because of (A.3.25), while $\mathbf{Q} = \mathbf{L}_m$ in (A.4.16) maximizes $\mathrm{tr}\mathbf{Q'X'XQ}$ subject to $\mathbf{Q'Q} = \mathbf{I}_m$ because of Theorem A.4.4. The substitution of (A.4.16) in $\mathbf{M} = \mathbf{P\Omega Q'}$ leads to (A.4.11). $\quad\square$

Matrix $\mathbf{M}$ in Theorem A.4.5 can be replaced by

$$
\mathbf{M} = \mathbf{FA'},
\tag{A.4.17}
$$

with $\mathbf{F}$ and $\mathbf{A}$ being $n \times m$ and $p \times m$ matrices, respectively. This replacement gives the formulation of principal component analysis in Chap. 5.

Though Theorem A.4.5 is referred to as Eckart and Young's (1936) theorem in some of the literature, other roots are detailed in Takane (2014, p. 61).

A.4.4 Modified Reduced Rank Approximation

In this section, we treat the *reduced rank approximation* problems for *generalized canonical correlation analysis* (*GCCA*) and *multiple correspondence analysis* (*MCA*). In this and the following sections, we use

$$
\mathrm{rank}(\mathbf{PQR}) = \mathrm{rank}(\mathbf{Q}) \text{ if } \mathbf{P} \text{ and } \mathbf{R} \text{ are nonsingular.}
\tag{A.4.18}
$$

Theorem A.4.6. GCCA Problems

For a given $n \times p$ block matrix $\mathbf{X} = [\mathbf{X}_1, \ldots, \mathbf{X}_J]$ with its jth block $\mathbf{X}_j$ ($n \times p_j$), we consider the following problem:

$$\text{Minimize } \eta(\mathbf{F}, \mathbf{C}) = \sum_{j=1}^{J} \left\| \mathbf{F} - \mathbf{X}_j \mathbf{C}_j \right\|^2 \text{ over } \mathbf{F} \text{ and } \mathbf{C}$$

$$\text{s.t. } \frac{1}{n}\mathbf{F}'\mathbf{F} = \mathbf{I}_m \text{ with } m \leq r = \text{rank}(\mathbf{X}). \qquad (A.4.19)$$

Here, $\mathbf{C} = \begin{bmatrix} \mathbf{C}_1 \\ \vdots \\ \mathbf{C}_J \end{bmatrix}$ is the $p \times m$ block matrix with its jth block, $\mathbf{C}_j$, being $p_j \times m$.

Problem (A.4.19) is equivalent to

$$\text{Minimize } f(\mathbf{F}, \mathbf{C}) = \left\| \mathbf{X}\mathbf{D}_{\mathbf{X}}^{-1/2} - \frac{1}{n}\mathbf{F}\mathbf{C}'\mathbf{D}_{\mathbf{X}}^{1/2} \right\|^2 \text{over } \mathbf{F} \text{ and } \mathbf{C}$$

$$\text{s.t. } \frac{1}{n}\mathbf{F}'\mathbf{F} = \mathbf{I}_m, \text{ with } m \leq r = \text{rank}(\mathbf{X}). \qquad (A.4.20)$$

Here, $\mathbf{D}_{\mathbf{X}} = \begin{bmatrix} \mathbf{X}_1'\mathbf{X}_1 & & \\ & \ddots & \\ & & \mathbf{X}_J'\mathbf{X}_J \end{bmatrix}$ is the $p \times p$ nonsingular block diagonal matrix.

Those problems are solved through the SVD of $\mathbf{X}\mathbf{D}_{\mathbf{X}}^{-1/2}$, defined as

$$\mathbf{X}\mathbf{D}_{\mathbf{X}}^{-1/2} = \mathbf{N}\boldsymbol{\Phi}\mathbf{M}', \qquad (A.4.21)$$

with $\mathbf{N}'\mathbf{N} = \mathbf{M}'\mathbf{M} = \mathbf{I}_r$ and $\boldsymbol{\Phi}$ a diagonal matrix whose diagonal elements are arranged in descending order. The minimization in (A.4.19) and (A.4.20) is attained for

$$\mathbf{F} = \sqrt{n}\,\mathbf{N}_m\mathbf{T} \quad \text{and} \quad \mathbf{C} = \sqrt{n}\,\mathbf{D}_{\mathbf{X}}^{-1/2}\mathbf{M}_m\boldsymbol{\Phi}_m\mathbf{T}, \qquad (A.4.22)$$

where $\mathbf{M}_m$ and $\mathbf{N}_m$ contain the first m columns of $\mathbf{M}$ and $\mathbf{N}$, respectively, $\boldsymbol{\Phi}_m$ is the first $m \times m$ diagonal block of $\boldsymbol{\Phi}$, and $\mathbf{T}$ is an $m \times m$ orthonormal matrix.

Proof. The loss function in (A.4.19) can be expanded as

$$\eta(\mathbf{F}, \mathbf{C}) = J\text{tr}\mathbf{F}'\mathbf{F} - 2\text{tr}\mathbf{F}' \sum_{j=1}^{J} \mathbf{X}_j\mathbf{C}_j + \text{tr} \sum_{j=1}^{J} \mathbf{C}_j'\mathbf{X}_j'\mathbf{X}_j\mathbf{C}_j$$

$$= nmJ - 2\text{tr}\mathbf{F}'\mathbf{X}\mathbf{C} + \text{tr}\mathbf{C}'\mathbf{D}_{\mathbf{X}}\mathbf{C}, \qquad (A.4.23)$$

and the function in (A.4.20) multiplied by n is expanded as

$$
\begin{aligned}
n \times f(\mathbf{F}, \mathbf{C}) &= n\mathrm{tr}\mathbf{X}\mathbf{D}_X^{-1}\mathbf{X}' - 2\mathrm{tr}\mathbf{D}_X^{-1/2'}\mathbf{X}'\mathbf{F}\mathbf{C}'\mathbf{D}_X^{1/2} + \frac{1}{n}\,\mathrm{tr}\mathbf{D}_X^{1/2}\mathbf{C}\mathbf{F}'\mathbf{F}\mathbf{C}'\mathbf{D}_X^{1/2} \\
&= n\mathrm{tr}\mathbf{X}\mathbf{D}_X^{-1}\mathbf{X}' - 2\mathrm{tr}\mathbf{X}'\mathbf{F}\mathbf{C}' + \mathrm{tr}\mathbf{C}'\mathbf{D}_X\mathbf{C},
\end{aligned}
$$

$$(A.4.24)$$

where the constraint $n^{-1}\mathbf{F}'\mathbf{F} = \mathbf{I}_m$ has been used. Since the parts relevant to $\mathbf{F}$ and $\mathbf{C}$ in (A.4.23) are the same as those in (A.4.24), problems (A.4.19) and (A.4.20) with the same constraints are equivalent.

Because of (A.4.18), $r = \mathrm{rank}(\mathbf{X}) = \mathrm{rank}(\mathbf{X}\mathbf{D}_X^{-1/2})$, while $\mathrm{rank}(n^{-1}\mathbf{F}\mathbf{C}'\mathbf{D}_X^{1/2}) \le m \le r$. Thus, problem (A.4.20) is the reduced rank approximation of $\mathbf{X}\mathbf{D}_X^{-1/2}$ by $n^{-1}\mathbf{F}\mathbf{C}'\mathbf{D}_X^{1/2}$ as the approximation of $\mathbf{X}$ by $\mathbf{M}$ in Theorem A.4.5; the minimization in (A.4.20) is attained for

$$
\frac{1}{n}\mathbf{F}\mathbf{C}'\mathbf{D}_X^{1/2} = \mathbf{N}_m\mathbf{\Phi}_m\mathbf{M}_m'. \tag{A.4.25}
$$

Matrices $\mathbf{F}$ and $\mathbf{C}$ in (A.4.22) satisfy (A.4.25) and the constraints in (A.4.19) and (A.4.20). $\qquad\square$

The constraint of $\mathbf{F}$ being *centered* is added to the above problems in those that follow:

Theorem A.4.7. MCA Problems

For a given $n \times p$ block matrix $\mathbf{G} = [\mathbf{G}_1, \ldots, \mathbf{G}_J]$ with its jth block $\mathbf{G}_j$ $(n \times p_j)$, we consider the following problem:

$$
\text{Minimize } \eta(\mathbf{F}, \mathbf{C}) = \sum_{j=1}^{J} \left\| \mathbf{F} - \mathbf{G}_j\mathbf{C}_j \right\|^2 \quad \text{over } \mathbf{F} \text{ and } \mathbf{C}
$$

$$
\text{s.t. } \mathbf{F}'\mathbf{F} = \mathbf{I}_m, \mathbf{J}\mathbf{F} = \mathbf{F}, \text{ and } m \le r = \mathrm{rank}(\mathbf{J}\mathbf{G}). \tag{A.4.26}
$$

Here, $\mathbf{J}$ is defined as (2.10) and $\mathbf{C} = \begin{bmatrix} \mathbf{C}_1 \\ \vdots \\ \mathbf{C}_J \end{bmatrix}$ is the $p \times m$ block matrix with its jth block, $\mathbf{C}_j$ $(p_j \times m)$.

Problem (A.4.26) is equivalent to

$$
\text{Minimize } f(\mathbf{F}, \mathbf{C}) = \left\| \mathbf{J}\mathbf{G}\mathbf{D}_G^{-1/2} - \frac{1}{n}\mathbf{F}\mathbf{C}'\mathbf{D}_G^{1/2} \right\|^2 \quad \text{over } \mathbf{F} \text{ and } \mathbf{C}
$$

$$
\text{s.t. } \frac{1}{n}\mathbf{F}'\mathbf{F} = \mathbf{I}_m, \mathbf{J}\mathbf{F} = \mathbf{F}, \text{ and } m \le r = \mathrm{rank}(\mathbf{J}\mathbf{G}), \tag{A.4.27}
$$

with $\mathbf{D}_G = \begin{bmatrix} \mathbf{G}_1'\mathbf{G}_1 & & \\ & \ddots & \\ & & \mathbf{G}_J'\mathbf{G}_J \end{bmatrix}$ the $p \times p$ nonsingular block diagonal

matrix.

Those problems are solved through the SVD of $\mathbf{JGD}_G^{-1/2}$, defined as

$$\mathbf{JGD}_G^{-1/2} = \mathbf{S\Theta P'}, \tag{A.4.28}$$

with $\mathbf{S'S} = \mathbf{P'P} = \mathbf{I}_r$ and $\mathbf{\Theta}$ a diagonal matrix whose diagonal elements are arranged in descending order. The minimization in (4.26) and (A.4.27) is attained for

$$\mathbf{F} = \sqrt{n}\,\mathbf{S}_m\mathbf{T} \quad \text{and} \quad \mathbf{C} = \sqrt{n}\,\mathbf{D}_G^{-1/2}\mathbf{P}_m\mathbf{\Theta}_m\mathbf{T}, \tag{A.4.29}$$

where $\mathbf{S}_m$ and $\mathbf{P}_m$ contain the first m columns of $\mathbf{S}$ and $\mathbf{P}$, respectively, $\mathbf{\Theta}_m$ is the first $m \times m$ diagonal block of $\mathbf{\Theta}$, and $\mathbf{T}$ is an $m \times m$ orthonormal matrix.

Proof. The loss function in (A.4.26) can be expanded as

$$\eta(\mathbf{F}, \mathbf{C}) = J\mathrm{tr}\mathbf{F'F} - 2\mathrm{tr}\mathbf{F'}\sum_{j=1}^{J}\mathbf{X}_j\mathbf{C}_j + \mathrm{tr}\sum_{j=1}^{J}\mathbf{C}_j'\mathbf{G}_j'\mathbf{G}_j\mathbf{C}_j$$

$$= nmJ - 2\mathrm{tr}\mathbf{F'JXC} + \mathrm{tr}\mathbf{C'D}_G\mathbf{C}, \tag{A.4.30}$$

where we have used $n^{-1}\mathbf{F'F} = \mathbf{I}_m$ and $\mathbf{JF} = \mathbf{F}$. On the other hand, (A.4.27) multiplied by n is expanded as

$$n \times f(\mathbf{F}, \mathbf{C}) = n\mathrm{tr}\mathbf{JGD}_G^{-1}\mathbf{GJ'} - 2\mathrm{tr}\mathbf{D}_G^{-1/2}\mathbf{G'JFC'D}_G^{1/2} + \frac{1}{n}\mathrm{tr}\mathbf{D}_G^{1/2}\mathbf{CF'FC'D}_C^{1/2}$$

$$= n\mathrm{tr}\mathbf{JGD}_G^{-1}\mathbf{GJ'} - 2\mathrm{tr}\mathbf{G'JFC'} + \mathrm{tr}\mathbf{C'D}_G\mathbf{C}, \tag{A.4.31}$$

where the constraint $n^{-1}\mathbf{F'F} = \mathbf{I}_m$ has been used. Since the parts relevant to $\mathbf{F}$ and $\mathbf{C}$ in (A.4.30) are the same as those in (A.4.31), problems (A.4.26) and (A.4.27) with the same constraints are equivalent.

Because of (A.4.18), with $\mathbf{D}_G$ being nonsingular, $\mathrm{rank}(\mathbf{JG}) = \mathrm{rank}(\mathbf{JGD}_G^{-1/2})$, while $\mathrm{rank}(n^{-1}\mathbf{FC'D}_G^{1/2}) \leq m$. Thus, problem (A.4.27) is the reduced rank approximation of $\mathbf{JGD}_G^{-1/2}$ by $n^{-1}\mathbf{FC'D}_G^{1/2}$ as the approximation of $\mathbf{X}$ by $\mathbf{M}$ in Theorem A.4.5; the minimization in (A.4.27) is attained for

$$\frac{1}{n}\mathbf{FC'D}_G^{1/2} = \mathbf{S}_m\mathbf{\Theta}_m\mathbf{P}_m'. \tag{A.4.32}$$

The $\mathbf{F}$ and $\mathbf{C}$ in (A.4.29) satisfy (A.4.32), $n^{-1}\mathbf{F'F} = \mathbf{I}_m$, and $\mathbf{JF} = \mathbf{F}$. The last identity follows from the fact that $\mathbf{F} = \sqrt{n}\,\mathbf{S}_m\mathbf{T}$, in (A.4.29), can be rewritten as $\mathbf{F} = \sqrt{n}\,\mathbf{S\Theta P'}\mathbf{P}_m\mathbf{\Theta}_m^{-1}\mathbf{T} = \sqrt{n}\,\mathbf{JGD}_G^{-1/2}\mathbf{P}_m\mathbf{\Theta}_m^{-1}\,\mathbf{T}$ with (2.12). □

The GCCA and MCA solutions (A.4.22) and (A.4.29) show that they have rotational indeterminacy. This can be avoided, if the constraint

$$\mathbf{C}'\mathbf{D}_G\mathbf{C} \text{ being a diagonal matrix whose}$$
$$\text{diagonal elements are arranged in descending order} \qquad (A.4.33)$$

is added to (A.4.26) and (A.4.27) for the MCA solution. The indeterminacy of the GCCA solution can also be avoided, by adding the constraint (A.4.33) with $\mathbf{D}_G$ replaced by $\mathbf{D}_X$ to (A.4.19) and (A.4.20), so that the GCCA solution is unique. Then, $\mathbf{T}$ in (A.4.22) and (A.4.29) is fixed to $\mathbf{I}_m$.

A.4.5 Modified Versions of Maximizing Trace Functions

In A.4.2, the parameter matrix $\mathbf{C}$ was constrained as $\mathbf{C}'\mathbf{C}$ being the identity matrix. In this section, $\mathbf{C}$ is constrained rather as $\mathbf{C}'\mathbf{V}\mathbf{C}$ being the identity matrix with $\mathbf{V}$ a *given positive definite matrix* (Note 8.2), and the *symmetric square roots* $\mathbf{V}^{1/2}$ and $\mathbf{V}^{-1/2}$ are used that satisfy $\mathbf{V}^{1/2}\mathbf{V}^{1/2} = \mathbf{V}$ and $\mathbf{V}^{-1/2}\mathbf{V}^{-1/2} = \mathbf{V}^{-1}$, respectively. How to obtain $\mathbf{V}^{1/2}$ and $\mathbf{V}^{-1/2}$ from $\mathbf{V}$ is described in Section A.4.6.

Theorem A.4.8

Let us define matrices as $\mathbf{V}_{11}$ ($p_1 \times p_1$), $\mathbf{V}_{22}$ ($p_2 \times p_2$), and $\mathbf{V}_{12}$ ($p_1 \times p_2$), with $\mathbf{V}_{11}$ and $\mathbf{V}_{22}$ symmetric and positive definite. We consider the following problem:

$$\text{Maximize } \mathrm{tr}\,\mathbf{C}_1'\mathbf{V}_{12}\mathbf{C}_2 \text{ over } \mathbf{C}_1(p_1 \times m) \text{ and } \mathbf{C}_2(p_2 \times m)$$
$$\text{s.t. } \mathbf{C}_1'\mathbf{V}_{11}\mathbf{C}_1 = \mathbf{C}_2'\mathbf{V}_{22}\mathbf{C}_2 = \mathbf{I}_m \text{ with } m \le r = \mathrm{rank}(\mathbf{V}_{12}). \qquad (A.4.34)$$

It is solved through the SVD of $\mathbf{V}_{11}^{-1/2}\mathbf{V}_{12}\mathbf{V}_{22}^{-1/2}$ defined as

$$\mathbf{V}_{11}^{-1/2}\mathbf{V}_{12}\mathbf{V}_{22}^{-1/2} = \mathbf{H}\mathbf{\Omega}\mathbf{R}', \qquad (A.4.35)$$

with $\mathbf{H}'\mathbf{H} = \mathbf{R}'\mathbf{R} = \mathbf{I}_r$ and $\mathbf{\Omega}$ the diagonal matrix whose diagonal elements are arranged in descending order. The maximization in (A.4.34) is attained for

$$\mathbf{C}_1 = \mathbf{V}_{11}^{-1/2}\mathbf{H}_m\mathbf{T} \quad \text{and} \quad \mathbf{C}_2 = \mathbf{V}_{22}^{-1/2}\mathbf{R}_m\mathbf{T}, \qquad (A.4.36)$$

where $\mathbf{H}_m$ and $\mathbf{R}_m$ contain the first m columns of $\mathbf{H}$ and those of $\mathbf{R}$, respectively, and $\mathbf{T}$ is an $m \times m$ orthonormal matrix.

Proof. By defining $\mathbf{A}$, $\mathbf{B}$, and $\mathbf{Y}$ as

$$\mathbf{A} = \mathbf{V}_{11}^{1/2}\mathbf{C}_1, \ \mathbf{B} = \mathbf{V}_{22}^{1/2}\mathbf{C}_2, \qquad (A.4.37)$$

$$\mathbf{Y} = \mathbf{V}_{11}^{-1/2}\mathbf{V}_{12}\mathbf{V}_{22}^{-1/2}, \qquad (A.4.38)$$

(A.4.34) can be transformed into the equivalent problem:

Maximize tr $\mathbf{A}'\mathbf{Y}\mathbf{B}$ over $\mathbf{A}(p_1 \times m)$ and $\mathbf{B}(p_2 \times m)$

$$\text{s.t. } \mathbf{A}'\mathbf{A} = \mathbf{B}'\mathbf{B} = \mathbf{I}_m \text{ with } m \leq r = \text{rank}(\mathbf{Y}), \qquad (A.4.39)$$

where we have used $r = \text{rank}(\mathbf{V}_{12}) = \text{rank}(\mathbf{Y})$, following from (A.4.18). Since problem (A.4.39) is equivalent to (A.4.6) in Theorem A.4.3, the solution for (A.4.39) is given by

$$\mathbf{A} = \mathbf{H}_m\mathbf{T} \quad \text{and} \quad \mathbf{B} = \mathbf{R}_m\mathbf{T}, \qquad (A.4.40)$$

when the SVD of (A.4.38) is defined as (A.4.35). Using (A.4.37) in (A.4.40), we have (A.4.36). $\qquad \square$

A related theorem is given next:

Theorem A.4.9.
Let $\mathbf{V}$ be a $p \times p$ symmetric positive definite matrix and $\mathbf{M} = \mathbf{U}\mathbf{U}'$ be the $p \times p$ symmetric and nonnegative definite matrix with its rank r. We consider the following problem:

Maximize tr$\mathbf{B}'\mathbf{M}\mathbf{B}$ over $\mathbf{B}(p \times m)$ s.t. $\mathbf{B}'\mathbf{V}\mathbf{B} = \mathbf{I}_m$ with $m \times r = \text{rank}(\mathbf{M})$.
$$\qquad (A.4.41)$$

This is solved through the SVD of $\mathbf{U}\mathbf{V}^{-1/2}$ defined as $\mathbf{U}'\mathbf{V}^{-1/2} = \mathbf{P}\mathbf{\Theta}\mathbf{Q}'$, which implies

$$\mathbf{V}^{-1/2}\mathbf{M}\mathbf{V}^{-1/2} = \mathbf{Q}\mathbf{\Theta}^2\mathbf{Q}', \qquad (A.4.42)$$

with $\mathbf{Q}'\mathbf{Q} = \mathbf{I}_r$ and $\mathbf{\Theta}$ the diagonal matrix whose diagonal elements are arranged in descending order. The maximization in (A.4.41) is attained for

$$\mathbf{B} = \mathbf{V}^{-1/2}\mathbf{Q}_m\mathbf{T}, \qquad (A.4.43)$$

where $\mathbf{Q}_m$ contains the first m columns of $\mathbf{Q}$, and $\mathbf{T}$ is an $m \times m$ orthonormal matrix.

Proof. By defining $\mathbf{W}$ and $\mathbf{Y}$ as

$$\mathbf{W} = \mathbf{V}^{1/2}\mathbf{B}, \tag{A.4.44}$$

$$\mathbf{Y} = \mathbf{V}^{-1/2}\mathbf{M}\mathbf{V}^{-1/2}, \tag{A.4.45}$$

(A.4.41) can be transformed into the equivalent problem:

Maximize $\mathrm{tr}\mathbf{W}'\mathbf{Y}\mathbf{W}$ over $\mathbf{W}(p \times m)$ s.t. $\mathbf{W}'\mathbf{W} = \mathbf{I}_m$ with $m \times \mathrm{rank}(\mathbf{Y})$, (A.4.46)

where we have used $r = \mathrm{rank}(\mathbf{M}) = \mathrm{rank}(\mathbf{Y})$, following from (A.4.18). Since (A.4.46) is equivalent to (A.4.8) in Theorem A.4.4, the solution for (A.4.46) is given by

$$\mathbf{W} = \mathbf{Q}_m\mathbf{T}, \tag{A.4.47}$$

when the SVD of (A.4.45) is defined as (A.4.42). Using (A.4.44) in (A.4.47), we have (A.4.43). $\qquad\square$

The solution of (A.4.36) is found to have rotational indeterminacy. Also, it is possessed by (A.4.43). Its indeterminacy is avoided by adding the following constraint to (A.4.41): $\mathbf{B}'\mathbf{M}\mathbf{B}$ is a diagonal matrix whose diagonal elements are arranged in descending order. Then, the solution is restricted to $\mathbf{B} = \mathbf{V}^{-1/2}\mathbf{Q}_m$. This solution has been used for the canonical discriminant analysis in Chap. 15.

A.4.6 Obtaining Symmetric Square Roots of Matrices

Let $\mathbf{V} = \mathbf{U}\mathbf{U}'$ be a $p \times p$ positive definite symmetric matrix. As in (A.3.13), the SVD of $\mathbf{V} = \mathbf{U}\mathbf{U}'$ can be defined as $\mathbf{V} = \boldsymbol{\Gamma}\boldsymbol{\Delta}^2\boldsymbol{\Gamma}'$ with $\boldsymbol{\Gamma}'\boldsymbol{\Gamma} = \boldsymbol{\Gamma}\boldsymbol{\Gamma}' = \mathbf{I}_p$ and $\boldsymbol{\Delta}^2$ a $p \times p$ diagonal matrix whose p diagonal elements are all positive.

The *symmetric square root* of $\mathbf{V}$ is given by

$$\mathbf{V}^{1/2} = \boldsymbol{\Gamma}\boldsymbol{\Delta}\boldsymbol{\Gamma}', \tag{A.4.48}$$

with each diagonal element of $\boldsymbol{\Delta}$ being the square root of the corresponding one of $\boldsymbol{\Delta}^2$. We can easily verify that $\mathbf{V}^{1/2}\mathbf{V}^{1/2} = \boldsymbol{\Gamma}\boldsymbol{\Delta}\boldsymbol{\Gamma}'\boldsymbol{\Gamma}\boldsymbol{\Delta}\boldsymbol{\Gamma}' = \boldsymbol{\Gamma}\boldsymbol{\Delta}\boldsymbol{\Delta}\boldsymbol{\Gamma}' = \boldsymbol{\Gamma}\boldsymbol{\Delta}^2\boldsymbol{\Gamma}' = \mathbf{V}$.

The *inverse* matrix of $\mathbf{V}$ is expressed as $\mathbf{V}^{-1} = \boldsymbol{\Gamma}\boldsymbol{\Delta}^{-2}\boldsymbol{\Gamma}'$. Its symmetric square root is given by

$$\mathbf{V}^{-1/2} = \boldsymbol{\Gamma}\boldsymbol{\Delta}^{-1}\boldsymbol{\Gamma}', \tag{A.4.49}$$

with each diagonal element of $\boldsymbol{\Lambda}^{-1}$ being the reciprocal of the square root of the corresponding element in $\boldsymbol{\Lambda}^2$. We can easily verify that $\mathbf{V}^{-1/2}\mathbf{V}^{-1/2} = \boldsymbol{\Gamma}\boldsymbol{\Lambda}^{-1}\boldsymbol{\Gamma}'\boldsymbol{\Gamma}\boldsymbol{\Lambda}^{-1}\boldsymbol{\Gamma}' = \boldsymbol{\Gamma}\boldsymbol{\Lambda}^{-1}\boldsymbol{\Lambda}^{-1}\boldsymbol{\Gamma}' = \boldsymbol{\Gamma}\boldsymbol{\Lambda}^{-2}\boldsymbol{\Gamma}' = \mathbf{V}^{-1}$.

Next, we consider the *symmetric square root* of the *block diagonal* matrix

$$\mathbf{D} = \begin{bmatrix} \mathbf{V}_1 & & \\ & \ddots & \\ & & \mathbf{V}_J \end{bmatrix} = \begin{bmatrix} \mathbf{U}_1'\mathbf{U}_1 & & \\ & \ddots & \\ & & \mathbf{U}_J'\mathbf{U}_J \end{bmatrix}, \text{ which is symmetric and positive}$$

definite. These properties imply that the diagonal blocks $\mathbf{V}_j = \mathbf{U}_j'\mathbf{U}_j$ $(p_j \times p_j)$ $(j = 1, \ldots, J)$ are also symmetric and positive definite. Thus, the SVD of $\mathbf{V}_j$ can be defined as $\mathbf{V}_j = \boldsymbol{\Gamma}_j\boldsymbol{\Lambda}_j^2\boldsymbol{\Gamma}_j'$, with $\boldsymbol{\Gamma}_j'\boldsymbol{\Gamma}_j = \boldsymbol{\Gamma}_j\boldsymbol{\Gamma}_j' = \mathbf{I}_{p_j}$ and $\boldsymbol{\Lambda}_j^2$ the $p_j \times p_j$ diagonal matrix whose p_j diagonal elements are positive.

The symmetric square root of $\mathbf{D}$ is given by

$$\mathbf{D}^{1/2} = \begin{bmatrix} \mathbf{V}_1^{1/2} & & \\ & \ddots & \\ & & \mathbf{V}_J^{1/2} \end{bmatrix} = \begin{bmatrix} \boldsymbol{\Gamma}_1\boldsymbol{\Lambda}_1\boldsymbol{\Gamma}_1' & & \\ & \ddots & \\ & & \boldsymbol{\Gamma}_J\boldsymbol{\Lambda}_J\boldsymbol{\Gamma}_J' \end{bmatrix} \tag{A.4.50}$$

and the root of $\mathbf{D}^{-1}$ is given by

$$\mathbf{D}^{-1/2} = \begin{bmatrix} \mathbf{V}_1^{-1/2} & & \\ & \ddots & \\ & & \mathbf{V}_J^{-1/2} \end{bmatrix} = \begin{bmatrix} \boldsymbol{\Gamma}_1\boldsymbol{\Lambda}_1^{-1}\boldsymbol{\Gamma}_1' & & \\ & \ddots & \\ & & \boldsymbol{\Gamma}_J\boldsymbol{\Lambda}_J^{-1}\boldsymbol{\Gamma}_J' \end{bmatrix}. \tag{A.4.51}$$

We can verify $\mathbf{D}^{1/2}\mathbf{D}^{1/2} = \mathbf{D}$ and $\mathbf{D}^{-1/2}\mathbf{D}^{-1/2} = \mathbf{D}^{-1}$ from the fact that $\boldsymbol{\Gamma}_j\boldsymbol{\Lambda}_j\boldsymbol{\Gamma}_j'\boldsymbol{\Gamma}_j\boldsymbol{\Lambda}_j\boldsymbol{\Gamma}_j' = \boldsymbol{\Gamma}_j\boldsymbol{\Lambda}_j^2\boldsymbol{\Gamma}_j' = \mathbf{V}_j$ and $\boldsymbol{\Gamma}_j\boldsymbol{\Lambda}_j^{-1}\boldsymbol{\Gamma}_j'\boldsymbol{\Gamma}_j\boldsymbol{\Lambda}_j^{-1}\boldsymbol{\Gamma}_j' = \boldsymbol{\Gamma}_j\boldsymbol{\Lambda}_j^{-2}\boldsymbol{\Gamma}_j' = \mathbf{V}_j^{-1}$.

Since $\mathbf{D}_G = \begin{bmatrix} \mathbf{G}_1'\mathbf{G}_1 & & \\ & \ddots & \\ & & \mathbf{G}_J'\mathbf{G}_J \end{bmatrix}$ in Theorem A.4.7 is diagonal, its square

root $\mathbf{D}_G^{1/2}$ is simply the diagonal matrix whose diagonal elements are the square roots of the corresponding ones in $\mathbf{D}_G$. On the other hand, the root of $\mathbf{D}_G^{-1}$ is given by $\mathbf{D}_G^{-1/2}$, whose diagonal elements are the reciprocals of the square roots of the corresponding elements in $\mathbf{D}_G$.

For the $p \times p$ symmetric positive definite matrix $\mathbf{V} = \mathbf{U}\mathbf{U}'$ appeared first in this section, $\mathbf{U}$ is called the *square root* of $\mathbf{V}$. It is given by $\mathbf{U} = \boldsymbol{\Gamma}\boldsymbol{\Lambda}$, using $\mathbf{V} = \boldsymbol{\Gamma}\boldsymbol{\Lambda}^2\boldsymbol{\Gamma}'$. The root $\mathbf{U}$ can also be used for solving the problems in the previous appendices. However, we must be careful about whether $\mathbf{U}$ or $\mathbf{U}'$ is used in solutions, as $\mathbf{U} = \boldsymbol{\Gamma}\boldsymbol{\Lambda}$ is *not symmetric*, which differs from the symmetric matrix in (A.4.48). Therefore, we chose to use the symmetric roots in this book.

A.5 Normal Maximum Likelihood Estimates

We derive the *maximum likelihoods* of *mean* vectors and *covariance* matrices for the *multivariate normal distributions*, which are used in Chaps. 8 and 15.

A.5.1 Estimates of Means and Covariances

Log likelihood (8.20) is presented again here:

$$l(\mathbf{\mu}, \mathbf{\Sigma}) = -\frac{n}{2}\log|\mathbf{\Sigma}| - \frac{1}{2}\sum_{i=1}^{n}(\mathbf{x}_i - \mathbf{\mu})'\mathbf{\Sigma}^{-1}(\mathbf{x}_i - \mathbf{\mu}). \tag{A.5.1}$$

In this appendix, it is shown that the maximum likelihood estimates (MLE) of $\mathbf{\mu}$ and $\mathbf{\Sigma}$ maximizing (A.5.1) are given by (8.21) and (8.22), i.e.,

$$\hat{\mathbf{\mu}} = \bar{\mathbf{x}} = \frac{1}{n}\sum_{i=1}^{n}\mathbf{x}_i, \tag{A.5.2}$$

$$\hat{\mathbf{\Sigma}} = \mathbf{V} = \frac{1}{n}\sum_{i=1}^{n}(\mathbf{x}_i - \bar{\mathbf{x}})(\mathbf{x}_i - \bar{\mathbf{x}})', \tag{A.5.3}$$

respectively, on the supposition of $\mathbf{\Sigma}$ being positive definite with $\mathbf{\Sigma}^{-1} = \mathbf{U}\mathbf{U}'$

For proving that (A.5.2) is the MLE of $\mathbf{\mu}$, we can use an extension of the *decomposition of the sum of squares* treated in Appendix A.2.1, as follows: in the right-hand side of (A.5.1), only the second term is relevant to $\mathbf{\mu}$, thus, its maximum likelihood estimate is the $\mathbf{\mu}$ minimizing that term multiplied by -2:

$$\sum_{i=1}^{n}(\mathbf{x}_i - \mathbf{\mu})'\mathbf{\Sigma}^{-1}(\mathbf{x}_i - \mathbf{\mu}) = \sum_{i=1}^{n}(\mathbf{x}_i - \bar{\mathbf{x}} + \bar{\mathbf{x}} - \mathbf{\mu})'\mathbf{\Sigma}^{-1}(\mathbf{x}_i - \bar{\mathbf{x}} + \bar{\mathbf{x}} - \mathbf{\mu})$$

$$= \sum_{i=1}^{n}(\mathbf{x}_i - \bar{\mathbf{x}})'\mathbf{\Sigma}^{-1}(\mathbf{x}_i - \bar{\mathbf{x}}) + n(\bar{\mathbf{x}} - \mathbf{\mu})'\mathbf{\Sigma}^{-1}(\bar{\mathbf{x}} - \mathbf{\mu}) + 2c.$$

$$\tag{A.5.4}$$

Here, c is found to be zero as

$$c = \sum_{i=1}^{n}(\mathbf{x}_i - \bar{\mathbf{x}})'\mathbf{\Sigma}^{-1}(\bar{\mathbf{x}} - \mathbf{\mu})$$

$$= \sum_{i=1}^{n}\mathbf{x}_i'\mathbf{\Sigma}^{-1}\bar{\mathbf{x}} - \sum_{i=1}^{n}\mathbf{x}_i'\mathbf{\Sigma}^{-1}\mathbf{\mu} - \sum_{i=1}^{n}\bar{\mathbf{x}}'\mathbf{\Sigma}^{-1}\mathbf{x} + \sum_{i=1}^{n}\bar{\mathbf{x}}'\mathbf{\Sigma}^{-1}\mathbf{\mu}$$

$$= n\bar{\mathbf{x}}'\mathbf{\Sigma}^{-1}\bar{\mathbf{x}} - n\bar{\mathbf{x}}'\mathbf{\Sigma}^{-1}\mathbf{\mu} - n\bar{\mathbf{x}}'\mathbf{\Sigma}^{-1}\bar{\mathbf{x}} + n\bar{\mathbf{x}}'\mathbf{\Sigma}^{-1}\mathbf{\mu} = 0. \tag{A.5.5}$$

This implies that the term relevant to $\boldsymbol{\mu}$ in (A.5.4) is only $n(\bar{\mathbf{x}} - \boldsymbol{\mu})'\boldsymbol{\Sigma}^{-1}(\bar{\mathbf{x}} - \boldsymbol{\mu}) = n||\boldsymbol{\Sigma}^{-1/2}(\bar{\mathbf{x}} - \boldsymbol{\mu})||^2$, which attains the lower limit, zero, for (A.5.2); it gives the MLE for $\boldsymbol{\mu}$.

Substituting (A.5.2) in (A.5.1), it is rewritten as

$$
\begin{aligned}
\log l(\boldsymbol{\Sigma}) &= -\frac{n}{2}\log|\boldsymbol{\Sigma}| - \frac{1}{2}\sum_{i=1}^{n}(\mathbf{x}_i - \bar{\mathbf{x}})'\boldsymbol{\Sigma}^{-1}(\mathbf{x}_i - \bar{\mathbf{x}}) \\
&= -\frac{n}{2}\log|\boldsymbol{\Sigma}| - \frac{n}{2}\operatorname{tr}\boldsymbol{\Sigma}^{-1}\left\{\frac{1}{n}\sum_{i=1}^{n}(\mathbf{x}_i - \bar{\mathbf{x}})(\mathbf{x}_i - \bar{\mathbf{x}})'\right\} \\
&= -\frac{n}{2}\log|\boldsymbol{\Sigma}| - \frac{n}{2}\operatorname{tr}\boldsymbol{\Sigma}^{-1}\mathbf{V} = -\frac{n}{2}(\log|\boldsymbol{\Sigma}| + \operatorname{tr}\boldsymbol{\Sigma}^{-1}\mathbf{V}). \quad (A.5.6)
\end{aligned}
$$

This shows that our remaining task is to minimize $\log|\boldsymbol{\Sigma}| + \operatorname{tr}\boldsymbol{\Sigma}^{-1}\mathbf{V} = \operatorname{tr}\boldsymbol{\Sigma}^{-1}\mathbf{V} - \log|\boldsymbol{\Sigma}^{-1}|$ over $\boldsymbol{\Sigma}$, which is equivalent to minimizing

$$
\begin{aligned}
g(\boldsymbol{\Sigma}) = \operatorname{tr}\boldsymbol{\Sigma}^{-1}\mathbf{V} - \log|\boldsymbol{\Sigma}^{-1}| - \log|\mathbf{V}| &= \operatorname{tr}\boldsymbol{\Sigma}^{-1}\mathbf{V} - \log|\boldsymbol{\Sigma}^{-1}\mathbf{V}| \\
&= \operatorname{tr}\mathbf{V}^{1/2}\boldsymbol{\Sigma}^{-1}\mathbf{V}^{1/2} - \log|\mathbf{V}^{1/2}\boldsymbol{\Sigma}^{-1}\mathbf{V}^{1/2}|,
\end{aligned}
$$
$$(A.5.7)$$

where we have used (8.11) and (8.12).

Let the SVD of $\mathbf{U}'\mathbf{V}^{1/2}$ be defined as $\mathbf{U}'\mathbf{V}^{1/2} = \boldsymbol{\Xi}\boldsymbol{\Omega}\boldsymbol{\Gamma}'$, which implies

$$
\mathbf{V}^{1/2}\boldsymbol{\Sigma}^{-1}\mathbf{V}^{1/2} = \boldsymbol{\Gamma}\boldsymbol{\Omega}^2\boldsymbol{\Gamma}' \quad (A.5.8)
$$

with $\boldsymbol{\Gamma}\boldsymbol{\Gamma}' = \boldsymbol{\Gamma}'\boldsymbol{\Gamma} = \mathbf{I}_p$ and $\boldsymbol{\Omega}^2$ a diagonal matrix whose jth diagonal element is $\omega_j > 0$. Using (A.5.8) in (A.5.7), it can be rewritten as

$$
\begin{aligned}
g(\boldsymbol{\Sigma}) &= \operatorname{tr}\boldsymbol{\Omega}^2 - \log(|\boldsymbol{\Omega}^2| \times |\boldsymbol{\Gamma}| \times |\boldsymbol{\Gamma}'|) \\
&= \operatorname{tr}\boldsymbol{\Omega}^2 - \log|\boldsymbol{\Omega}^2| = \sum_{j=1}^{p}\omega_j - \sum_{j=1}^{p}\log\omega_j = \sum_{j=1}^{p}g_j(\omega_j), \quad (A.5.7')
\end{aligned}
$$

with $g(\omega_j) = \omega_j - \log\omega_j$. Here, we have used the fact that $\boldsymbol{\Gamma}' = \boldsymbol{\Gamma}^{-1}$ and (8.12) leads to $|\boldsymbol{\Gamma}| \times |\boldsymbol{\Gamma}'| = 1$. It is known that the *differentiation* of $g_j(\omega_j)$ with respect to ω_j is given by $g'_j(\omega_j) = dg_j(\omega_j)/d\omega_j = 1 - 1/\omega_j$, which is found to satisfy

$$
\begin{aligned}
g'_j(\omega_j) &< 0 \text{ for } 0 < \omega_j < 1, \\
g'_j(\omega_j) &= 0 \text{ for } \omega_j = 1, \\
g'_j(\omega_j) &> 0 \text{ for } \omega_j > 1. \quad (A.5.9)
\end{aligned}
$$

This shows that (A.5.7′) is minimized for $\omega_j = 1$ $(j = 1, \ldots, p)$, i.e., $\mathbf{\Omega}^2 = \mathbf{I}_p$. Using this, (A.5.8) is rewritten as

$$\mathbf{V}^{1/2}\mathbf{\Sigma}^{-1}\mathbf{V}^{1/2} = \mathbf{\Gamma}\mathbf{\Gamma}' = \mathbf{I}_p, \tag{A.5.10}$$

which leads to (A.5.3).

A.5.2 Multiple Groups with Homogeneous Covariances

Let us consider an $n \times p$ block data matrix $\mathbf{X} = \begin{bmatrix} \mathbf{X}_1 \\ \vdots \\ \mathbf{X}_K \end{bmatrix}$, whose kth block is an $n_k \times p$ matrix $\mathbf{X}_k$ $(k = 1, \ldots, K)$ with its ith row being $\mathbf{x}'_{ki}$. We suppose $\mathbf{x}_{ki} \sim N_p(\mathbf{\mu}_k, \mathbf{\Sigma})$, i.e., that the probability density of $\mathbf{x}_{ki}$ observed is given by

$$P(\mathbf{x}_{ki}|\mathbf{\mu}_k,\mathbf{\Sigma}) = \frac{1}{(2\pi)^{p/2}|\mathbf{\Sigma}|^{1/2}}\exp\left\{-\frac{1}{2}(\mathbf{x}_{ki} - \mathbf{\mu}_k)'\mathbf{\Sigma}^{-1}(\mathbf{x}_{ki} - \mathbf{\mu}_k)\right\}. \tag{A.5.11}$$

Here, it should be noted that $\mathbf{\mu}_k$ has subscript k, but $\mathbf{\Sigma}$ does not, which implies that the *mean* vectors of the distributions *differ* across K blocks, while their *covariance* matrices are *homogeneous* across them.

We further suppose the rows of $\mathbf{X}$ to be mutually independently observed. Then, the likelihood for $\mathbf{X}$ is expressed as the product of (A.5.11) over $k = 1, \ldots, K$, and $i = 1, \ldots, n_k$:

$$\begin{aligned} P(\mathbf{X}|\mathbf{\mu}_1,\ldots,\mathbf{\mu}_K, \mathbf{\Sigma}) &= \prod_{k=1}^{K}\prod_{i=1}^{n_k}\left\{\frac{1}{(2\pi)^{p/2}|\mathbf{\Sigma}|^{1/2}}\exp\left\{-\frac{1}{2}(\mathbf{x}_{ki} - \mathbf{\mu}_k)'\mathbf{\Sigma}^{-1}(\mathbf{x}_{ki} - \mathbf{\mu}_k)\right\}\right\} \\ &= \frac{1}{(2\pi)^{np/2}|\mathbf{\Sigma}|^{n/2}}\exp\left\{-\frac{1}{2}\sum_{k=1}^{K}\sum_{i=1}^{n_k}(\mathbf{x}_{ki} - \mathbf{\mu}_k)'\mathbf{\Sigma}^{-1}(\mathbf{x}_{ki} - \mathbf{\mu}_k)\right\}. \end{aligned} \tag{A.5.12}$$

This leads to

$$\log l(\mathbf{\mu}_1,\ldots,\mathbf{\mu}_K, \mathbf{\Sigma}) = -\frac{n}{2}\log|\mathbf{\Sigma}| - \frac{1}{2}\sum_{k=1}^{K}\sum_{i=1}^{n_k}(\mathbf{x}_{ki} - \mathbf{\mu}_k)'\mathbf{\Sigma}^{-1}(\mathbf{x}_{ki} - \mathbf{\mu}_k), \tag{A.5.13}$$

where the terms irrelevant to $\boldsymbol{\mu}_1$, ..., $\boldsymbol{\mu}_K$, and $\boldsymbol{\Sigma}$ have been omitted. Log likelihood (A.5.13) is maximized for

$$\hat{\mu}_k = \bar{\mathbf{x}}_k = \frac{1}{n_k}\sum_{i=1}^{n_k}\mathbf{x}_{ki}, \qquad (A.5.14)$$

$$\widehat{\boldsymbol{\Sigma}} = \mathbf{W} = \frac{1}{n}\sum_{k=1}^{K}\sum_{i=1}^{n_k}(\mathbf{x}_{ki} - \bar{\mathbf{x}}_k)(\mathbf{x}_{ki} - \bar{\mathbf{x}}_k)', \qquad (A.5.15)$$

which is proved in the following paragraphs.

As (A.5.4) and (A.5.5) are derived, the term relevant to $\boldsymbol{\mu}_k$ in (A.5.13) can be decomposed as

$$\sum_{i=1}^{n}(\mathbf{x}_{ki} - \boldsymbol{\mu}_k)'\boldsymbol{\Sigma}^{-1}(\mathbf{x}_{ki} - \boldsymbol{\mu}_k)$$

$$= \sum_{i=1}^{n}(\mathbf{x}_{ki} - \bar{\mathbf{x}}_k)'\boldsymbol{\Sigma}^{-1}(\mathbf{x}_{ki} - \bar{\mathbf{x}}_k) + n(\bar{\mathbf{x}}_k - \boldsymbol{\mu}_k)'\boldsymbol{\Sigma}^{-1}(\bar{\mathbf{x}}_k - \boldsymbol{\mu}_k), \qquad (A.5.16)$$

which shows (A.5.14) is MLE for $\boldsymbol{\mu}_k$.

Substituting (A.5.14) into $\boldsymbol{\mu}_k$ in (A.5.13), we can rewrite it as

$$\log l(\boldsymbol{\Sigma}) = -\frac{n}{2}\log|\boldsymbol{\Sigma}| - \frac{1}{2}\sum_{k=1}^{K}\sum_{i=1}^{n_k}(\mathbf{x}_{ki} - \bar{\mathbf{x}}_k)'\boldsymbol{\Sigma}^{-1}(\mathbf{x}_{ki} - \bar{\mathbf{x}}_k)$$

$$= -\frac{n}{2}\log|\boldsymbol{\Sigma}| - \frac{n}{2}\mathrm{tr}\boldsymbol{\Sigma}^{-1}\mathbf{W} = -\frac{n}{2}(\log|\boldsymbol{\Sigma}| + \mathrm{tr}\boldsymbol{\Sigma}^{-1}\mathbf{W}), \qquad (A.5.17)$$

which is *equivalent* to (A.5.6) if $\mathbf{W}$ is replaced by $\mathbf{V}$. Thus, MLE of $\boldsymbol{\Sigma}$ is given by (A.5.15), i.e., (A.5.3) with $\mathbf{V}$ replaced by $\mathbf{W}$.

A.6 Iterative Algorithms

In this appendix, *iterative algorithms* used in statistical data analysis are first outlined, followed by an illustration of a gradient algorithm. Though the descriptions in this book are very elementary, more advanced and exhaustive descriptions of a variety of iterative algorithms used in statistical computing are found in Lange (2010). Further, matrix-intensive descriptions of the algorithms are found in Hansen, Pereyra, and Scherer (2013) and Absil et al. (2008).

A.6.1   General Methodology

Let us use $\phi(\boldsymbol{\theta})$ for a function of parameter vector $\boldsymbol{\theta} = [\theta_1, \ldots, \theta_q]'$ to be *minimized* and $\hat{\boldsymbol{\theta}}$ for the solution, i.e., the vector $\boldsymbol{\theta}$ minimizing $\phi(\boldsymbol{\theta})$. For log likelihood $l(\boldsymbol{\theta})$, we can set $\phi(\boldsymbol{\theta}) = -l(\boldsymbol{\theta})$ so that the *maximum likelihood method* for maximizing $l(\boldsymbol{\theta})$ is equivalent to *minimizing* $\phi(\boldsymbol{\theta})$. The following stories hold for any optimization including least squares and maximum likelihood methods.

If the solution $\hat{\boldsymbol{\theta}}$ is not explicitly given, we must find $\hat{\boldsymbol{\theta}}$ by using an iterative algorithm in which the update of $\boldsymbol{\theta}$ is iterated. By expressing the vector $\boldsymbol{\theta}$ at the *t*th iteration as $\boldsymbol{\theta}_{[t]}$, any iterative algorithm can be described with the following steps:

Step 1. Set $\boldsymbol{\theta}$ to an initial value vector $\boldsymbol{\theta}_{[t]}$ with $t = 0$.
Step 2. Update $\boldsymbol{\theta}_{[t]}$ to $\boldsymbol{\theta}_{[t+1]}$ so that $\phi(\boldsymbol{\theta}_{[t+1]}) \leq \phi(\boldsymbol{\theta}_{[t]})$.
Step 3. Regard $\boldsymbol{\theta}_{[t+1]}$ as $\hat{\boldsymbol{\theta}}$ if convergence is reached; otherwise, increase t by one and go back to Step 2.

The convergence in Step 3 can be defined as $\phi(\boldsymbol{\theta}_{[t]}) - \phi(\boldsymbol{\theta}_{[t+1]})$, $\|\boldsymbol{\theta}_{[t]} - \boldsymbol{\theta}_{[t+1]}\|$, or the maximum of $|\theta_k^{[t]} - \theta_k^{[t+1]}|$ over k is small enough to be ignored, with $\theta_k^{[t]}$ the *k*th element of $\boldsymbol{\theta}_{[t]}$.

There are various types of iterative algorithms. They can be roughly classified into three groups: parameter partition, auxiliary function, and gradient algorithms. They differ with respect to the way in which the update in Step 2 is performed.

In the *parameter partition algorithms*, the elements of $\boldsymbol{\theta}$ are partitioned into subsets as $\boldsymbol{\theta}' = [\boldsymbol{\theta}_1', \ldots, \boldsymbol{\theta}_s', \ldots, \boldsymbol{\theta}_S']$ with the *s*th subset vector $\boldsymbol{\theta}_s'$ at the *t*th iteration expressed as $\boldsymbol{\theta}_s^{[t]}$. Then, Step 2 is divided into the sub-steps, in each of which $\phi(\boldsymbol{\theta})$ is minimized over only $\boldsymbol{\theta}_s$, with the other parameter sets kept fixed, and the resulting $\boldsymbol{\theta}_s$ gives $\boldsymbol{\theta}_s^{[t+1]}$. In the most simple case, with $S = 2$ and $\boldsymbol{\theta} = [\boldsymbol{\theta}_1, \boldsymbol{\theta}_2]$, Step 2 consists of the following sub-steps:

Step 2.1.   Minimize $\phi(\boldsymbol{\theta}_1, \boldsymbol{\theta}_2)$ over $\boldsymbol{\theta}_1$ with $\boldsymbol{\theta}_2$ fixed at $\boldsymbol{\theta}_2^{[t]}$ and the resulting $\boldsymbol{\theta}_1$ (that minimizes $\phi(\boldsymbol{\theta}_1, \boldsymbol{\theta}_2^{[t]})$) gives $\boldsymbol{\theta}_1^{[t+1]}$.
Step 2.2.   Minimize $\phi(\boldsymbol{\theta}_1, \boldsymbol{\theta}_2)$ over $\boldsymbol{\theta}_2$ with $\boldsymbol{\theta}_1$ fixed at $\boldsymbol{\theta}_1^{[t+1]}$ and the resulting $\boldsymbol{\theta}_2$ (that minimizes $\phi(\boldsymbol{\theta}_1^{[t+1]}, \boldsymbol{\theta}_2)$) gives $\boldsymbol{\theta}_2^{[t+1]}$.

This approach is useful for cases in which it is easy to minimize $\phi(\boldsymbol{\theta})$ over a subset of parameters with the other parameters being fixed. In particular, the parameter partition algorithms for least squares problems are known as *alternating least squares algorithms (ALS)* (e.g., Young 1981). An example of them has been shown in Sects. 7.4–7.7.

In *auxiliary function algorithms*, a different function $\eta(\boldsymbol{\theta})$ is used, which satisfies $\phi(\boldsymbol{\theta}_{[t]}) = \eta(\boldsymbol{\theta}_{[t]}) \geq \eta(\boldsymbol{\theta}_{[t+1]}) \geq \phi(\boldsymbol{\theta}_{[t+1]})$ with $\eta(\boldsymbol{\theta})$ being easier to handle than $\phi(\boldsymbol{\theta})$. Here, the update of $\boldsymbol{\theta}_{[t]}$ leading to $\eta(\boldsymbol{\theta}_{[t]}) \geq \eta(\boldsymbol{\theta}_{[t+1]})$ implies $\phi(\boldsymbol{\theta}_{[t]}) \geq \phi(\boldsymbol{\theta}_{[t+1]})$. One of the auxiliary function algorithms is the *EM algorithm* leading to the formulas in Note 12.1. The EM algorithm is originally presented by Dempster et al. (1977) and its book-length description is found in McLachlan and Krishnan (2008). The auxiliary function algorithm also includes the *majorization algorithm* introduced in Chap. 16. Majorization algorithms useful for some multivariate analysis procedures can be found in Kiers (2002).

In *gradient algorithms*, the differential of $\phi(\mathbf{\theta})$ with respect to $\mathbf{\theta}$ is used. This type of algorithm is illustrated in the remaining sections.

A.6.2 Gradient Algorithm for Single Parameter Cases

For introducing the gradient algorithms, we consider an example of $\phi(\mathbf{\theta})$ to be minimized:

$$\phi(\theta) = 16\theta^4 - 192\theta^3 + 880\theta^2 - 1824\theta + 1444, \tag{A.6.1}$$

which is a function of a *single* parameter $\mathbf{\theta} = [\theta]$. Figure A.6 shows $\phi(\theta)$ values against θ, where (A.6.1) is found to attain its minimum at $\theta = 3$, i.e., the solution $\hat{\theta} = 3$. However, we suppose that only formula (A.6.1) is given and $\hat{\theta}$ is unknown. Then, a *gradient algorithm* can be used, in which the *derivative* of $\phi(\theta)$ with respect to θ,

$$\begin{aligned} \frac{d\phi(\theta)}{d\theta} &= 16 \times 4\theta^3 - 192 \times 3\theta^2 + 880 \times 2\theta - 1824 \\ &= 64\theta^3 - 576\theta^2 + 1760\theta - 1824, \end{aligned} \tag{A.6.2}$$

is noted. The value of (A.6.2) with θ set to a specific value, $\theta_{[t]}$, that is,

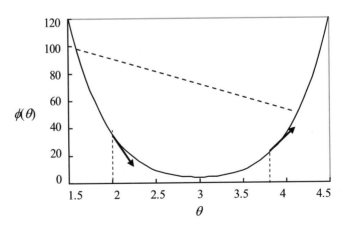

Fig. A.6 Function $\phi(\theta)$ against θ with *arrows* expressing the gradients of $\phi(\theta)$ and a *dotted line* for illustrating the convexity of $\phi(\theta)$

$$\nabla\phi(\theta_{[t]}) = 64\theta_{[t]}^3 - 576\theta_{[t]}^2 + 1760\,\theta_{[t]} - 1824 \qquad \text{(A.6.3)}$$

is called the *gradient* of $\phi(\theta)$ at $\theta = \theta_{[t]}$.

For example,

$$\text{if } \theta_{[t]} = 2, \text{ then } \nabla\phi(2) = 64 \times 2^3 - 576 \times 2^2 + 1760 \times 2 - 1824 = -96,$$
$$\text{(A.6.4)}$$

$$\text{if } \theta_{[t]} = 3.8, \text{ then } \nabla\phi(3.8) = 64 \times 3.8^3 - 576 \times 3.8^2 + 1760 \times 3.8 - 1824$$
$$= 58.4.$$

$$\text{(A.6.5)}$$

These values show the directions of the tangential lines represented as the arrows in Fig. A.6. Let us note that $\theta_{[t]} = 2$, giving the negative value (=−96) as (A.6.4), is less than the solution $\hat{\theta}$ (=3), while $\theta_{[t]} = 3.8$, leading to the positive value (A.6.5), is greater than $\hat{\theta}$. These relationships of $\theta_{[t]}$ to the solution $\hat{\theta}$ generally hold for (A.6.3); $\nabla\phi(\theta_t)$ is *negative* when $\theta_{[t]} > \hat{\theta}$, thus, $\theta_{[t]}$ should be updated to a *larger* value so as to approach the solution $\hat{\theta}$, while $\nabla\phi(\theta_{[t]})$ is *positive* for $\theta_{[t]} > \hat{\theta}$, thus, $\theta_{[t]}$ is to be updated to a *smaller* value to approach $\hat{\theta}$. This implies that $\theta_{[t]}$ is to be updated in the direction of $-1 \times \nabla(\theta_{[t]})$, i.e., in the *opposite direction of the sign of* $\nabla\phi(\theta_{[t]})$. This update is formally expressed as

$$\theta_{[t+1]} = \theta_{[t]} - s\nabla\phi(\theta_{[t]}), \qquad \text{(A.6.6)}$$

with s a suitable positive value. The resulting $\theta_{[t+1]}$ can be closer to $\hat{\theta}$ than $\theta_{[t]}$, with $\phi(\theta_{[t+1]}) \leq \phi(\theta_{[t]})$, if s is suitably chosen.

We find that whether update (A.6.6) is *successful or not* depends on which s is chosen. One unsuccessful example is if $s = 1$ is chosen for $\theta_{[t]} = 2$. Then, (A.6.4) and (A.6.6) show $\theta_{[t+1]} = \theta_{[t]} - s\nabla\phi(\theta_{[t]}) = 2 - (-96) = 98$; the updated $\theta_{[t+1]}$ far exceeds $\hat{\theta}$ and $\phi(\theta_{[t+1]}) > \phi(\theta_{[t]})$. However, such cases can be avoided by choosing s with the following steps:

Step 2.1. Set s to 1.

Step 2.2. Obtain $\theta_{[t+1]}$ with (A.6.6).

Step 2.3. Finish if $\phi(\theta_{[t+1]}) \leq \phi(\theta_{[t]})$; otherwise, set $s := s/2$ and go back to Step 2.2.

Here, "$s := s/2$" stands for reduce the s value to half; s is reduced as 1, 1/2, 1/2², 1/2³, … . When $\theta_{[t]} = 2$, returning to Step 2.2 seven times leads to $\theta_{[t+1]} = \theta_{[t]} - s\nabla\phi(\theta_{[t]}) = 2 - (1/2^7) \times 98 = 2 - 1/128 \times (-96) = 2.75$, which is close to $\hat{\theta}$.

The three steps in Sect. A.6.1, with Step 2 replaced by the above Steps 2.1, 2.2, and 2.3, allow us to find $\hat{\theta}$ if $\phi(\theta)$ is *convex*. Here, the adjective "convex", roughly speaking, stands for the fact that the curve of $\phi(\theta)$ is not a zigzag. The exact definition, with multiple parameters considered, is as follows: $\phi(\mathbf{\theta})$ is said to be *convex*,

$$\phi(\alpha\mathbf{\theta}_1 + (1 - \alpha)\mathbf{\theta}_2) \leq \alpha\phi(\mathbf{\theta}_1) + (1 - \alpha)\phi(\mathbf{\theta}_2), \tag{A.6.7}$$

for every pair of $q \times 1$ vectors $\mathbf{\theta}_1$ and $\mathbf{\theta}_2$, and every α taking a value within the range from 0 to 1. This implies that, as a dotted line in Fig. A.6, the line connecting the two points of $\phi(\mathbf{\theta})$ is not lower than $\phi(\mathbf{\theta})$.

Although more efficient procedures than the one in the above Step 2.3 have been developed for choosing s (for example, Boyed and Vandenberghe 2004), they are beyond the scope in this book.

A.6.3   Gradient Algorithm for Multiple Parameter Cases

For *multiple* parameter cases with $\mathbf{\theta} = [\theta_1, \ldots, \theta_q]'$, update formula (A.6.6) is extended as

$$\mathbf{\theta}_{[t+1]} = \mathbf{\theta}_{[t]} - s\nabla\phi(\mathbf{\theta}_{[t]}), \tag{A.6.8}$$

with $\nabla\phi(\mathbf{\theta}_{[t]})$ the $q \times 1$ gradient vector, which is the vector $\partial\phi(\mathbf{\theta})/\partial\mathbf{\theta}$ at $\mathbf{\theta} = \mathbf{\theta}_{[t]}$. Here, $\partial\phi(\mathbf{\theta})/\partial\mathbf{\theta}$ denotes the *partial derivative* of $\phi(\mathbf{\theta})$ with respect to $\mathbf{\theta}$. That is, $\partial\phi(\mathbf{\theta})/\partial\mathbf{\theta}$ is the $q \times 1$ vector, and its kth element is the derivative of $\phi(\mathbf{\theta})$ with respect to θ_k (the kth element of $\mathbf{\theta}$), where $\phi(\mathbf{\theta})$ is regarded as a function of *only* θ_k with $\phi(\mathbf{\theta}) = \phi(\theta_k)$ and θ_l ($l \neq k$) treated as a *fixed constant*. For example, when $q = 3$ and

$$\phi(\mathbf{\theta}) = 3\theta_1^2 + 6\theta_2^2 - 4\theta_1\theta_3 + 5\theta_2\theta_3 - 7\theta_2 + 9\theta_3, \tag{A.6.9}$$

its partial derivative is

$$\frac{\partial\phi(\mathbf{\theta})}{\partial\mathbf{\theta}} = \begin{bmatrix} d\phi(\theta_1)/d\theta_1 \\ d\phi(\theta_2)/d\theta_2 \\ d\phi(\theta_3)/d\theta_3 \end{bmatrix} = \begin{bmatrix} 6\theta_1 - 4\theta_3 \\ 12\theta_2 + 5\theta_3 - 7 \\ -4\theta_1 + 5\theta_2 + 9 \end{bmatrix}. \tag{A.6.10}$$

Note its second element. There, (A.6.9) has been regarded as a function of only θ_2, i.e., $\phi(\theta_2) = (6)\theta_2^2 + (5\theta_3 - 7)\theta_2 + (3\theta_1^2 - 4\theta_1\theta_3 + 9\theta_3)$, with the parenthesized terms being treated as fixed constants.

In multiple parameter cases, the three steps in the last section are simply replaced by their vector versions:

Step 2.1. Set s to 1.
Step 2.2. Obtain $\mathbf{\theta}_{[t+1]}$ with (A.6.8).
Step 2.3. Finish if $\phi(\mathbf{\theta}_{[t+1]}) \leq \phi(\mathbf{\theta}_{[t]})$; otherwise, set $s := s/2$ and go back to Step 2.2.

Fig. A.7 Illustration of $\boldsymbol{\theta}_{[t]}$ approaching the solution $\hat{\theta}$ with an increase in t, where the horizontal axis represents the q-dimensional space for $\boldsymbol{\theta} = [\theta_1, \ldots, \theta_q]'$ and subscript t is attached to s, as the s value chosen for t differs from the one for $t - 1$

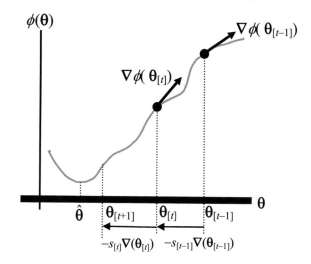

The three steps in Section A.6.1, with Step 2 replaced by the above steps, allow us to find $\hat{\theta}$, if $\phi(\boldsymbol{\theta})$ is convex with (A.6.7). The algorithm with (A.6.8) is illustrated in Fig. A.7

Though we have only introduced a procedure using the (first) derivative, more effective procedures, including one in which first and *second* derivatives are used, have been developed, which are beyond the scope of this book. Advanced theories for gradient and related algorithms are detailed in Absil et al. (2008), and Boyed and Vandenberghe (2004). One republication of a classical book dealing with such theories is Ortega and Rheinboldt (2000).

References

Absil, P.-A., Mahony, R., & Sepulchre, R. (2008). *Optimization algorithms on matrix manifolds.* Princeton University Press: Princeton, NJ.

Adachi, K. (2006). *Multivariate data analysis*: An introduction for psychology, pedagogy, and sociology courses. Kyoto, Nakanishiya-Shuppan (in Japanese).

Adachi, K. (2004). Correct classification rates in multiple correspondence analysis. *Journal of the Japanese Society of Computational Statistics, 17,* 1–20.

Adachi, K. (2009). Joint Procrustes analysis for simultaneous nonsingular transformation of component score and loading matrices. *Psychometrika, 74,* 667–683.

Adachi, K. (2011). Constrained principal component analysis of standardized data for biplots with unit-length variable vectors. *Advances in Data Analysis and Classification, 5,* 23–36.

Adachi, K. (2016). Three-way principal component analysis with its applications to psychology. In T. Sakata (Ed.), *Applied matrix and tensor variate data analysis* (pp. 1–21). Tokyo: Springer.

Adachi, K. (2013). Factor analysis with EM algorithm never gives improper solutions when sample covariance and initial parameter matrices are proper. *Psychometrika, 78,* 380–394.

Adachi, K., & Murakami, T. (2011). *Nonmetric multivariate analysis: From principal component analysis to multiple correspondence analysis.* Tokyo: Asakura-Shoten. (in Japanese).

Adachi, K. & Trendafilov, N.T. (2015a). Sparse orthogonal factor analysis. In E. Carpita, E. Brentari, & Qannari, E. M. (Eds.), *Advances in latent variable: Methods, models, and Applications* (pp. 227–239). Cham, Switzerland: Springer.

Adachi, K. & Trendafilov, N. T., (2015b). Sparse principal component analysis subject to prespecified cardinality of loadings. *Computational Statistics.* doi: 10.1007/s00180-015-0608-4

Akaike, H. (1974). A new look at the statistical model identification. *IEEE Transactions on Automatic Control, 19,* 716–723.

Anderson, T. W. (2003). *An introduction to multivariate statistical analysis* (3rd ed.). New York: Wiley.

Banerjee, S., & Roy, A. (2014). *Linear algebra and matrix analysis for statistics.* Boca Raton, FL: CRC Press.

Bartholomew, D., Knott, M., & Moustaki, I. (2011). *Latent variable models and factor analysis*: A unified approach (Third Edition). Chichester: West Sussex: Wiley.

Bentler, P. M. (1985). *Theory and implementation of EQS: A structural equation program* (manual for program version 2.0). Los Angels: BMDP Statistical Software.

Bishop, C. M. (2006). *Pattern recognition and machine learning.* New York: Springer.

Bollen, K. A. (1989). *Structural equations with latent variables.* New York: Wiley.

Borg, I., & Groenen, P. J. F. (2005). *Modern multidimensional scaling: Theory and applications* (2nd ed.). New York: Springer.

Borg, I., Groenen, P. J. F., & Mair, P. (2013). *Applied multidimensional scaling.* Heidelberg: Springer.

© Springer Nature Singapore Pte Ltd. 2016
K. Adachi, *Matrix-Based Introduction to Multivariate Data Analysis,*
DOI 10.1007/978-981-10-2341-5

Boyed, S., & Vandenberghe, L. (2004). *Convex optimization*. Cambridge: Cambridge University Press.

Browne, M. (2001). An overview of analytic rotation in exploratory factor analysis. *Multivariate Behavioral Research, 36*, 111–150.

Carroll, J. D., Green, P. E., & Chaturvedi, A. (1997). *Mathematical tools for applied multivariate analysis* (Revised ed.). San Diego, California: Academic Press.

Committee for Guiding Psychological Experiments. (1985). *Experiments and tests: The foundations of psychology* (Explanation version). Tokyo: Bifukan (in Japanese).

Cox, T. F., & Cox, M. A. A. (2000). *Multidimensional scaling* (2nd ed.). London: Chapman & Hall.

de Leeuw, J. (1977). Application of convex analysis to multidimensional scaling. In J. R. Barra, F. Brodeau, G. Rominer, & B. Van Custem (Eds.), *Recent developments in statistics* (pp. 133–145). Amsterdam: North-Holland.

Dempster, A. P., Laird, N. M., & Rubin, D. B. (1977). Maximum likelihood from incomplete data via the EM algorithm. *Journal of the Royal Statistical Society, Series B, 39*, 1–38.

De Soete, G., & Carroll, J. D. (1994). K-means clustering in a low-dimensional Euclidean space. In E. Diday, Y. Lechevallier, M. Schader., P. Bertrand, & B. Burtschy (Eds), *New approaches in classification and data analysis* (pp. 212–219). Berlin: Springer.

Eckart, C., & Young, G. (1936). The approximation of one matrix by another of lower rank. *Psychometrika, 1*, 211–218.

Eldén, L. (2007). *Matrix methods in data mining and pattern recognition*. Philadelphia, PA, Society of Industrial and Applied Mathematics (SIAM).

Everitt, B. S. (1993). *Cluster analysis* (3rd edition ed.). London: Edward Arnold.

Fisher, R. A. (1936). The use of multiple measurements in taxonomic problems. *Annals of Eugenics, 7*, 179–188.

Gan, G., Ma, C., & Wu, J. (2007) *Data clustering: Theory, algorithms, and applications*. Philadelphia, PA: Society of Industrial and Applied Mathematics (SIAM).

Gifi, A. (1990). *Nonlinear multivariate analysis*. Chichester: Wiley.

Gentle, J. E. (2007). *Matrix algebra: Theory, computations and applications in statistics*. New York: Springer.

Golub, G. H., & Van Loan, C. F. (1996). *Matrix computations* (3rd ed.). Baltimore: Johns Hopkins University Press.

Gower, J. C., Lubbe, S., & le Roux, N. (2011). *Understanding biplots*. Chichester: Wiley.

Gower, J. C., & Dijksterhuis, G. B. (2004). *Procrustes problems*. Oxford: Oxford University Press.

Greenacre, M. J. (1984). *Theory and applications of correspondence analysis*. London: Academic Press.

Greenacre, M. J. (2007). *Correspondence analysis in practice* (2nd ed.). Boca Raton, FL: CRC Press/Taylor & Francis Group.

Groenen, P. J. F. (1993). *The majorization approach to multidimensional scaling: Some problems and extensions*. Leiden, The Netherlands: DSWO.

Haavelmo, T. (1943). The statistical implications of a system of simultaneous equations. *Econometrika, 11*, 1–12.

Hand, D. J. (1997). *Construction and assessment of classification rules*. Chichester: Wiley.

Hansen, P. C., Pereyra, V., & Scherer G. (2013). *Least square data fitting with applications*. Baltimore, Maryland: The John Hopkins University Press.

Harman, H. H. (1976). *Modern factor analysis* (3rd ed.). Chicago: The University of Chicago Press.

Harman, H. H., & Jones, W. H. (1966). Factor analysis by minimizing residuals (Minres). *Psychomerika, 31*, 351–369.

Hartigan, J. A., & Wang, M. A. (1979). *A k-means clustering algorithm. Applied Statistics, 28*, 100–108.

Harville, D. A. (1997). *Matrix algebla from a statistician's perspective*. New York: Springer.

Hastie, T., Tibshirani, R., & Friedman, J. (2009). *The elements of statistical learning: Data mining, inference, and prediction* (2nd ed.). New York: Springer.

Hastie, T., Tibshirani, R., & Wainwright, M. (2015). *Statistical learning with sparsity: The lasso and generalizations*. Boca Raton, FL: CRC Press/Taylor & Francis Group.

Harshman, R.A. (1970). Foundations of the PARAFAC procedure: Models and conditions for an "exploratory" multi-mode factor analysis. *UCLA Working Papers in Phonetics, 16,* 1–84.

Hayashi, C. (1952). On the prediction of phenomena from qualitative data and the quantification of qualitative data from mathematico-statistical point of view. *Annals of the Institute of Mathematical Statistics, 3,* 69–98.

Heiser, W. J. (1991). A generalized majorization method for least squares multidimensional scaling of pseudo distances that may be negative. *Psychometrika, 56,* 7–27.

Hempel, C. (1966). *Philosophy in natural science*. Englewood Cliffs, New Jersey: Prentice Hall.

Hitchcock, F. L. (1927). Multiple invariants and generalized rank of a p-way matrix or tensor. *Journal of Mathematics and Physics, 7,* 39–79.

Hirose, K., & Yamamoto, M. (2014). Estimation of an oblique structure via penalized likelihood factor analysis. *Computational Statistics and Data Analysis, 79,* 120–132.

Hoerl, A. E., & Kennard, R. (1970). Ridge regression: Biased estimation for non-orthogonal roblems. *Technometrics, 12,* 55-67. (Reprinted in *Technometrics, 42* (2000), 80–86).

Hogg, R. V., McKean, J. W., & Craig, A. T. (2005). *Introduction to mathematical statistics* (5th ed.). Upper Saddle River, New Jersey: Prentice Hall.

Holzinger K. J. & Swineford F. (1939). *A study in factor analysis: The stability of a bi-factor solution*. University of Chicago: Supplementary Educational Monographs No. 48.

Horn, R. A., & Johonson, C. R. (2013). *Matrix analysis* (2nd ed.). Cambridge: Cambridge University Press.

Hotelling, H. (1933). Analysis of a complex of statistical variables into principal components. *Journal of Educational Statistics, 24,* 417–441.

Hotelling, H. (1936). Relations between sets of variables. *Biometrika, 28,* 321–377.

Huley, J. R., & Cattell, R. B. (1962). The Procrustes program: Producing direct rotation to test a hypothesized factor structure. *Behavioral Science, 7,* 258–262.

Izenman, A. J. (2008). *Modern multivariate statistical techniques: Regression, classification, and manifold learning*. New York, NY: Springer.

Jennrich, R. I. (2001). A simple general method for orthogonal rotation. *Psychometrika, 66,* 289–306.

Jennrich, R. I. (2002). A simple general method for oblique rotation. *Psychometrika, 67,* 7–20.

Jennrich, R. I., & Sampson, P. F. (1966). Rotation for simple loadings. *Psychometrika, 31,* 313–323.

Jolliffe, I. T. (2002). *Principal component analysis* (2nd ed.). New York: Springer.

Jolliffe, I. T., Trendafilov, N. T., & Uddin, M. (2003). A modified principal component technique based on the LASSO. *Journal of Computational and Graphical Statistics, 12,* 531–547.

Jöreskog, K. G. (1969). A general approach to confirmatory maximum likelihood factor analysis. *Psychometrika, 34,* 183–202.

Jöreskog, K. G. (1970). A general method for analysis of covariance structures. *Biometrika, 57,* 239–251.

Kaiser, H. F. (1958). The varimax criterion for analytic rotation in factor analysis. *Psychometrika, 23,* 187–200.

Kaplan, D. (2000). *Structural equation modeling: Foundations and extensions*. Thousand Oaks, California: Sage Publications.

Kettenring, J. R. (1971). Canonical analysis of several sets of variables. *Biometrika, 58,* 433–460.

Kiers, H. A. L. (1994). Simplimax: Oblique rotation to an optimal target with simple structure. *Psychometrika, 59,* 567–579.

Kiers, H. A. L. (2002). Setting up alternating least squares and iterative majorization algorithms for solving various matrix optimization problems. *Computational Statistics and Data Analysis, 41,* 157–170.

Koch, I. (2014). *Analysis of multivariate and high-dimensional data*. Cambridge: Cambridge University Press.

Konishi, S. (2014). *Introduction to multivariate analysis: Linear and nonlinear modeling*. Boca Raton, FL: CRC Press.

Konishi, S., & Kitagawa, G. (2007). *Information criteria and statistical modeling*. New York: Springer.

Kroonenberg, P. M. (2008). *Applied multiway data analysis*. Hoboken: Wiley.

Lange, K. (2010). *Numerical analysis for statisticians* (2nd ed.). New York: Springer.

Lattin, J., Carroll, J. D., & Green, P. E. (2003). *Analyzing multivariate data*. Pacific Grove: CA, Thomson Learning Inc.

Lütkepohl, H. (1996). *Handbook of matrices*. Chichester: Wiley.

McLachlan, G. J., & Krishnan, T. (2008). *The EM algorithm and extensions* (2nd ed.). New York: Wiley.

McLachlan, G. J. (1992). *Discriminant analysis and statistical pattern recognition*. New York: Wiley.

MacQueen, J.B. (1967). Some methods for classification and analysis of multivariate observations. *Proceedings of the 5th Berkeley Symposium, 1*, 281–297.

Magnus, J. R., & Neudecker, H. (1991). *Matrix differential calculus with an applications in statistics and econometrics* (2nd ed.). Chichester: Wily.

Montgomery, D. C., Peck, E. A., & Vining, G. G. (2012). *Introduction to regression analysis* (5th ed.). Hoboken, New Jersey: Wiley.

Mosier, C. I. (1939). Determining a simple structure when loadings for certain tests are known. *Psychometrika, 4*, 149–162.

Mulaik, S. A. (2010). *Foundations of factor analysis* (2nd ed.). Boca Raton: CRC Press.

Nishisato, S. (1980). *Analysis of categorical data: Dual scaling and its applications*. Toronto: University of Toronto Press.

Ortega, J. M. & Rheinboldt, W. C. (2000). *Iterative solution of nonlinear equations in several variables*. Philadelphia, PA: Society of Industrial and Applied Mathematics (SIAM).

Pearson, K. (1901). On lines and planes of closest fit to systems of points in space. *Philosophical Magazines, 2*, 559–572.

Rao, C. R. (1973). *Linear statistical inference and its applications*. New York: Wiley.

Rao, C. R. (2001). *Linear statistical inference and its applications* (2nd ed.). New York: Wiley.

Rao, C. R., & Mitra, S. K. (1971). *Generalized inverse of matrices and its applications*. New York: Wiley.

Reyment, R., & Jöreskog, K. G. (1996). *Applied factor analysis in the natural sciences*. Cambridge: Cambridge University Press.

Schott, J. R. (2005). *Matrix analysis for statistics* (2nd ed.). Hoboken, New Jersey: Wiley.

Rencher, A. C., & Christensen, W. F. (2012). *Methods of multivariate analysis* (3rd ed.). Hoboken, New Jersey: Wiley.

Rubin, D. B., & Thayer, D. T. (1982). EM algorithms for ML factor analysis. *Psychometrika, 47*, 69–76.

Schwarz, G. (1978). Estimating the dimension of a model. *Annals of Statistics, 6*, 461–464.

Seber, G. A. F. (1984). *Multivariate observations*. New York: Wiley.

Seber, G. A. F. (2008). *A matrix handbook for statisticians*. Hoboken, New Jersey: Wiley.

Smilde, A., Bro, R., & Geladi, P. (2004). *Multi-way analysis: Applications in the chemical sciences*. Chichester: Wiley.

Spearman, C. (1904). "General intelligence" objectively determined and measured. *American Journal of Psychology, 15*, 201–293.

Takane, Y. (2014). *Constrained principal component analysis and related techniques*. Boca Raton, FL: CRC Press.

Takane, Y., Young, F. W., & de Leeuw, J. (1977). Nonmetric individual differences multidimensional scaling: An alternating least squares method with optimal scaling features. *Psychometrika, 42*, 7–67.

ten Berge, J. M. F. (1983). A generalization of Kristof's theorem on the trace of certain matrix products. *Psychometrika, 48,* 519–523.

ten Berge, J. M. F. (1993). *Least squares optimization in multivariate analysis.* Leiden: DSWO Press.

ten Berge, J. M. F., & Kiers, H. A. L. (1996). Optimality criteria for principal component analysis and generalizations. *British Journal of Mathematical and Statistical Psychology, 49,* 335–345.

ten Berge, J. M. F., Knol, D. L., & Kiers, H. A. L. (1988). A treatment of the orthomax rotation family in terms of diagonalization, and a re-examination of a singular value approach to varimax rotation. *Computational Statistics Quarterly, 3,* 207–217.

Thurstone, L. L. (1935). *The vectors of mind.* Chicago: University of Chicago Press.

Thurstone, L. L. (1947). *Multiple factor analysis.* Chicago: University of Chicago Press.

Tipping, M. E., & Bishop, C. M. (1999). Probabilistic principal component analysis. *Journal of the Royal Statistical Society: Series B (Statistical Methodology), 61,* 611–622.

Torgerson, W. S. (1952). Multidimensional scaling: I. Theory and method. *Psychometrika, 17,* 401–419.

Toyoda, H. (1988). *Covariance structure analysis (structural equation modeling): Introductory part.* Tokyo: Asakura-Shoten (in Japanese).

Trendafilov, N. T. (2014). From simple structure to sparse components: a review. *Computational Statistics, 29,* 431–454.

Tucker, L. R. (1966). Some mathematical notes on three-mode factor analysis. *Psychometrika, 31,* 279–311.

van de Geer, J. P. (1984). Linear relations among k sets of variables. *Psychometrika, 49,* 79–94.

Vichi, M., & Kiers, H. A. L. (2001). Factorial k-means analysis for two-way data. *Computational Statistics and Data Analysis, 37,* 49–64.

Wright, S. (1918). On the nature of size factors. *Genetics, 3,* 367–374.

Unkel, S., & Trendafilov, N. T. (2010). Simultaneous parameter estimation in exploratory factor analysis: An expository review. *International Statistical Review, 78,* 363–382.

Wang, J., & Wang, X. (2012). *Structural equation modeling: Applications using Mplus.* Chichester, West Sussex: Wiley.

Wright, S. (1960). Path coefficients and path regressions: Alternative or complementary cocepts? *Biometrics, 16,* 189–202.

Yanai, H., & Ichikawa, M. (2007). Factor analysis. In C. R. Rao & S. Sinharay (Eds.), *Handbook of statistics* (Vol. 26, pp. 257–296)., Amsterdam: Elsevier.

Yanai, H., Takeuchi, K., & Takane, Y. (2011). *Projection matrices, generalized inverse matrices, and singular value decomposition.* New York: Springer.

Yates, A. (1987). *Multivariate exploratory data analysis: A perspective on exploratory factor analysis.* Albany: State University of New York Press.

Yeung, K. Y., & Ruzzo, W. L. (2001). Principal component analysis for clustering gene expression data. *Bioinformatics, 17,* 763–774.

Young, F. W. (1981). Quantitative analysis of qualitative data. *Psychometrika, 46,* 357–388.

Zou, D. M., Hastie, T., & Tibshirani, R. (2006). Sparse principal component analysis. *Journal of Computational and Graphical Statistics, 15,* 265–286.

Index

A

Akaike's information criterion (AIC), 120
Alternating least squares (ALS) algorithm, 104,
   285
Alternating least squares scaling (ALSCAL),
   243
Angle of vectors, 34, 255
Auxiliary function algorithm, 285
Average, 17, 20

B

Bayes classification rule, 230
Bayesian estimation method, 241
Bayesian information criterion (BIC), 120
Bayes theorem, 230
Best fitting plane, 85, 88
Biplot, 75
Block diagonal matrix, 213, 214
Block matrix, 187, 207, 208, 217
Blocks of a matrix, 207

C

Canonical correlation analysis (CCA), 207, 210
Canonical correlation coefficient, 212
Canonical discriminant analysis (CDA),
   225–227, 279
Cauchy–Schwarz inequality, 42, 249
Causal model, 127
Centered matrix, 38
Centered score, 20, 23–25, 27, 28
Centering matrix, 22, 23, 31, 38, 42, 43, 51,
   236, 248
Centroid, 228, 229
Classical scaling, 243
Cluster, 93–95, 97, 98, 100–103
Cluster analysis, 93, 227
Cluster center matrix, 97

Cluster feature matrix, 94, 100
Clustering, 93, 94, 103, 109, 110, 144, 226
Coefficient of determination, 55
Column, 3–5, 27, 39, 41–43, 50, 61, 64, 69, 71,
   73, 75, 81–83, 87, 88, 91, 94, 105, 115,
   167, 184, 187, 194, 195, 199, 203, 204,
   209, 216–219, 222, 223, 225, 227, 244,
   253, 256, 258–260, 264–266, 269, 271,
   272, 274, 276–278
Column vector, 7, 8, 218, 256, 257
Column–orthonormal matrix, 276
Common factor, 153, 176, 182
Compact version of SVD, 265
Complexity of loadings, 69, 200
Component loading, 63
Confirmatory factor analysis (CFA), 146
Confirmatory principal component analysis,
   159
Constraint, 68–70, 74, 76, 79, 80, 91, 105, 109,
   110, 135, 168, 184, 196, 199, 200, 202,
   204, 210, 217, 218, 226, 252, 253,
   270–272, 275–277, 279
Contingency table, 222
Continuous variable, 113, 117, 124
Convergence, 98–100, 102, 104, 105, 144,
   159, 187, 204, 224, 240, 246, 285
Convex function, 287, 289
Coordinates of object, 7, 29, 84, 85, 88, 195,
   197, 198, 219, 232, 243, 244, 251
Correlation coefficient, 29, 32, 34, 35, 37, 42,
   56, 61, 71, 137, 168, 187, 212
Correspondence analysis, 221–223, 225
Cosine theorem, 255
Covariance, 29–33, 35–37, 42, 54, 126, 136,
   140
Covariance matrix, 31, 36–38, 42, 56, 61, 69,
   73, 81, 91, 116, 117, 119, 133, 134, 136,

© Springer Nature Singapore Pte Ltd. 2016
K. Adachi, *Matrix-Based Introduction to Multivariate Data Analysis*,
DOI 10.1007/978-981-10-2341-5